The Evolution Debate
1813–1870

Edited by David Knight

The Evolution Debate
1813–1870
Edited and with new introductions by David Knight

The Evolution Debate
1813–1870

Volume II

Geology and Mineralogy,
Considered with Reference to Natural Theology,
Volume I
William Buckland

with a new introduction by
David Knight

Published in Association with the Natural History Museum

Routledge
Taylor & Francis Group
LONDON AND NEW YORK

THE
NATURAL
HISTORY
MUSEUM

First published 1836

This edition reprinted 2003
by Routledge
2 Park Square, Milton Park, Abingdon, Oxon, OX14 4RN

Simultaneously published in the USA and Canada
by Routledge
270 Madison Ave, New York, NY 10016

Transferred to Digital Printing 2006

Routledge is an imprint of the Taylor & Francis Group

Editorial material and selection © 2003 David Knight

Typeset in Times by Keystroke, Jacaranda Lodge, Wolverhampton
Printed and bound in Great Britain by TJI Digital, Padstow, Cornwall

British Library Cataloguing in Publication Data
A catalogue record for this book is available from the British Library

Library of Congress Cataloging in Publication Data
A catalog record for this book has been requested

ISBN 0–415–28922–X (set)
ISBN 0–415–28924–6 (volume II)

William Buckland (1784–1856)
© The Natural History Museum, London

INTRODUCTION TO
VOLUMES II AND III

David Knight

Buckland (1784–1856) was a parson's son from Devonshire who went up to Oxford in 1801, and after graduation was ordained and elected to a Fellowship of Corpus Christi College there, later becoming Reader in Mineralogy.[1] There were no courses in that or other sciences, and the post was poorly paid, but members of the university were welcome to attend his lectures if they chose. He was an attractive and informal lecturer, occasionally getting down on hands and knees to demonstrate how some extinct creature walked. In 1818 he was elected a Fellow of the Royal Society, and in 1821 a government-funded Oxford Professorship was obtained for him. In that same year he investigated a cave at Kirkdale in Yorkshire, where there were the bones of hyenas and their prey.[2] His research earned him the Copley Medal of the Royal Society, its highest award – never before given to a geologist;[3] and it seemed to him clear that the hyenas had been drowned in Noah's Flood.[4] He fed bones to living hyenas to establish that they crunched them in the same way as the extinct Yorkshire ones had, and that their faeces were similar. The science professor in a crusty, conservative and clerical university had become a star.[5]

But his pupil Charles Lyell argued in his *Principles of Geology*[6] [1830–3] that the geologist should invoke only causes operating now, rather than inexplicable catastrophes. Geological processes worked slowly over hundreds of millions of years. Like the best professors, Buckland was convinced by a good student, and recanted his ideas about the universal Flood. By then he had been commissioned to write a treatise, funded by a legacy from the Earl of Bridgewater, demonstrating God's wisdom and goodness using geology. His magnificent fold-out table of the geological epochs demonstrates his commitment to a long history; and he argued that the dinosaurs were not only prehistoric, but in a sense pre-Biblical. After the first sentence, the book of Genesis is concerned with our epoch and the appearance of humans. His book was very successful, with an edition as late as 1858; and it convinced educated church people that there was a vast range of extinct creatures, and that the Earth had a very long history indeed: it had not all begun in 4004 BC. What had happened over these aeons was that the Earth was prepared for mankind,

needing a cooler climate and coal deposits. God had loved the dinosaurs and designed them appropriately for the state of things in their time: they became extinct not because of design faults, but because circumstances changed. God thereupon designed a new fauna and flora, in a progressive series of new creations coming up to our era.

Buckland was particularly interested in dinosaurs as the magnificent plates, done by lithography, in this book show: it was said that he had spent all his Bridgewater legacy money (£1,000) on having them done. He was fascinated by the footprints of dinosaurs, and from their fossilised turds he worked out features of their digestive systems. His book was a great success, and he was twice President of the Geological Society of London. The 1830s and early 1840s were a time of tremen-dous religious excitement and bitterness in Oxford, orchestrated by John Henry Newman;[7] Buckland as a liberal man was unhappy, and undergraduates caught up in this enthusiasm had no time for voluntary science lectures. In 1845 Buckland moved to London as Dean of Westminster, succeeding Samuel Wilberforce who became Bishop of Oxford. In taking up a post in the church rather in academe, Buckland demonstrated that his new understanding of Genesis had not undermined his faith. While they believed that God had created each kind of living creature, it would be completely wrong to see Buckland (or Wilberforce) as anything like modern 'Creationists': fundamentalism is a twentieth-century response to twentieth-century problems.[8] But they could not accept that chance lay behind the exquisite adaptations of animals and plants to their environments, and rejected evolutionary ideas.

The copy reproduced here was presented to Charles König at the British Museum, where he looked after fossils, and kept there until the Natural History collections were transferred to the new Natural History Museum in 1881. König has cross-referenced some of these illustrations to the museum in Oxford where the original fossils described by Buckland are kept.

Durham, October 2002

1   E.O. Gordon, *The Life and Correspondence of William Buckland, DD, FRS,* London: Murray, 1894.

2   W. Buckland, *Reliquiæ Diluviance: or, Observations on the Organic Remains contained in Caves, Fissures, and Diluvial Gravel, and other Geological Phenomena, attesting the Action of an Universal Deluge,* London: Murray, 1823, 2nd edn, 1824.

3   N.A. Rupke, *The Great Chain of History: William Buckland and the English School of Geology, 1814–49,* Oxford: Oxford University Press., 1983.

4   N. Cohn, *Noah's Flood: the Genesis Story in Western Thought,* New Haven, Conn.: Yale University Press, 1996, pp.113–26.

5   D.M. Knight, 'Genesis and Geology: a very English Compromise', *Nuncius,* 15 (2000), 639–64.

6   C. Lyell, *Principles of Geology, being an Attempt to Explain the Former Changes of the Earth's Surface, by Reference to Causes now in Operation,* intr. M.J.S. Rudwick, Lehre: Cramer, 1970.

7   F. Turner, *John Henry Newman: the Challenge to Evangelical Religion,* New Haven, Conn.: Yale University Press, 2002.

8   K. Armstrong, *The Battle for God: Fundamentalism in Judaism, Christianity and Islam*, London: HarperCollins, 2000.

Chas Hones
British Museum

"From the Author"

THE BRIDGEWATER TREATISES

ON THE POWER WISDOM AND GOODNESS OF GOD

AS MANIFESTED IN THE CREATION

TREATISE VI

GEOLOGY AND MINERALOGY CONSIDERED WITH REFERENCE

TO NATURAL THEOLOGY

BY THE REV. WILLIAM BUCKLAND, D.D.

IN TWO VOLUMES

VOL I

THOU LORD IN THE BEGINNING HAST LAID THE FOUNDATION OF THE EARTH.
PSALM CII. 25.

" Let us take a Survey of the principal Fabrick, viz. the Terraqueous Globe itself; a most stupendous work in every particular of it, which doth no less aggrandize its Maker than every curious complete work doth its Workman. Let us cast our eyes here and there, let us ransack all the Globe, let us with the greatest accuracy inspect every part thereof, search out the inmost secrets of any of the creatures, let us examine them with all our gauges, measure them with our nicest rules, pry into them with our microscopes and most exquisite instruments, still we find them to bear testimony to their infinite Workman."

DERHAM'S PHYSICO-THEOLOGY, BOOK II. P. 38.

" Could the body of the whole Earth - - be submitted to the Examination of our Senses, were it not too big and disproportioned for our Enquiries, too unwieldy for the Management of the Eye and Hand, there is no question but it would appear to us as curious and well-contrived a frame as that of an human body. We should see the same Concatenation and Subserviency, the same Necessity and Usefulness, the same Beauty and Harmony in all and every of its Parts, as what we discover in the Body of every single Animal."

SPECTATOR, NO. 543.

GEOLOGY AND MINERALOGY

CONSIDERED WITH REFERENCE TO

NATURAL THEOLOGY

BY

THE REV. WILLIAM BUCKLAND, D.D.

CANON OF CHRIST CHURCH AND READER IN GEOLOGY AND MINERALOGY
IN THE UNIVERSITY OF OXFORD

VOL I

ALDI
DISCIP.
ANGLVS

LONDON
WILLIAM PICKERING
1836

C. WHITTINGHAM, TOOKS COURT, CHANCERY LANE.

DAVIES GILBERT, ESQ.

D.C.L. (BY DIPLOMA), F.R.S., HON. M.R.S.E., HON. M.R.I.A.,
F.S.A., F.L.S., F.G.S., F.R.A.S., ETC. ETC.

MY DEAR SIR,

I ONLY fulfil a gratifying duty in dedicating to you the present Essay, which owes its existence principally to your favourable opinion of my ability to discharge the trust confided to me.

To have been thus selected for such a service, is a distinction which I prize, as one of the most honourable results of my devotion of many years to the study of the mineral structure of the Earth. I fear, however, that your estimate of my qualifications has been raised above my deserts, by your affectionate regard for the University, with which it has been our common happiness to be so long connected.

G.   b

Whatever other results may have attended my public exertions in this place, I assure you that it is a source of much satisfaction to me, to find them thus rewarded by the approbation of a Philosopher, whose attainments placed him in the chair once occupied by Newton, and who is endeared by his urbanity to all, who have ever enjoyed the happiness of communication with him, either as the President of the Royal Society of London, or in that more familiar intercourse of private friendship to which it has been my privilege to be admitted.

Believe me to remain,

My dear Sir,

Your much obliged and faithful Servant,

WILLIAM BUCKLAND.

Christ Church, Oxford,
May 30, 1836.

PREFACE.

THREE important subjects of enquiry in Natural Theology come under consideration in the present Treatise.

The first regards the inorganic Elements of the Mineral Kingdom, and the actual dispositions of the Materials of the Earth : many of these, although produced or modified by the agency of violent and disturbing forces, afford abundant proofs of wise and provident Intention, in their adaptations to the uses of the Vegetable and Animal Kingdoms, and especially to the condition of Man.

The second relates to Theories which have been entertained respecting the Origin of the World; and the derivation of existing systems of organic Life, by an eternal succession, from preceding individuals of the same species; or by gradual transmutation of one species into another. I have endeavoured to show, that to all these Theories the phenomena of Geology are decidedly opposed.

The third extends into the Organic Remains of a former World the same kind of investigation, which Paley has pursued with so much success in his examination of the evidences of Design in the mechanical structure of the corporeal frame of Man, and of the inferior Animals which are placed with him on the present surface of the Earth.

The myriads of petrified Remains which are disclosed by the researches of Geology all tend to prove, that our Planet has been occupied in times preceding the Creation of the Human Race, by extinct species of Animals and Vegetables, made up, like living Organic Bodies, of " Clusters of Contrivances," which demonstrate the exercise of stupendous Intelligence and Power. They further show that these extinct forms of Organic Life were so closely allied, by Unity in the principles of their construction, to Classes, Orders, and Families, which make up the existing Animal and Vegetable Kingdoms, that they not only afford an argument of surpassing force, against the doctrines of the Atheist and the Polytheist; but supply a chain of connected evidence, amounting to demonstration, of the continuous Being, and of many of the highest Attributes of the One Living and True God.

The scientific Reader will feel that much value has been added to the present work, from the whole of the Palæontology, during its progress through the Press, having had the great advantage of passing under the revision of Mr. Broderip, and from the botanical part having being submitted to Mr. Robert Brown. I have also to acknowledge my obligations to Mr. Clift for his important assistance in the anatomy of the Megatherium; to Professor Agassiz of Neuchatel for his unreserved communications of his discoveries relating to Fossil Fishes; to Mr. Owen for his revision of some parts of my Chapter on Mollusks; and to Mr. James Sowerby for his assistance in engraving most of my figures of radiated animals, and some of those of Mollusks.

To all these Gentlemen I feel it my duty thus to offer my public acknowledgments.

Many obligations to other scientific friends are also acknowledged in the course of the work.

The Wood-cuts have been executed by Mr. Fisher and Mr. Byfield, and most of the Steel plates of Mollusks by Mr. Zeitter.

NOTICE.

THE series of Treatises, of which the present is one, is published under the following circumstances:

The RIGHT HONOURABLE and REVEREND FRANCIS HENRY, EARL of BRIDGEWATER, died in the month of February, 1829; and by his last Will and Testament, bearing date the 25th of February, 1825, he directed certain Trustees therein named to invest in the public funds the sum of Eight thousand pounds sterling; this sum, with the accruing dividends thereon, to be held at the disposal of the President, for the time being, of the Royal Society of London, to be paid to the person or persons nominated by him. The Testator further directed, that the person or persons selected by the said President should be appointed to write, print, and publish one thousand copies of a work *On the Power, Wisdom, and Goodness of God, as manifested in the Creation; illustrating such work by all reasonable arguments, as for instance the variety and formation of God's creatures in the animal, vegetable, and mineral kingdoms; the effect of digestion, and thereby of conversion; the construction of the hand of man, and an infinite variety of other arguments; as also by discoveries ancient and modern, in arts, sciences, and the whole extent of literature.* He desired, moreover, that the profits arising from the sale of the works so published should be paid to the authors of the works.

The late President of the Royal Society, Davies Gilbert, Esq. requested the assistance of his Grace the Archbishop of Canterbury and of the Bishop of London, in determining upon the best mode of carrying into effect the intentions of the Testator. Acting with their advice, and with the concurrence of a nobleman immediately connected with the deceased, Mr. Davies Gilbert appointed the following eight gentlemen to write separate Treatises on the different branches of the subject as here stated:

THE REV. THOMAS CHALMERS, D.D.
PROFESSOR OF DIVINITY IN THE UNIVERSITY OF EDINBURGH.
ON THE POWER, WISDOM, AND GOODNESS OF GOD AS MANIFESTED IN THE ADAPTATION OF EXTERNAL NATURE TO THE MORAL AND INTELLECTUAL CONSTITUTION OF MAN.

JOHN KIDD, M.D. F.R.S.
REGIUS PROFESSOR OF MEDICINE IN THE UNIVERSITY OF OXFORD.
ON THE ADAPTATION OF EXTERNAL NATURE TO THE PHYSICAL CONDITION OF MAN.

THE REV. WILLIAM WHEWELL, M.A. F.R.S.
FELLOW OF TRINITY COLLEGE, CAMBRIDGE.
ASTRONOMY AND GENERAL PHYSICS CONSIDERED WITH REFERENCE TO NATURAL THEOLOGY.

SIR CHARLES BELL, K.G.H. F.R.S. L. & E.
THE HAND: ITS MECHANISM AND VITAL ENDOWMENTS AS EVINCING DESIGN.

PETER MARK ROGET, M.D.
FELLOW OF AND SECRETARY TO THE ROYAL SOCIETY.
ON ANIMAL AND VEGETABLE PHYSIOLOGY.

THE REV. WILLIAM BUCKLAND, D. D. F. R. S.

CANON OF CHRIST CHURCH, AND READER IN GEOLOGY AND MINERALOGY
IN THE UNIVERSITY OF OXFORD.

ON GEOLOGY AND MINERALOGY.

THE REV. WILLIAM KIRBY, M. A. F. R. S.

ON THE HISTORY, HABITS, AND INSTINCTS OF ANIMALS.

WILLIAM PROUT, M. D. F. R. S.

CHEMISTRY, METEOROLOGY, AND THE FUNCTION OF DIGESTION, CONSIDERED WITH REFERENCE TO NATURAL THEOLOGY.

His Royal Highness the Duke of Sussex, President of the Royal Society, having desired that no unnecessary delay should take place in the publication of the above mentioned treatises, they will appear at short intervals, as they are ready for publication.

CONTENTS

OF THE FIRST VOLUME.

Page

INTRODUCTION.

CHAPTER I.

Extent of the Province of Geology.

IF a stranger, landing at the extremity of England, were to traverse the whole of Cornwall and the North of Devonshire; and crossing to St. David's, should make the tour of all North Wales; and passing thence through Cumberland, by the Isle of Man, to the south-western shore of Scotland, should proceed either through the hilly region of the Border Counties, or, along the Grampians, to the German Ocean; he would conclude from such a journey of many hundred miles, that Britain was a thinly peopled sterile region, whose principal inhabitants were miners and mountaineers.

Another foreigner, arriving on the coast of Devon, and crossing the Midland Counties, from the mouth of the Exe, to that of the Tyne, would find a continued succession of fertile

G.   B

hills and valleys, thickly overspread with towns
and cities, and in many parts crowded with a
manufacturing population, whose industry is
maintained by the coal with which the strata of
these districts are abundantly interspersed.*

A third foreigner might travel from the
coast of Dorset to the coast of Yorkshire, over
elevated plains of oolitic limestone, or of chalk;
without a single mountain, or mine, or coal-pit,
or any important manufactory, and occupied by
a population almost exclusively agricultural.

Let us suppose these three strangers to meet at
the termination of their journeys, and to compare their respective observations; how different
would be the results to which each would have
arrived, respecting the actual condition of Great
Britain. The first would represent it as a thinly
peopled region of barren mountains; the second, as a land of rich pastures, crowded with

* It may be seen, in any correct geological map of England,
that the following important and populous towns are placed
upon strata belonging to the single geological formation of the
new red sandstone:—Exeter, Bristol, Worcester, Warwick, Birmingham, Lichfield, Coventry, Leicester, Nottingham, Derby,
Stafford, Shrewsbury, Chester, Liverpool, Warrington, Manchester, Preston, York, and Carlisle. The population of these
nineteen towns, by the census of 1830, exceeded a million.

The most convenient small map to which I can refer my
readers, in illustration of this and other parts of the present
essay, is the single sheet, reduced by Gardner from Mr.
Greenough's large map of England, published by the Geological
Society of London.

a flourishing population of manufacturers; the third, as a great corn field, occupied by persons almost exclusively engaged in the pursuits of husbandry.

These dissimilar conditions of three great divisions of our country, result from differences in the geological structure of the districts through which our three travellers have been conducted. The first will have seen only those north-western portions of Britain, that are composed of rocks belonging to the primary and transition series : the second will have traversed those fertile portions of the new red sandstone formation which are made up of the detritus of more ancient rocks, and have beneath, and near them, inestimable treasures of mineral coal : the third will have confined his route to wolds of limestone, and downs of chalk, which are best adapted for sheep-walks, and the production of corn.*

Hence it appears that the numerical amount

* The road from Bath through Cirencester and Oxford to Buckingham, and thence by Kettering and Stamford to Lincoln, affords a good example of the unvaried sameness in the features and culture of the soil, and in the occupations of the people, that attends the line of direction, in which the oolite formation crosses England from Weymouth to Scarborough.

. The road from Dorchester, by Blandford and Salisbury, to Andover and Basingstoke, or from Dunstable to Royston, Cambridge, and Newmarket, affords similar examples of the dull uniformity that we observe in a journey along the line of bearing

of our population, their varied occupations, and the fundamental sources of their industry and wealth, depend, in a great degree, upon the geological character of the strata on which they live. Their physical condition also, as indicated by the duration of life and health, depending on the more or less salubrious nature of their employments; and their moral condition, as far as it is connected with these employments, are directly affected by the geological causes in which their various occupations originate.

From this example of our own country, we learn that the same constituent materials of the

of the chalk, from near Bridport on the coast of Dorset, to Flamborough Head on the coast of Yorkshire.

In the same line of direction, or line of *bearing* of the strata across England, a journey might be made from Lyme Regis to Whitby, almost entirely upon the lias formation; and from Weymouth to the Humber, without once leaving the Oxford clay. Indeed almost any route, taking a north-east and south-west direction across England, will for the most part pass continuously along the same formation; whilst a line from south-east to north-west, at right angles to the former, will nowhere continue on the same stratum beyond a few miles. Such a line will give the best information of the order of superposition, and various conditions of the very numerous strata, that traverse our island in a succession of narrow belts, the main direction of which is nearly north-east and south-west. This line has afforded to Mr. Conybeare the instructive section, from Newhaven near Brighton, to Whitehaven, published in his Geology of England and Wales; along which nearly seventy changes in the character of the strata take place.

earth are not uniformly continuous in all directions over large superficial areas. In one district, we trace the course of crystalline and granitic rocks; in another, we find mountains of slate; in a third, alternating strata of sandstone, shale, and limestone ; in a fourth, beds of conglomerate rock; in a fifth, strata of marl and clay; in a sixth, gravel, loose sand, and silt. The subordinate mineral contents of these various formations are also different; in the more ancient, are veins of gold and silver, tin, copper, lead and zinc; in another series, we find beds of coal; in others, salt and gypsum; many are composed of freestone, fit for the purposes of architecture ; or of limestone, useful both for building and cement; others of clay, convertible by fire into materials for building, and pottery : in almost all we find that most important of mineral productions, iron.

Again, if we look to the great phenomena of physical geography, the grand distributions of the solids and fluids of the globe; the disposition of continents and islands above and amidst the waters ; the depth and extent of seas, and lakes, and rivers ; the elevation of hills and mountains; the extension of plains ; and the excavation, depression, and fractures of valleys ; we find them all originating in causes which it is the province of Geology to investigate.

A more minute examination traces the progress of the mineral materials of the earth, through various stages of change and revolution, affecting the strata which compose its surface; and discloses a regular order in the superposition of these strata; recurring at distant intervals, and accompanied by a corresponding regularity in the order of succession of many extinct races of animals and vegetables, that have followed one after another during the progress of these mineral formations; arrangements like these could not have originated in chance, since they afford evidence of law and method in the disposition of mineral matter; and still stronger evidence of design in the structure of the organic remains with which the strata are interspersed.

How then has it happened that a science thus important, comprehending no less than the entire physical history of our planet, and whose documents are co-extensive with the globe, should have been so little regarded, and almost without a name, until the commencement of the present century ?

Attempts have been made at various periods, both by practical observers and by ingenious speculators, to establish theories respecting the formation of the earth; these have in great part failed, in consequence of the then imperfect state of those subsidiary sciences, which, within the last half century, have enabled the geologist to return from the region of fancy to that of facts, and to establish his conclusions on the firm basis of philosophical induction. We now approach the study of the natural history of the globe, aided not only by the higher branches of Physics, but by still more essential recent discoveries, in Mineralogy, and Chemistry, in Botany, Zoology, and Comparative Anatomy. By the help of these sciences, we are enabled to extract from the archives of the interior of the earth, intelligible records of former conditions of our planet, and to decipher documents, which were a sealed book to all our predecessors in the attempt to illustrate subterranean history. Thus enlarged in its views, and provided with fit means of pursuing them, Geology extends its researches into regions more vast and remote, than come within the scope of any other physical science except Astronomy. It not only comprehends the entire range of the mineral kingdom, but includes also the history of innumerable extinct races of animals and vegetables; in each of which it exhibits evidences of design and contrivance, and of adaptations to the varying condition of the lands and waters on which they were placed; and besides all these, it discloses an ulterior prospective accommodation of the mineral elements, to existing tribes of plants and animals, and more especially to the uses of man. Evidences like these make up a history of a high and ancient order, unfolding records of the operations of the Almighty Author of the Universe, written by the finger of God himself, upon the foundations of the everlasting hills.

CHAPTER II.

Consistency of Geological Discoveries with Sacred History.

IT may seem just matter of surprise, that many learned and religious men should regard with jealousy and suspicion the study of any natural phenomena, which abound with proofs of some of the highest attributes of the Deity ; and should receive with distrust, or total incredulity, the announcement of conclusions, which the geologist deduces from careful and patient investigation of the facts which it is his province to explore. These doubts and difficulties result from the disclosures made by geology, respecting the lapse of very long periods of time, before the creation of man. Minds which have been long accustomed to date the origin of the universe, as well as that of the human race, from an era of about six thousand years ago, receive reluctantly any information, which if true, demands some new modification of their present ideas of cosmogony; and, as in this respect, Geology has shared the fate of other infant sciences, in being for a while considered hostile to revealed religion; so like them, when fully understood, it will be found a potent and consistent auxiliary to it, exalting our conviction of the Power, and Wisdom, and Goodness of the Creator.*

No reasonable man can doubt that all the phenomena of the natural world derive their origin from God; and no one who believes the Bible to be the word of God, has cause to fear any discrepancy between this, his word, and the results of any discoveries respecting the nature of his works; but the early and deliberative stages of scientific discovery are always those of perplexity and alarm, and during these stages the human mind is naturally circumspect, and slow to admit new conclusions in any department of knowledge. The prejudiced persecutors of Galileo apprehended danger to religion, from

---

* Hæc et hujusmodi cœlorum phænomena, ad Epocham sexmillennem, salvis naturæ legibus, ægrè revocari possunt. Quin fatendum erit potius non eandem fuisse originem, neque coævam, Telluris nostræ et totius Universi: sive Intellectualis, sive Corporei. Neque mirum videri debet hæc non distinxisse Mosem, aut Universi originem non tractâsse seorsim ab illâ mundi nostri sublunaris : Hæc enim non distinguit populus, aut separatim æstimat.—Rectè igitur Legislator sapientissimus philosophis reliquit id negotii, ut ubi maturuerit ingenium humanum, per ætatem, usum, et observationes, opera Dei alio ordine digererent, perfectionibus divinis atque rerum naturæ adaptato.— *Burnet's Archæologiæ Philosophicæ.* C. viii. p. 306. 4to. 1692.

the discoveries of a science, in which a Kepler,* and a Newton found demonstration of the most sublime and glorious attributes of the Creator A Herschel has pronounced that "Geology, in the magnitude and sublimity of the objects of which it treats, undoubtedly ranks in the scale of sciences next to astronomy;" and the history of the structure of our planet, when it shall be fully understood, must lead to the same great moral results that have followed the study of the mechanism of the heavens; Geology has already

* Kepler concludes one of his astronomical works with the following prayer, which is thus translated in the Christian Observer, Aug. 1834, p. 495.

"It remains only that I should now lift up to heaven my eyes and hands from the table of my pursuits, and humbly and devoutly supplicate the Father of lights. O thou, who by the light of nature dost enkindle in us a desire after the light of grace, that by this thou mayst translate us into the light of glory; I give thee thanks, O Lord and Creator, that thou hast gladdened me by thy creation, when I was enraptured by the work of thy hands. Behold, I have here completed a work of my calling, with as much of intellectual strength as thou hast granted me. I have declared the praise of thy works to the men who will read the evidences of it, so far as my finite spirit could comprehend them in their infinity. My mind endeavoured to its utmost to reach the truth by philosophy; but if any thing unworthy of thee has been taught by me—a worm born and nourished in sin—do thou teach me that I may correct it. Have I been seduced into presumption by the admirable beauty of thy works, or have I sought my own glory among men, in the construction of a work designed for thine honour? O then graciously and mercifully forgive me; and finally grant me this favour, that this work may never be injurious, but may conduce to thy glory and the good of souls."

proved by physical evidence, that the surface of the globe has not existed in its actual state from eternity, but has advanced through a series of creative operations, succeeding one another at long and definite intervals of time; that all the actual combinations of matter have had a prior existence in some other state; and that the ultimate atoms of the material elements, through whatever changes they may have passed, are, and ever have been, governed by laws, as regular and uniform, as those which hold the planets in their course.   All these results entirely accord with the best feelings of our nature, and with our rational conviction of the greatness and goodness of the Creator of the universe; and the reluctance with which evidences, of such high importance to natural theology, have been admitted by many persons, who are sincerely zealous for the interests of religion, can only be explained by their want of accurate information in physical science; and by their ungrounded fears lest natural phenomena should prove inconsistent with the account of creation in the book of Genesis.

It is argued unfairly against Geology, that because its followers are as yet agreed on no complete and incontrovertible theory of the earth; and because early opinions advanced on imperfect evidence have yielded, in succession, to more extensive discoveries; therefore nothing certain is known upon the whole subject; and that all geological deductions must be crude, unauthentic, and conjectural.

It must be candidly admitted that the season has not yet arrived, when a perfect theory of the whole earth can be fixedly and finally established, since we have not yet before us all the facts on which such a theory may eventually be founded; but, in the mean while, we have abundant evidence of numerous and indisputable phenomena, each establishing important and undeniable conclusions; and the aggregate of these conclusions, as they gradually accumulate, will form the basis of future theories, each more and more nearly approximating to perfection; the first, and second, and third story of our edifice may be soundly and solidly constructed; although time must still elapse before the roof and pinnacles of the perfect building can be completed. Admitting therefore, that we have yet much to learn, we contend that much sound knowledge has been already acquired; and we protest against the rejection of established parts, because the whole is not yet made perfect.

It was assuredly prudent, during the infancy of Geology, in the immature state of those physical sciences which form its only sure foundation, not to enter upon any comparison of the Mosaic account of creation with the structure of the earth, then almost totally un-known ; the time was not then come when the knowledge of natural phenomena was suffici-ently advanced to admit of any profitable in-vestigation of this question ; but the discoveries of the last half century have been so extensive in this department of natural knowledge, that, whether we will or not, the subject is now forced upon our consideration, and can no longer es-cape discussion. The truth is, that all observers, however various may be their speculations, re-specting the secondary causes by which geolo-gical phenomena have been brought about, are now agreed in admitting the lapse of very long periods of time to have been an essential condi-tion to the production of these phenomena.

It may therefore be proper, in this part of our enquiry, to consider how far the brief account of creation, contained in the Mosaic narrative, can be shown to accord with those natural phe-nomena, which will come under consideration in the course of the present essay. Indeed some examination of this question seems in-dispensable at the very threshold of an inves-tigation, the subject matter of which will be derived from a series of events, for the most part, long antecedent to the creation of the human species. I trust it may be shown, not only that there is no inconsistency between our interpreta-tion of the phenomena of nature and of the Mosaic narrative, but that the results of geological enquiry throw important light on parts of this history, which are otherwise involved in much obscurity.

If the suggestions I shall venture to propose require some modification of the most commonly received and popular interpretation of the Mosaic narrative, this admission neither involves any impeachment of the authenticity of the text, nor of the judgment of those who have formerly interpreted it otherwise, in the absence of information as to facts which have but recently been brought to light; and if, in this respect, geology should seem to require some little concession from the literal interpreter of scripture, it may fairly be held to afford ample compensation for this demand, by the large additions it has made to the evidences of natural religion, in a department where revelation was not designed to give information.

The disappointment of those who look for a detailed account of geological phenomena in the Bible, rests on a gratuitous expectation of finding therein historical information, respecting all the operations of the Creator in times and places with which the human race has no concern; as reasonably might we object that the Mosaic history is imperfect, because it makes no specific mention of the satellites of Jupiter, or the rings of Saturn, as feel disappointment at

not finding in it the history of geological phenomena, the details of which may be fit matter
for an encyclopedia of science, but are foreign
to the objects of a volume intended only to be
a guide of religious belief and moral conduct.

We may fairly ask of those persons who consider physical science a fit subject for revelation,
what point they can imagine short of a communication of Omniscience, at which such a revelation might have stopped, without imperfections
of omission, less in degree, but similar in kind, to
that which they impute to the existing narrative
of Moses? A revelation of so much only of
astronomy, as was known to Copernicus, would
have seemed imperfect after the discoveries of
Newton; and a revelation of the science of Newton would have appeared defective to La Place:
a revelation of all the chemical knowledge of
the eighteenth century would have been as
deficient in comparison with the information of
the present day, as what is now known in this
science will probably appear before the termination of another age; in the whole circle of
sciences, there is not one to which this argument
may not be extended, until we should require
from revelation a full developement of all the
mysterious agencies that uphold the mechanism of the material world. Such a revelation might indeed be suited to beings of a
more exalted order than mankind, and the at-

tainment of such knowledge of the works as
well as of the ways of God, may perhaps form
some part of our happiness in a future state ; but
unless human nature had been constituted otherwise than it is, the above supposed communication
of omniscience would have been imparted to creatures, utterly incapable of receiving it, under any
past or present moral or physical condition of
the human race; and would have been also at
variance with the design of all God's other disclosures of himself, the end of which has uniformly been, not to impart intellectual but moral
knowledge.

Several hypotheses have been proposed, with a
view of reconciling the phenomena of Geology,
with the brief account of creation which we find
in the Mosaic narrative. Some have attempted
to ascribe the formation of all the stratified
rocks to the effects of the Mosaic Deluge ; an
opinion which is irreconcileable with the enormous thickness and almost infinite subdivisions
of these strata, and with the numerous and regular
successions which they contain of the remains
of animals and vegetables, differing more and
more widely from existing species, as the
strata in which we find them are placed at
greater depths. The fact that a large proportion of these remains belong to extinct genera, and almost all of them to extinct species,
that lived and multiplied and died on or near

the spots where they are now found, shows that the strata in which they occur were deposited slowly and gradually, during long periods of time, and at widely distant intervals. These extinct animals and vegetables could therefore have formed no part of the creation with which we are immediately connected.

It has been supposed by others, that these strata were formed at the bottom of the sea, during the interval between the creation of man and the Mosaic deluge ; and that, at the time of that deluge, portions of the globe which had been previously elevated above the level of the sea, and formed the antediluvian continents, were suddenly submerged ; while the ancient bed of the ocean rose to supply their place. To this hypothesis also, the facts I shall subsequently advance offer insuperable objections.

A third opinion has been suggested, both by learned theologians and by geologists, and on grounds independent of one another ; viz. that the Days of the Mosaic creation need not be understood to imply the same length of time which is now occupied by a single revolution of the globe ; but successive periods, each of great extent : and it has been asserted that the order of succession of the organic remains of a former world, accords with the order of creation recorded in Genesis. This assertion, though to a certain degree apparently correct, is not entirely

G.   C

supported by geological facts; since it appears
that the most ancient marine animals occur in
the same division of the lowest transition strata
with the earliest remains of vegetables; so that
the evidence of organic remains, as far as it
goes, shows the origin of plants and animals to
have been contemporaneous: if any creation of
vegetables preceded that of animals, no evidence
of such an event has yet been discovered by the
researches of geology. Still there is, I believe,
no sound critical, or theological objection, to the
interpretation of the word " day," as meaning a
long period; but there will be no necessity for
such extension, in order to reconcile the text of
Genesis with physical appearances, if it can be
shown that the time indicated by the phenomena
of Geology* may be found in the undefined
interval, following the announcement of the first
verse.

In my inaugural lecture, published at Oxford,
1820, pp. 31, 32, I have stated my opinion in

* A very interesting treatise on the Consistency of Geology
with Sacred History has recently been published at Newhaven,
1833, by Professor Silliman, as a supplement to an American
edition of Bakewell's Geology, 1833. The author contends that
the period alluded to in the first verse of Genesis, " In the beginning," is not necessarily connected with the first day, and that
it may be regarded as standing by itself, and admitting of any
extension backward in time which the facts may seem to require.

He is further disposed to consider the six days of creation as
periods of time of indefinite length, and that the word " day" is
not of necessity limited to twenty-four hours.

favour of the hypothesis, " which supposes the word ' *beginning*,' as applied by Moses in the first verse of the book of Genesis, to express an undefined period of time, which was antecedent to the last great change that affected the surface of the earth, and to the creation of its present animal and vegetable inhabitants; during which period a long series of operations and revolutions may have been going on; which, as they are wholly unconnected with the history of the human race, are passed over in silence by the sacred historian, whose only concern with them was barely to state, that the matter of the universe is not eternal and self-existent, but was originally created by the power of the Almighty."

I have great satisfaction in finding that the view of this subject, which I have here expressed, and have long entertained, is in perfect accordance with the highly valuable opinion of Dr. Chalmers, recorded in the following passages of his Evidence of the Christian Revelation, chap. vii. :—" Does Moses ever say, that when God created the heavens and the earth he did more, at the time alluded to, than transform them out of previously existing materials? Or does he ever say that there was not an interval of many ages between the first act of creation described in the first verse of the Book of Genesis, and said to have been performed at the *beginning*, and those more detailed operations, the account of which commences at the second verse, and which are described to us as having been performed in so many days? Or, finally, does he ever make us to understand that the genealogies of man went any farther than to fix the antiquity of the species, and, of consequence, that they left the antiquity of the globe a free subject for the speculation of philosophers?"

It has long been matter of discussion among learned theologians, whether the first verse of Genesis should be considered prospectively, as containing a summary announcement of that new creation, the details of which follow in the record of the operations of the six successive days; or as an abstract statement that the heaven and earth were made by God, without limiting the period when that creative agency was exerted. The latter of these opinions is in perfect harmony with the discoveries of Geology.

The Mosaic narrative commences with a declaration, that " In the beginning God created the heaven and the earth." These few first words of Genesis may be fairly appealed to by the geologist, as containing a brief statement of the creation of the material elements, at a time distinctly preceding the operations of the first day: it is nowhere affirmed that God created the heaven and the earth in the *first day*, but in the *beginning;* this beginning may have been an epoch at an unmeasured distance, followed by periods of undefined duration, during which all the physical operations disclosed by Geology were going on.

The first verse of Genesis, therefore, seems explicitly to assert the creation of the Universe; " the heaven," including the sidereal systems ;* " and the earth," more especially specifying our own planet, as the subsequent scene of the operations of the six days about to be described : no information is given as to events which may have occurred upon this earth, unconnected with the history of man, between the creation of its component matter recorded in the first verse, and the era at which its history is resumed in the second verse; nor is any limit fixed to the time during which these intermediate events may have been going on : millions of millions of years may have occupied the indefinite interval, between the beginning in which God created the heaven and the earth, and the

* The Hebrew plural word, *shamaim*, Gen. i. 1, translated heaven, means etymologically, the higher regions, all that seems above the earth : as we say, God above, God on high, God in heaven ; meaning thereby to express the presence of the Deity in space distinct from this earth.—E. B. Pusey.

evening or commencement of the first day of the
Mosaic narrative.*

The second verse may describe the condition
of the earth on the evening of this first day; (for in
the Jewish mode of computation used by Moses,

---

* I have much satisfaction in subjoining the following note
by my friend, the Regius Professor of Hebrew in Oxford, as
it enables me to advance the very important sanction of Hebrew
criticism, in support of the interpretations, by which we may reconcile the apparent difficulties arising from geological phenomena,
with the literal interpretation of the first chapter of Genesis.—
" Two opposite errors have, I think, been committed by critics,
with regard to the meaning of the word *bara*, created; the one,
by those who asserted that it *must* in itself signify " created out
of nothing;" the other, by those who endeavoured, by aid of
etymology, to show that it *must* in itself signify " formation out
of existing matter."   In fact, neither is the case; nor am I aware
of any language in which there is a word signifying *necessarily*
" created out of nothing;" as of course, on the other hand, no
word when used of the agency of God would, *in itself*, imply the
previous existence of matter.   Thus the English word, create, by
which *bara* is translated, expresses that the thing created received
its existence from God, without in itself conveying whether God
called that thing into existence *out of nothing*, or no; for our
very addition of the words " out of nothing," shows that the word
creation has not, in itself, that force: nor indeed, when we speak
of ourselves as creatures of God's hand, do we at all mean that
we were *physically* formed out of nothing.   In like manner,
whether *bara* should be paraphrased by " created out of nothing"
(as far as we can comprehend these words), or, " gave a new
and distinct state of existence to a substance already existing,"
must depend upon the context, the circumstances, or what God
has elsewhere revealed, not upon the mere force of the word.
This is plain, from its use in Gen. i. 27, of the creation of man,
who, as we are instructed, chap. ii. 7, was formed out of previously existing matter, the ' dust of the ground.'   The word *bara*

each day is reckoned from the beginning of one evening to the beginning of another evening). This first evening may be considered as the termination of the indefinite time which followed the primeval creation announced in the first

is indeed so far stronger than *asah*, "made," in that *bara* can only be used with reference to God, whereas *asah* may be applied to man. The difference is exactly that which exists in English between the words by which they are rendered, "created" and "made." But this seems to me to belong rather to our mode of conception than to the subject itself; for making, when spoken of with reference to God, is equivalent to creating.

The words accordingly, bara, *created*—asah, *made*—yatsar, *formed*, are used repeatedly by Isaiah, and are also employed by Amos, as equivalent to each other. *Bara* and *asah* express alike a formation of something new (de novo), something whose existence in this new state originated in, and depends entirely upon the will of its creator or maker. Thus God speaks of Himself as the Creator " *boree*" of the Jewish people, *e. g.* Isaiah xliii. 1, 15; and a new event is spoken of under the same term as " a creation," Numb. xvi. 30, English version, " If the Lord make a new thing :" in the margin, Heb. " create a creature."   Again, the Psalmist uses the same word, Ps. civ. 30, when describing the renovation of the face of the earth through the successive generations of living creatures, " Thou sendest forth thy spirit, they are *created;* and thou renewest the face of the earth." The question is popularly treated by Beausobre, Hist. de Manicheisme, tom. ii. lib. 5, c. 4; or, in a better spirit, by Petavius Dogm. Theol. tom. iii. de opificio sex dierum, lib. 1, c. 1, § 8.

After having continually re-read and studied this account, I can come to no other result than that the words "created" and "made" are synonymous, (although the former is to us the stronger of the two), and that, because they are so constantly interchanged; as, Gen. i. ver. 21, " God *created* great whales :" ver. 25, " God *made* the beast of the earth ;" ver. 26; " Let us *make* man ;" ver. 27, " So God *created* man."   At the same time it is

verse, and as the commencement of the first of the six succeeding days, in which the earth was to be fitted up, and peopled in a manner fit for the reception of mankind. We have in this second verse, a distinct mention of earth and waters, as

very probable that bara, "*created*," as being the stronger word, was selected to describe the first production of the heaven and the earth.

The point, however, upon which the interpretation of the first chapter of Genesis appears to me *really* to turn, is, whether the two first verses are merely a summary statement of what is related in detail in the rest of the chapter, and a sort of introduction to it, or whether they contain an account of an act of creation. And this last seems to me to be their true interpretation, first, because there is no other account of the creation of the earth ; secondly, the second verse describes the condition of the earth when so created, and thus prepares for the account of the work of the six days; but if they speak of any creation, it appears to me that this creation " in the beginning" was previous to the six days, because, as you will observe, the creation of each day is preceded by the declaration that God said, or willed, that such things should be (" and God said"), and therefore the very form of the narrative seems to imply that the creation of the first day began when these words are first used, i.e. with the creation of light in ver. 3. The time then of the creation in ver. 1 appears to me not to be defined : we are told only what alone we are concerned with, that all things were made by God. Nor is this any new opinion. Many of the fathers (they are quoted by Petavius, *l. c.* c. 11, § i.—viii.) supposed the two first verses of Genesis to contain an account of a distinct and prior act of creation ; some, as Augustine, Theodoret, and others, that of the creation of matter ; others, that of the elements ; others again (and they the most numerous) imagine that, not these visible heavens, but what they think to be called elsewhere " the highest heavens," the " heaven of heavens," are here spoken of, our visible heavens being related to have been created on the second day. Petavius himself regards the light as the only act of creation of the first day (c. vii. " de opere primæ diei, i. e.

already existing, and involved in darkness;
their condition also is described as a state of
confusion and emptiness, (*tohu bohu*), words
which are usually interpreted by the vague and
indefinite Greek term, "chaos," and which may

luce"), considering the two first verses as a summary of the account of creation which was about to follow, and a general declaration that all things were made by God.

Episcopius again, and others, thought that the creation and
fall of the bad angels took place in the interval here spoken of:
and misplaced as such speculations are, still they seem to show
that it is natural to suppose that a considerable interval may have
taken place between the creation related in the first verse of Genesis and that of which an account is given in the third and following verses. Accordingly, in some old editions of the English
Bible, where there is no division into verses, you actually find a
break at the end of what is now the second verse; and in Lu-
ther's Bible (Wittenburg, 1557) you have in addition the figure 1
placed against the third verse, as being the beginning of the
account of the creation on the first day.

This then is just the sort of confirmation which one wished for,
because, though one would shrink from the impiety of bending
the language of God's book to any other than its obvious meaning, we can not help fearing lest we might be unconsciously
influenced by the floating opinions of our own day, and therefore
turn the more anxiously to those who explained Holy Scripture,
before these theories existed. You must allow me to add that I
would not define further. We know nothing of creation, nothing
of ultimate *causes*, nothing of space, except what is bounded by
actual existing bodies, nothing of time, but what is limited by the
revolution of those bodies. I should be very sorry to appear to
dogmatize upon that, of which it requires very little reflection, or
reverence, to confess that we are necessarily ignorant. "Hardly
do we guess aright of things that are upon earth, and with labour
do we find the things that are before us; but the things that are in
heaven who hath searched out?"—Wisdom, ix. 16.—E. B. Pusey.

be geologically considered as designating the wreck and ruins of a former world. At this intermediate point of time, the preceding undefined geological periods had terminated, a new series of events commenced, and the work of the first morning of this new creation was the calling forth of light from a temporary darkness, which had overspread the ruins of the ancient earth.*

We have further mention of this ancient earth and ancient sea in the ninth verse, in which the waters are commanded to be *gathered together* into one place, and the dry land to *appear;* this dry land being the same earth whose material creation had been announced in the first verse, and whose temporary submersion and temporary darkness are described in the second verse; the *appearance* of the land and the *gathering together* of the waters are the only facts affirmed respecting them in the ninth verse, but neither land nor waters are said to have been *created* on the third day.

A similar interpretation may be given of the fourteenth and four succeeding verses; what is

---

* I learn from Professor Pusey that the words " *let there be light*," *yehi or*, Gen. i. 3, by no means necessarily imply, any more than the English words by which they are translated, that light had *never* existed before. They may speak only of the substitution of light for darkness upon the surface of this, our planet: whether light had existed before in other parts of God's creation, or had existed upon this earth, before the darkness described in v. 2, is foreign to the purpose of the narrative.

herein stated of the celestial luminaries seems to
be spoken solely with reference to our planet,
and more especially to the human race, then
about to be placed upon it. We are not told that
the substance of the sun and moon were first
called into existence upon the fourth day :*
the text may equally imply that these bodies were
then prepared, and appointed to certain offices,
of high importance to mankind; " to give light
upon the earth, and to rule over the day,
and over the night," " to be for signs, and for
seasons, and for days, and for years."   The fact
of their creation had been stated before in the
first verse. The stars also are mentioned (Gen. i.
16) in three words only, almost parenthetically ;
as if for the sole purpose of announcing, that
they also were made by the same Power, as
those luminaries which are more important to
us, the sun and moon.† This very slight
notice of the countless host of celestial bodies,
all of which are probably suns, the centres
of other planetary systems, whilst our little
satellite, the moon, is mentioned as next in importance to the sun, shows clearly that astronomical phenomena are here spoken of only
according to their relative importance to our
earth, and to mankind, and without any regard

* See notes, p. 22 and p. 26.

† The literal translation of the words *veeth haccocabim*, is,
" And the stars."—E. B. Pusey.

to their real importance in the boundless universe. It seems impossible to include the fixed stars among those bodies which are said (Gen. i. v. 17.) to have been set in the firmament of the heaven to give light upon the earth; since without the aid of telescopes, by far the greater number of them are invisible. The same principle seems to pervade the description of creation which concerns our planet: the creation of its component matter having been announced in the first verse, the phenomena of Geology, like those of astronomy, are passed over in silence, and the narrative proceeds at once to details of the actual creation which have more immediate reference to man.*

* The following observations by Bishop Gleig (though, at the time of writing them, he was not entirely convinced of the reality of facts announced by geological discoveries) show his opinion of the facility of so interpreting the Mosaic account of creation, as to admit of an indefinite lapse of time prior to the existence of the human race.

" I am indeed strongly inclined to believe that the *matter* of the corporeal universe was all created at once, though different portions of it may have been reduced to *form* at very different periods ; *when* the universe was created, or how long the solar system remained in a chaotic state are vain enquiries, to which no answer can be given. Moses records the history of the earth only in its present state ; he affirms, indeed, that it was *created*, and that it was without form and void, when the spirit of God began to move on the surface of the fluid mass ; but, he does not say how long that mass had been in the state of chaos, or whether it was, or was not the wreck of some former system, which had been inhabited by living creatures of different kinds from those which occupy the present. I say this, not to meet

The interpretation here proposed seems more-over to solve the difficulty, which would otherwise attend the statement of the appearance of light upon the first day, whilst the sun and moon and stars are not made to appear until the fourth. If we suppose all the heavenly bodies, and the earth, to have been created at the indefi-nitely distant time, designated by the word be-ginning, and that the darkness described on the evening of the first day, was a temporary dark-ness, produced by an accumulation of dense va-

the objection which has sometimes been urged against the Mo-saic cosmogony, from its representing the works of creation as being no more than six or seven thousand years old, for Moses gives no such representation of the age of those works. How-ever distant the period may be, and it is probably very distant, when God created the heavens and the earth ; there has been a time when it was not distant one year, one day, or one hour. Those, therefore, who contend that the glory of the Almighty God manifested in his works, cannot be limited to the short period of six or seven thousand years, are not aware that the same objection may be made to the longest period which can possibly be conceived by the mind of man. No assignable quantity of successive duration bears any proportion to eternity, and though we should suppose the corporeal universe to have been created six millions or six hundred millions of years ago, a caviller might still say, and with equal reason, that the glory of Almighty God manifested in his works cannot be so limited. It is not to silence such objections as this, that I have admitted the existence of a former earth and visible heavens to be not incon-sistent with the cosmogony of Moses, or indeed with any other part of scripture, but only to prevent the faith of the pious reader from being unsettled by the discoveries, whether real or pretended, of our modern geologists. If these philosophers have

pours "upon the face of the deep;" an incipient dispersion of these vapours may have readmitted light to the earth, upon the first day, whilst the exciting cause of light was still obscured ; and the further purification of the atmosphere, upon the fourth day, may have caused the sun and moon and stars to reappear in the firmament of heaven, to assume their new relations to the newly modified earth, and to the human race.*

We have evidence of the presence of light during long and distant periods of time, in which

really discovered fossil bones that must have belonged to species and genera of animals, which now no where exist, either on the earth or in the ocean, and if the destruction of these genera or species cannot be accounted for by the general deluge, or any other catastrophe to which we know, from authentic history, that our globe has been actually subjected, or if it be a fact, that towards the surface of the earth are found strata, which could not have been so disposed as they are, but by the sea, or at least some watery mass remaining over them in a state of tranquillity, for a much longer period than the duration of Noah's flood ; if these things be indeed well ascertained, of which I am however by no means convinced, there is nothing in the sacred writings forbidding us to suppose that they are the ruins of a former earth, deposited in the chaotic mass of which Moses informed us that God formed the present system. His history, as far as it comes down, is the history of the present earth, and of the primeval ancestors of its present inhabitants ; and one of the most scientific and ingenious of geologists has clearly proved,† that the human race cannot be much more ancient than it appears to be in the writings of the Hebrew lawgiver."—*Stackhouse's Bible, by Bishop Gleig*, p. 6, 7, 1816.

  * See note, p. 26.
  † See Cuvier's Essay on the Theory of the Earth.

the many extinct fossil forms of animal life
succeeded one another upon the early surface of
the globe : this evidence consists in the petrified
remains of eyes of animals, found in geological
formations of various ages. In a future chapter
I shall show, that the eyes of Trilobites, which
are preserved in strata of the transition formation, (Pl. 45, Figs. 9, 10, 11), were constructed
in a manner so closely resembling those of existing crustacea ; and that the eyes of Ichthyosauri, in the lias, (Pl. 10, Figs. 1, 2), contained
an apparatus, so like one in the eyes of many
birds, as to leave no doubt that these fossil eyes
were optical instruments, calculated to receive, in
the same manner, impressions of the same light,
which conveys the perception of sight to living
animals. This conclusion is further confirmed
by the general fact, that the heads of all fossil
fishes and fossil reptiles, in every geological formation, are furnished with cavities for the reception of eyes, and with perforations for the
passage of optic nerves, although the cases are
rare, in which any part of the eye itself has been
preserved. The influence of light is also so
necessary to the growth of existing vegetables,
that we cannot but infer, that it was equally
essential to the development of the numerous
fossil species of the vegetable kingdom, which
are coextensive and coeval with the remains of
fossil animals.

It appears highly probable from recent discoveries,* that light is not a material substance, but
only an effect of undulations of ether; that this
infinitely subtle and elastic ether pervades all
space, and even the interior of all bodies; so
long as it remains at rest, there is total darkness; when it is put into a peculiar state of
vibration, the sensation of light is produced:
this vibration may be excited by various causes;
e. g. by the sun, by the stars, by electricity, combustion, &c. If then light be not a substance,
but only a series of vibrations of ether, i. e.
an effect produced on a subtile fluid, by the
excitement of one or many extraneous causes, it
can hardly be said, nor is it said, in Gen. i. 3, to
have been *created*,† though it may be literally
said to be called into action.

Lastly, in the reference made in the Fourth
Commandment, Exod. xx. 11, to the six days of
the Mosaic creation, the word *asah*, " made," is
the same which is used in Gen. i. 7, and Gen. i.
16, and which has been shown to be less strong
and less comprehensive than *bara*, " created;"
and as it by no means necessarily implies creation out of nothing, it may be here employed

* For a general statement of the undulatory theory of light,
see Sir John Herschell, art. Light, part iii. sec. 2. Encyc.
Metropol. See also Professor Airy's Mathematical Tracts, 2nd
edit. 1831, p. 249; and Mrs. Somerville's Connexion of the
Physical Sciences, 1834, p. 185.

† See Note, p. 26.

to express a new arrangement of materials that existed before.*

After all, it should be recollected that the question is not respecting the correctness of the Mosaic narrative, but of our interpretation of it; and still further, it should be borne in mind that the object of this account was, not to state *in what manner*, but *by whom*, the world was made. As the prevailing tendency of men in those early days was to worship the most glorious objects of nature, namely, the sun and moon and stars; it should seem to have been one important point in the Mosaic account of creation, to guard the Israelites against the Polytheism and idolatry of the nations around them; by announcing that all these magnificent celestial bodies were no Gods, but the works of One Almighty Creator, to whom alone the worship of mankind is due.†

* See Note, p. 22.

† Having thus far ventured to enter into a series of explanations, which I think will reconcile even the letter of the text of Genesis with the phenomena of Geology, I forbear to say more on this important subject, and have much satisfaction in being able to refer my readers to some admirable articles in the Christian Observer (May, June, July, August, 1834) for a very able and comprehensive summary of the present state of this question; explaining the difficulties with which it is surrounded, and offering many temperate and judicious suggestions, as to the spirit in which investigations of this kind ought to be conducted. I would also refer to Bishop Horsley's Sermons, 8vo. 1816, vol. iii. ser. 39; to Bishop Bird Sumner's Records of Creation, vol. ii. p. 356; Douglas's Errors regarding Religion, 1830, p. 261-264,

G.   D

CHAPTER III.

Proper Subjects of Geological Enquiry.

THE history of the earth forms a large and
complex subject of enquiry, divisible at its out-
set, into two distinct branches; the first, comprehending the history of unorganized mineral

Higgins on the Mosaical and Mineral Geologies, 1832; and
more especially to Professor Sedgwick's eloquent and admirable
discourse on the Studies of the University of Cambridge, 1833, in
which he has most ably pointed out the relations which Geology
bears to natural religion, and thus sums up his valuable opinion
as to the kind of information we ought to look for in the Bible:
" The Bible instructs us that man, and other living things, have
been placed but a few years upon the earth; and the physical
monuments of the world bear witness to the same truth: if the
astronomer tells us of myriads of worlds not spoken of in the
sacred records; the geologist, in like manner, proves (not by
arguments from analogy, but by the incontrovertible evidence of
physical phenomena) that there were former conditions of our
planet, separated from each other by vast intervals of time,
during which man, and the other creatures of his own date, had
not been called into being.   Periods such as these belong not,
therefore, to the moral history of our race, and come neither
within the letter nor the spirit of revelation.   Between the first
creation of the earth and that day in which it pleased God
to place man upon it, who shall dare to define the interval?
On this question scripture is silent, but that silence destroys not
the meaning of those physical monuments of his power that God
has put before our eyes, giving us at the same time faculties
whereby we may interpret them and comprehend their meaning."

matter, and of the various changes through which it has advanced, from the creation of its component elements to its actual condition ; the second, embracing the past history of the animal and vegetable kingdoms, and the successive modifications which these two great departments of nature have undergone, during the chemical and mechanical operations that have affected the surface of our planet. As the study of both these branches forms the subject of the science of Geology, it is no less important to examine the nature and action of the physical forces, that have affected unorganized mineral bodies, than to investigate the laws of life, and varied conditions of organization, that prevailed while the crust of our globe was in process of formation.

Before we enter on the history of fossil animals and vegetables, we must therefore first briefly review the progressive stages of mineral formations; and see how far we can discover in the chemical constitution, and mechanical arrangement of the materials of the earth, proofs of general prospective adaptation to the economy of animal and vegetable life.

As far as our planet is concerned, the first act of creation seems to have consisted in giving origin to the elements of the material world, These inorganic elements appear to have received no subsequent addition to their number,

and to have undergone no alteration in their
nature and qualities; but to have been submitted at their creation to the self same laws that
regulate their actual condition, and to have continued subject to these laws during every succeeding period of geological change. The same
elements also which enter the composition of
existing animals and plants, appear to have
performed similar functions in the economy of
many successive animal and vegetable creations.

In tracing the history of these natural phenomena we enter at once into the consideration
of Geological Dynamics, including the nature
and mode of operation of all kinds of physical
agents, that have at any time, and in any
manner, affected the surface and interior of the
earth. In the foremost rank of these agents, we
find Fire and Water,—those two universal and
mighty antagonizing forces, which have most
materially influenced the condition of the globe;
and which man also has converted into the most
efficient instruments of his power, and obedient
auxiliaries of his mechanical and chemical and
culinary operations.

The state of the ingredients of crystalline
rocks has, in a great degree, been influenced
by chemical and electro-magnetic forces; whilst
that of stratified sedimentary deposits has resulted chiefly from the mechanical action of
moving water, and has occasionally been modi-

fied by large admixtures of animal and vegetable remains.

As the action of all these forces will be rendered most intelligible by examples of their effects, I at once refer my readers for a synoptic view of them, to the section which forms the first of my series of plates.* The object of this section is, first, to represent the order in which the successive series of stratified formations are piled on one another, almost like courses of masonry; secondly, to mark the changes that occur in their mineral and mechanical condition; thirdly, to show the manner in which all stratified rocks have at various periods been disturbed, by the intrusion of unstratified crystalline rocks; and variously affected by elevations, depressions, fractures, and dislocations; fourthly, to give examples of the alterations in the forms of animal and vegetable life, that have accompanied these changes of the mineral conditions of the earth.

From the above section it appears that there are eight distinct varieties of the crystalline unstratified rocks, and twenty-eight well defined divisions of the stratified formations. Taking the average maximum thickness of each of these divisions, at one thousand feet,† we should have

* The detailed explanation of this section is given in the description of the plates in vol. ii.

† Many formations greatly exceed, whilst others fall short, of the average here taken.

a total amount of more than five miles; but as
the transition and primary strata very much
exceed this average, the aggregate of all the
European stratified series may be considered to
be at least ten miles.

Chapter IV.

Relation of Unstratified to Stratified Rocks.

I SHALL enter into no further details respecting
the component members of each group of stratified rocks, than are represented by the lines of
division and colours upon the section.* They
are arranged under the old divisions of *primary,
transition, secondary,* and *tertiary* series, more

---

* For particular information respecting the mineral character
and organic remains of the strata composing each series, I must
refer to the numerous publications that have been devoted to
these subjects. A most convenient summary of the contents of
these publications will be found in De La Beche's Manual of
Geology, and in Von Meyer's Palæologia, (Frankfurt, 1832);
ample details respecting the English strata are given in Conybeare and Phillips's Geology of England and Wales. See also
Bakewell's Introduction to Geology, 1833; and Professor Phillips's article Geology, in the Encyclopædia Metropolitana; also
Professor Phillips's Guide to Geology, 8vo. 1834; and De La
Beche's Researches in Theoretical Geology, 8vo. 1834. The
history of the organic remains of the tertiary period has been
most ably elucidated in Lyell's Principles of Geology.

from a sense of the convenience of this long received arrangement, than from the reality of any strongly defined boundaries by which the strata, on the confines of each series, are separated from one another.

As the materials of stratified rocks are in great degree derived, directly or indirectly, from those which are unstratified,* it will be premature to enter upon the consideration of derivative strata, until we have considered briefly the history of the primitive formations. We therefore commence our inquiry at that most ancient period, when there is much evidence to render it probable that the entire materials of the globe were in a fluid state, and that the cause of this fluidity was heat. The form of the earth being that of an oblate spheroid, compressed at the poles, and enlarged at the equator, is that which a fluid mass would assume from revolution round its axis. The further fact, that the shortest diameter coincides with the existing axis

---

* In speaking of crystalline rocks of supposed igneous origin as unstratified, we adopt a distribution which, though not strictly accurate, has long been in general use among geologists. Ejected masses of granite, basalt, and lava have frequently horizontal partings, dividing them into beds of various extent and thickness, such as those which are most remarkable in what the Wernerians have called the Floetz trap formation, Pl. 1, section Fig. 6.; but they do not present that subdivision into successions of small beds, and still smaller laminæ, which usually exists in sedimentary strata that have been deposited by the action of water.

of rotation, shows that this axis has been the same ever since the crust of the earth attained its present solid form.

Assuming that the whole materials of the globe may have once been in a fluid, or even a nebular state,* from the presence of intense heat, the passage of the first consolidated portions of this fluid, or nebulous matter, to a solid state may have been produced by the radiation of heat from its surface into space; the gradual abstraction of such heat would allow the particles of matter to approximate and crystallize; and the first result of this crystallization might have been the formation of a shell or crust, composed of oxidated metals and metalloids, constituting various rocks of the granitic series, around an incandescent nucleus, of melted matter, heavier than granite; such as forms the more weighty substance of basalt and compact lava.

It is now unnecessary to dwell on controversies which have prevailed during the last half century, respecting the origin of this large and important class of unstratified crystalline rocks,

* The nebular hypothesis offers the most simple, and therefore the most probable theory, respecting the first condition of the material elements that compose our solar system. Mr. Whewell has shown how far this theory, supposing it to be established, would tend to exalt our conviction of the prior existence of some presiding Intelligence.—Bridgewater Treatises, No. III. Chap. vii.

which the common consent of nearly all modern geologists and chemists refers to the action of fire. The agency of central heat, and the admission of water to the metalloid bases of the earths and alkalis, offer two causes which, taken singly or conjointly, seem to explain the production and state of the mineral ingredients of these rocks; and to account for many of the grand mechanical movements that have affected the crust of the globe.

The gradations are innumerable, which connect the infinite varieties of granite, syenite, porphyry, greenstone, and basalt with the trachytic porphyries and lavas that are at this day ejected by volcanos. Although there still remain some difficulties to be explained, there is little doubt that the fluid condition in which all unstratified crystalline rocks originally existed, was owing to the solvent power of heat; a power whose effect in melting the most solid materials of the earth we witness in the fusion of the hardest metals, and of the flinty materials of glass.*

* The experiments of Mr. Gregory Watt on bodies cooled slowly after fusion; and of Sir James Hall, on reproducing artificial crystalline rocks, from the pounded ingredients of the same rocks highly heated under strong pressure; and the more recent experiments of Professor Mitscherlich, on the production of artificial crystals, by fusion of definite proportions of their component elements, have removed many of the objections, which were once urged against the igneous origin of crystalline rocks.

Beneath the whole series of stratified rocks that appear on the surface of the globe (see section Pl. 1), there probably exists a foundation of unstratified crystalline rocks; bearing an irregular surface, from the detritus of which the materials of stratified rocks have in great measure been derived,* amounting, as we have stated, to a thickness of many miles. This is indeed but a small depth, in comparison with the diameter of the globe; but small as it is, it affords certain evidence of a long series of changes and revolutions; affecting not only the mineral condition of the nascent surface of the earth, but attended also by important alterations in animal and vegetable life.

The detritus of the first dry lands, being drifted into the sea, and there spread out into extensive beds of mud and sand and gravel, would for ever have remained beneath the surface of the water, had not other forces been subsequently employed to raise them into dry land : these forces appear to have been the same expansive powers of heat and vapour which, having caused the elevation of the first raised portions of the fundamental crystalline rocks,

* Either directly, by the accumulation of the ingredients of disintegrated granitic rocks; or indirectly, by the repeated destruction of different classes of stratified rocks, the materials of which had, by prior operations, been derived from unstratified formations.

continued their energies through all succeeding geological epochs, and still exert them in producing the phenomena of active volcanoes; phenomena incomparably the most violent that now appear upon the surface of our planet.*

The evidence of design in the employment of forces, which have thus effected a grand general purpose, viz. that of forming dry land, by elevating strata from beneath the waters in which they were deposited, stands independent of the truth or error of contending theories, respecting the origin of that most ancient class of stratified rocks, which are destitute of organic remains (see pl. 1.—section 1, 2, 3, 4, 5, 6, 7). It is

* " The fact of great and frequent alteration in the relative level of the sea and land is so well established, that the only remaining questions regard the mode in which these alterations have been effected, whether by elevation of the land itself, or subsidence in the level of the sea? And the nature of the force which has produced them? The evidence in proof of great and frequent movements of the land itself, both by protrusion and subsidence, and of the connection of these movements with the operations of volcanos, is so various and so strong, derived from so many different quarters on the surface of the globe, and every day so much extended by recent enquiry, as almost to demonstrate that these have been the causes by which those great revolutions were effected; and that although the action of the inward forces which protrude the land has varied greatly in different countries, and at different periods, they are now and ever have been incessantly at work in operating present change and preparing the way for future alteration in the exterior of the globe."—Geological Sketch of the Vicinity of Hastings, by Dr. Fitton, pp. 85, 86.

immaterial to the present question, whether they were formed (according to the theory of Hutton) from the detritus of the earlier granitic rocks, spread forth by water into beds of clay and sand; and subsequently modified by heat: or whether they have been produced, (as was maintained by Werner) by chemical precipitation from a fluid, having other powers of solution than those possessed by the waters of the present ocean. It is of little importance to our present purpose, whether the non-appearance of animals and vegetables in these most ancient strata was caused by the high temperature of the waters of the ocean, in which they were mechanically deposited; or by the compound nature and uninhabitable condition of a primeval fluid, holding their materials in solution. All observers admit that the strata were formed beneath the water, and have been subsequently converted into dry land: and whatever may have been the agents that caused the movements of the gross unorganized materials of the globe; we find sufficient evidence of prospective wisdom and design, in the benefits resulting from these obscure and distant revolutions, to future races of terrestrial creatures, and more especially to Man.*

* In describing geological phenomena, it is impossible to avoid the use of theoretical terms, and the provisional adoption of many theoretical opinions as to the manner in which these phenomena have been produced. From among the various and

In unstratified crystalline rocks, wholly destitute of animal or vegetable remains, we search in vain for those most obvious evidences of contrivance, which commence with the first traces of organic life, in strata of the transition period ; the chief agencies which these rocks indicate, are those of fire and water; and yet even here we find proof of system and intention, in the purpose which they have accomplished, of supplying and accumulating at the bottom of the water the materials of stratified formations, which, in after times, were to be elevated into dry lands, in an ameliorated condition of fertility. Still more decisive are the evidences of design and method, which arise from the consideration of the

conflicting theories that have been proposed to explain the most difficult and complicated problems of Geology, I select those which appear to carry with them the highest degree of probability; but as results remain the same from whatever cause they have originated, the force of inferences from these results will be unaffected by changes that may arise in our opinions as to the physical causes by which these have been produced. As in estimating the merits of the highest productions of human art it is not requisite to understand perfectly the nature of the machinery by which the work has been effected in order to appreciate the skill and talent of the artist by whom it was contrived ; so our minds may be fully impressed with a perception of the magnificent results of creative intelligence, which are visible in the phenomena of nature, although we can but partially comprehend the mechanism that has been instrumental to their production; and although the full developement of the workings of the material instruments by which they were effected, has not yet been, and perhaps may never be, vouchsafed to the prying curiosity of man.

structure and composition of their crystalline mineral ingredients. In every particle of matter to which crystallization has been applied, we recognize the action of those undeviating laws of polar forces, and chemical affinity, which have given to all crystallized bodies a series of fixt definite forms and definite compositions. Such universal prevalence of law, method, and order assuredly attests the agency of some presiding and controlling mind.

A further argument, which will be more insisted on in speaking on the subject of metallic veins, may be founded on the dispensation whereby the primary and transition rocks are made the principal repositories of many valuable metals, which are of such peculiar and indispensable importance to mankind.

CHAPTER V.

Volcanic Rocks, Basalt, and Trap.

In the state of tranquil equilibrium which our planet has attained in the region we inhabit, we are apt to regard the foundation of the solid earth, as an emblem of duration and stability. Very different are the feelings of those whose lot is cast near the foci of volcanic eruptions ; to them the earth affords no stable resting place, but during the paroxysms of volcanic activity, reels to and fro, and vibrates beneath their feet; overthrowing cities, yawning with dreadful chasms, converting seas into dry lands, and dry lands into seas. (See Lyell's Geology, vol. i. Passim.)

To the inhabitants of such districts we speak a language which they fully comprehend, when we describe the crust of the globe as floating on an internal nucleus of molten elements; they have seen these molten elements burst forth in liquid streams of lava; they have felt the earth beneath them quivering and rolling, as if upon the billows of a subterranean sea; they have seen mountains raised and valleys depressed, almost in an instant of time; they can duly appreciate, from sensible experience, the force of the terms in which geologists describe the tremulous throes, and convulsive agitations of the earth; during the passage of its strata from the bottom of the seas, in which they received their origin, to the plains and mountains in which they find their present place of rest.

We see that the streams of earthy matter, which issue in a state of fusion from active volcanos, are spread around their craters in sheets of many kinds of lava; some of these so much resemble beds of basalt, and various trap rocks, that occur in districts remote from any existing volcanic vent, as to render it probable that the latter also have been poured forth from the interior of the earth. We further find the rocks adjacent to volcanic craters, intersected by rents and fissures, which have been filled with injections of more recent lava, forming transverse walls or dykes. Similar dykes occur not only in districts occupied by basalt and trap rocks, at a distance from the site of any modern volcanic activity; but also in strata of every formation, from the most ancient primary, to the most recent tertiary (see Plate 1. section f 1—f 8. h 1—h 2. i 1—i 5): and as the mineral characters of these dykes present insensible gradations, from a state of compact lava, through infinite varieties of greenstone, serpentine, and porphyry to granite, we refer them all to a common igneous origin.

The sources from which the matter of these ejected rocks ascends are deeply seated beneath the granite; but it is not yet decided whether the immediate cause of an eruption be the access of water to local accumulations of the metalloid bases of the earths and alkalies; or whether lava be derived directly from that general mass of incandescent elements, which may probably exist at a depth of about one hundred miles beneath the surface of our planet.*

* See chapter on the internal temperature of the earth.

Our section shows how closely the results of volcanic forces now in action are connected, both with the phenomena of basaltic formations, and also with the more ancient eruptions of greenstone, porphyry, syenite, and granite. The intrusion both of dykes and irregular beds of un-stratified crystalline matter, into rocks, of every age and every formation, all proceeding upwards from an unknown depth, and often accumulated into vast masses overlying the surface of strati-fied rocks, are phenomena coextensive with the globe.

Throughout all these operations, however tur-bulent and apparently irregular, we see ultimate proofs of method and design, evinced by the uniformity of the laws of matter and motion, which have ever regulated the chemical and mechanical forces by which such grand effects have been produced. If we view their aggre-gate results, in causing the elevation of land from beneath the sea, we shall find that volcanic forces assume a place of the highest importance, among the second causes which have influenced the past, as well as the present condition of the globe ; each individual movement has contri-buted its share towards the final object, of con-ducting the molten materials of an uninhabit-able planet, through long successions of change and of convulsive movements, to a tranquil state

G.   E

of equilibrium; in which it has become the
convenient and delightful habitation of man, and
of the multitudes of terrestrial creatures that are
his fellow tenants of its actual surface.*

Chapter VI.

Primary Stratified Rocks.

In the summary we have given of the leading
phenomena of unstratified and volcanic rocks, we
have unavoidably been led into theoretical speculations, and have seen that the most probable
explanation of these phenomena is found in the
hypothesis of the original fluidity of the entire
materials of the earth, caused by the presence of
intense heat. From this fluid mass of metals,
and metalloid bases of the earths, and alkalies,
the first granitic crust appears to have been
formed, by oxydation of these bases; and subsequently broken into fragments, disposed at
unequal levels above and below the surface of
the first formed seas.

Wherever solid matter arose above the water,
it became exposed to destruction by atmospheric

* See further details respecting the effects of volcanic forces
in the description of Pl. I. Vol. ii.

agents; by rains, torrents, and inundations; at that time probably acting with intense violence, and washing down and spreading forth, in the form of mud and sand and gravel, upon the bottom of the then existing seas, the materials of primary stratified rocks, which, by subsequent exposure to various degrees of subterranean heat, became converted into beds of gneiss, and mica slate, and hornblende slate, and clay slate. In the detritus thus swept from the earliest lands into the most ancient seas, we view the commencement of that enormous series of derivative strata which, by long continued repetition of similar processes, have been accumulated to a thickness of many miles.*

The total absence of organic remains through-

---

* Mr. Conybeare (in his admirable Report on Geology to the British Association for the Advancement of Science, 1832, p. 367) shows, that many of the most important principles of the igneous theory, which has been almost demonstrated by modern discoveries, had been anticipated by the universal Leibnitz. " In the fourth section of his Protogæa, Leibnitz presents us with a masterly sketch of his general views, and, perhaps, even in the present day, it would be difficult to lay down more clearly the fundamental positions which must be necessarily common to every theory, attributing geological phenomena in great measure to central igneous agency. He attributes the primary and fundamental rocks to the refrigeration of the crust of this volcanic nucleus; an assumption which well accords with the now almost universally admitted igneous origin of the fundamental granite, and with the structure of the primitive slates, for the insensible gradation of these formations appears to prove that gneiss must have undergone in a greater,

out those lowest portions of these strata, which
have been called primary, is a fact consistént

and mica slate in a less degree, the same action of which the
maximum intensity produced granite.

" The dislocations and deranged position of the strata he
attributes to the breaking in of vast vaults, which the vesicular
and cavernous structure assumed by masses, during their refrigeration from a state of fusion must necessarily have occasioned
in the crust, thus cooling down and consolidated. He assigns
the weight of the materials and the eruption of elastic vapours as
the concurrent causes of these disruptions; to which we should
perhaps add, that the oscillations of the surface of the still fluid
nucleus may, independently of any such cavities, have readily
shattered into fragments the refrigerated portion of the crust;
especially, as at this early period, it must have been necessarily
very thin, and resembling chiefly the scoriæ floating on a surface
of lava just beginning to cool. He justly adds, that these disruptions of the crust must, from the disturbances communicated
to the incumbent waters, have been necessarily attended with
diluvial action on the largest scale. When these waters had
subsequently, in the intervals of quiescence between these
convulsions, deposited the materials first acquired by their force
of attrition, these sediments formed, by their consolidation,
various stony and earthy strata. Thus, he observes, we may
recognise a double origin of the rocky masses, the one by
refrigeration from igneous fusion, (which, as we have seen, he
considered principally to be assignable to the primary and
fundamental rocks,) the other by concretion from aqueous
solution. We have here distinctly stated the great basis
of every scientific classification of rock formations. By the
repetition of similar causes (i. e. disruption of the crust and consequent inundations) frequent alternations of new strata were
produced, until at length these causes having been reduced to a
condition of quiescent equilibrium, a more permanent state of
things emerged. Have we not here clearly indicated the data
on which, what may be termed the chronological investigation of
the series of geological phenomena, must ever proceed?"

with the hypothesis which forms part of the theory of gradual refrigeration; viz. that the waters of the first formed oceans were too much heated to have been habitable by any kind of organic beings.*

In these most ancient conditions, both of land and water, Geology refers us to a state of things incompatible with the existence of animal and vegetable life; and thus on the evidence of natural phenomena, establishes the important fact that we find a starting point, on this side of which all forms, both of animal and vegetable beings, must have had a beginning.

As, in the consideration of other strata, we find abundant evidence in the *presence* of organic remains, in proof of the exercise of creative power, and wisdom, and goodness, attending the progress of life, through all its stages of advancement upon the surface of the globe; so, from the *absence* of organic remains in the primary strata, we may derive an important argument, showing that there was a point of time in the history of our planet, (which no other researches but those of geology can possibly approach,) antecedent to the beginning of either animal or vegetable life. This

---

* So long as the temperature of the earth continued intensely high, water.could have existed only in the state of steam or vapour, floating in the atmosphere around the incandescent surface.

conclusion is the more important, because it has
been the refuge of some speculative philosophers
to refer the origin of existing organizations, either
to an external succession of the same species,
or to the formation of more recent from more
ancient species, by successive developments,
without the interposition of direct and repeated
acts of creation ; and thus to deny the existence
of any first term, in the infinite series of successions which this hypothesis assumes. Against
this theory, no decisive evidence has been accessible, until the modern discoveries of geology
had established two conclusions of the highest
value in relation to this long disputed question :
the first proving, that existing species have had
a beginning ; and this at a period comparatively
recent in the physical history of our globe : the
second showing that they were preceded by
several other systems of animal and vegetable life, respecting each of which it may
no less be proved, that there was a time when
their existence had not commenced ; and that
to these more ancient systems also, the doctrine
of eternal succession, both retrospective and
prospective, is equally inapplicable.*

* Mr. Lyell, in the four first chapters of the second volume of
his Principles of Geology, has very ably and candidly examined
the arguments that have been advanced in support of the doctrine of transmutation of species, and arrives at the conclusion,
—" that species have a real existence in nature, and that each

Having this evidence both of the beginning and end of several systems of organized life, each affording internal proof of the repeated exercise of creative design, and wisdom, and power, we are at length conducted back to a period anterior to the earliest of these systems; a period in which we find a series of primary strata, wholly destitute of organic remains; and from this circumstance, we infer their deposition to have preceded the commencement of organic life. Those who contend that life may have existed during the formation of the primary strata, and the animal remains have been obliterated by the effects of heat, on strata nearest to the granite, do but remove to one point further backwards the first term of the finite series of organic beings; and there still remains beyond this point an antecedent period, in which a state of total fusion pervaded the entire materials of the fundamental granite; and one universal mass of incan-

was endowed, at the time of its creation, with the attributes and organization by which it is now distinguished."

Mr. Dela Beche also says (Geological Researches, 1834, p. 239, 1st edit. 8vo.) " There can be no doubt that many plants can adapt themselves to altered conditions, and many animals accommodate themselves to different climates; but when we view the subject generally, and allow full importance to numerous exceptions, terrestrial plants and animals seem intended to fill the situations they occupy, as these were fitted for them; they appear created as the conditions arose, the latter not causing a modification in previously existing forms productive of new species."

descent elements, wholly incompatible with any
condition of life, which can be shown to have
ever existed, formed the entire substance of the
globe.*

* In adopting the hypothesis that the primary stratified rocks
have been altered and indurated by subjacent heat, it should be
understood, that although heat is in this case referred to as one
cause of the consolidation of strata, there are other causes which
have operated largely to consolidate the secondary and tertiary
strata, which are placed at a distance above rocks of igneous
origin. Although many kinds of limestone may have been in
certain cases converted to crystalline marble, by the action of heat
under high pressure, there is no need for appealing to such
agency to explain the consolidation of ordinary strata of carbonate
of lime; beds of secondary and tertiary sandstone have often a
calcareous cement, which may have been precipitated from
water, like the substance of stalactites and ordinary limestone.
When their cement is siliceous, it may also have been supplied
by some humid process, analogous to that by which the siliceous
matter of chalcedony and of quartz is either suspended or dissolved in nature; a process, the existence of which we cannot
deny, although it has yet baffled all the art of chemistry to imi-
tate it. The beds of clay which alternate with limestone, and
sand, or sandstone, in secondary and tertiary formations, show no
indications of the action of heat; having undergone no greater
consolidation than may be referred to pressure, or to the admixture of certain proportions of carbonate of lime, where the clay
beds pass into marl and marlstone. Beds of soft unconsolidated
clay, or of loose unconsolidated sand, are very rarely if ever found
amongst any of the primary strata, or in the lower regions of the
transition formation; the effects of heat appear to have converted
the earlier deposits of sand into compact quartz rock, and beds
of clay into clay slate, or other forms of primary slate. The rock
which some authors have called primary grauwacke, seems
to be a mechanical deposit of coarse sandstone, in which the form
of the fragments has not been so entirely obliterated by heat, as
in the case of compact quartz rock.

It may be said we have no right to deny the possible existence of life and organization upon the surface, or in the interior of our planet, under a state of igneous fusion. " Who," says the ingenious and speculative Tucker, (Light of Nature, book iii. chap. 10), " can reckon up all the varieties that infinite wisdom can contrive, or show the impossibility of organizations dissimilar to any within our experience? Who knows what cavities lie within the earth, or what living creatures they may contain, endued with senses unknown to us, to whom the streams of magnetism may serve instead of light, and those of electricity affect them as sensibly as sounds and odours affect us? Why should we pronounce it impossible that there should be bodies formed to endure the burning sun, to whom fire may be the natural element, whose bones and muscles are composed of fixed earth, their blood and juices of molten metals? Or others made to live in the frozen regions of Saturn, having their circulation carried on by fluids more subtle than the highest rectified spirits raised by chemistry?"

It is not for us to meet questions of this kind by dogmatizing as to possible existences, or to presume to speculate on the bounds which creative Power may have been pleased to impose on its own operations. We can only assert, that as the laws that now regulate the movements and properties of all the material elements, can be shown to have undergone no change since matter was first created upon our planet; no forms of organization such as now exist, or such as Geology shows to have existed, during any stages of the gradual formation of the earth, could have supported, for an instant, the state of fusion here supposed.

We therefore conclude, that whatever beings of wholly different natures and properties may be imagined to be within the range of possible existences, not one of all the living or fossil species of animals or vegetables, could ever have endured the temperature of an incandescent planet. All these species must therefore have had a beginning, posterior to the state of universal fusion which Geology points out.

I know not how I can better sum up the conclusion of this argument, than in the words of my Inaugural Lecture, (Oxford, 1819, p. 20).

" The consideration of the evidences afforded by geological phenomena may enable us to lay more securely the very foundations of natural theology, inasmuch as they clearly point out to us a period antecedent to the habitable state of the earth, and consequently antecedent to the existence of its inhabitants. When our minds become thus familiarized with the idea of a beginning and first creation of the beings we see around us, the proofs of design, which the structure of those beings affords, carry with them a more forcible conviction of an intelligent Creator, and the hypothesis of an eternal succession of causes, is thus at once removed. We argue thus: it is demonstrable from Geology that there was a period when no organic beings had existence; these organic beings must therefore have had a beginning subsequently to this period; and where is that beginning to be found but in the will and *fiat* of an intelligent and all-wise Creator?"

The same conclusion is stated by Cuvier, to be the result of his observations on geological phenomena: "Mais ce qui étonne davantage encore, et ce qui n'est pas moins certain, c'est qui la vie n'a pas toujours existé sur la globe, et qu'il est facile à l'observateur de reconnoître le point où elle a commencé à déposer ses produits."—Cuvier, Ossemens Fossiles, Disc. Prelim. 1821, vol. i. p. ix.

CHAPTER VII.

Strata of the Transition Series.

THUS far we have been occupied with rocks, in which we trace chiefly the results of chemical and mechanical forces ; but, as soon as we enter on the examination of strata of the Transition Series, the history of organic life becomes associated with that of mineral phenomena.*

The mineral character of the transition formations presents alternations of slate and shale, with slaty sandstone, limestone, and conglomerate rocks ; the latter bearing evidence of the action of water in violent motion ; the former showing, by their composition and structure, and by the organic remains which they frequently contain, that they were for the most part deposited in the form of mud and sand, at the bottom of the sea.

Here, therefore, we enter on a new and no less curious than important field of enquiry,

---

* It is most convenient to include within the Transition series, all kinds of stratified rocks, from the earliest slates, in which we find the first traces of animal or vegetable remains, to the termination of the great coal formation. The animal remains in the more ancient portion of this series, viz. the Grauwacke group, though nearly allied in genera, usually differ in species from those in its more recent portion, viz. the Carboniferous group.

and commence our examination of the relics of a former world, with a view to ascertain how far the fossil members of the animal and vegetable kingdoms may, or may not, be related to existing genera and species, as parts of one great system of creation, all bearing marks of derivation from a common author.*

Beginning with the animal kingdom, we find the four great existing divisions of *Vertebrata, Mollusca, Articulata,* and *Radiata,* to have been coeval with the commencement of organic life upon our globe.†

---

* In Plate 1, I have attempted to convey some idea of the organic remains preserved in the several series of formations, by introducing over each, restored figures of a few of the most characteristic animals and vegetables that occupied the lands and waters, at the periods in which they were deposited.

† " It has not been found necessary, in discussing the history of fossil plants and animals, to constitute a single new class; they all fall naturally into the same great sections as the existing forms.—We are warranted in concluding that the older organic creations were formed upon the same general plan as at present. They cannot, therefore, be correctly described as entirely different systems of nature, but should rather be viewed as corresponding systems, composed of different details. The difference of these details arises mostly from minute specific distinctions; but sometimes, especially among terrestrial plants, certain crustacea, and reptiles, the differences are of a more general nature, and it is not possible to refer the fossil tribes to any known recent genus, or even family. Thus we find the problem of the resemblance of recent and fossil organic beings to resolve itself into a general analogy of system, frequent agreement in important points, but almost universal distinction of minute organization."—Phillips's Guide to Geology, p. 61-63, 1834.

No higher condition of Vertebrata has been yet discovered in the transition formation than that of fishes, whose history will be reserved for a subsequent chapter.

The Mollusca,* in the transition series, afford examples of several families, and many genera, which seem at that time to have been universally diffused over all parts of the world. Some of these, (e. g. the Orthoceratite, Spirifer, and Producta) became extinct at an early period in the history of stratification, whilst other genera (as the Nautilus and Terebratula) have continued through all formations unto the present hour.

The earliest examples of Articulated animals are those afforded by the extinct family of Trilobites, (see Plates 45 and 46) to the history of which we shall devote peculiar consideration under the head of Organic Remains. Although nearly fifty species of these Trilobites occur in strata of the transition period, they appear to have become extinct before the commencement of the secondary series.

The Radiated Animals are among the most frequent organic remains in the transition strata; they present numerous forms of great beauty, from which I shall select the family of Crinoidea,

---

* In this great division, Cuvier includes a vast number of animals having soft bodies, without any articulated skeleton or spinal marrow, such as the Cuttle-fish, and the inhabitants of univalve and bivalve shells.

or lily-shaped animals allied to Star-fish, for peculiar consideration in a future chapter. (See Pl. 47, Figs. 5, 6, 7.) Fossil corallines also abound among the radiata of this period, and show that this family had entered thus early upon the important geological functions of adding their calcareous habitations to the solid materials of the strata of the globe. Their history will also be considered in another chapter.

Remains of Vegetables in the Transition Series.

Some idea may be formed of the vegetation which prevailed during the deposition of the upper strata of the transition series, from the figures represented in our first plate (Fig. 1 to 13). In the inferior regions of this series plants are few in number, and principally marine;* but in its superior regions the remains of land plants are accumulated in prodigious quantities, and preserved in a state which gives them a high and two-fold importance; first, as illustrating the history of the earliest vegetation that appeared upon our planet, and the state of climate

* M. A. Brogniart mentions the occurrence of four species of fucoids in the transition strata of Sweden and Quebec; and Dr. Harlan has described another species found in the Alleghany Mountains.

and geological changes which then prevailed :*
secondly, as affecting, in no small degree, the
actual condition of the human race.

The strata in which these vegetable remains
have been collected together in such vast abundance have been justly designated by the name
of the carboniferous order, or great coal formation. (See Conybeare and Phillips' Geology of
England and Wales, book iii.) It is in this
formation chiefly, that the remains of plants
of a former world have been preserved and converted into beds of mineral coal ; having been
transported to the bottom of former seas and
estuaries, or lakes, and buried in beds of sand
and mud, which have since been changed into
sandstone and shale. (See Pl. 1, sec. 14.)†

---

* The nature of these vegetables, and their relations to existing species, will be considered in a future chapter.

  † The most characteristic type that exists in this country of the
general condition and circumstances of the strata composing the
great carboniferous order, is found in the north of England.   It
appears from Mr. Forster's section of the strata from Newcastle-
upon-Tyne to Cross Fell, in Cumberland, that their united thickness along this line exceeds 4,000 feet.   This enormous mass is
composed of alternating beds of shale or indurated clay, sandstone, limestone, and coal : the coal is most abundant in the
upper part of the series, near Newcastle and Durham, and the
limestone predominates towards the lower part ; the individual
strata enumerated by Forster are thirty-two beds of coal, sixty-
two of sandstone, seventeen of limestone, one intruding bed of
trap, and one hundred and twenty-eight beds of shale and clay.
The animal remains hitherto noticed in the limestone beds are
almost exclusively marine ; hence we infer that these strata were

Besides this coal, many strata of the carboniferous order contain subordinate beds of a rich argillaceous iron ore, which the near position of the coal renders easy of reduction to a metallic state; and this reduction is further facilitated by the proximity of limestone, which is requisite as a flux to separate the metal from the ore, and usually abounds in the lower regions of the carboniferous strata.

A formation that is at once the vehicle of two such valuable mineral productions as coal and iron, assumes a place of the first importance among the sources of benefit to mankind; and

deposited at the bottom of the sea. The fresh-water shells that occur occasionally in the upper regions of this great series show that these more recent portions of the coal formation were deposited in water that was either brackish or entirely fresh. It has lately been shown that fresh-water deposits occur also occasionally in the lower regions of the carboniferous series. (See Dr. Hibbert's account of the limestone of Burdie House, near Edinburgh; Transactions of the Royal Society of Edinburgh, vol. xiii.; and Professor Phillips's Notice of fresh-water shells of the genus Unio, in the lower part of the coal series of Yorkshire; London Phil. Mag. Nov. 1832, 349.) The causes which collected these vegetables in beds thus piled above each other, and separated by strata of vast thickness, composed of drifted sand and clay, receive illustration from the manner in which drifted timber from the existing forests of America is now accumulated in the estuaries of the great rivers of that continent, particularly in the estuary of the Mississippi, and on the River Mackenzie. See Lyell's Principles of Geology, 3rd edit. Vol. iii. Book iii. Ch. xv. and Prof. Phillips's Article Geology in Encyclopædia Metropolitana, Pt. 37, page 596.

this benefit is the direct result of physical changes which affected the earth at those remote periods of time, when the first forms of vegetable life appeared upon its surface.

The important uses of coal and iron in administering to the supply of our daily wants, give to every individual amongst us, in almost every moment of our lives, a personal concern, of which but few are conscious, in the geological events of these very distant eras. We are all brought into immediate connection with the vegetation that clothed the ancient earth, before one-half of its actual surface had yet been formed. The trees of the primeval forests have not, like modern trees, undergone decay, yielding back their elements to the soil and atmosphere by which they had been nourished; but, treasured up in subterranean storehouses, have been transformed into enduring beds of coal, which in these later ages have become to man the sources of heat, and light, and wealth. My fire now burns with fuel, and my lamp is shining with the light of gas, derived from coal that has been buried for countless ages in the deep and dark recesses of the earth. We prepare our food, and maintain our forges and furnaces, and the power of our steam-engines, with the remains of plants of ancient forms and extinct species, which were swept from the earth ere the formation of the transition strata was completed. Our instruments of cutlery, the tools of our mechanics, and the countless machines which we construct, by the infinitely varied applications of iron, are derived from ore, for the most part coeval with, or more ancient than the fuel, by the aid of which we reduce it to its metallic state, and apply it to innumerable uses in the economy of human life. Thus, from the wreck of forests that waved upon the surface of the primeval lands, and from ferruginous mud that was lodged at the bottom of the primeval waters, we derive our chief supplies of coal and iron; those two fundamental elements of art and industry, which contribute more than any other mineral production of the earth, to increase the riches, and multiply the comforts, and ameliorate the condition of mankind.

Chapter VIII.

Strata of the Secondary Series.

WE may consider the history of secondary, and also of tertiary strata, in two points of view : the one, respecting their actual state as dry land, destined to be the habitation of man ; the other, regarding their prior condition, whilst in progress of formation at the bottom of the waters, and occupied by crowds of organic beings in the enjoyment of life.*

With regard to their adaptation to human uses, it may be stated generally, that the greater number of the most populous and highly civilized assemblages of mankind inhabit those portions of the earth which are composed of secondary and tertiary formations. Viewed, therefore, in their relations to that agricultural stage of human society in which man becomes established in a settled habitation, and applies his industry to till the earth, we find in these formations which have been accumulated, in apparently accidental succession, an arrangement highly advantageous to the cultivation of their surface. The movements of the waters, by which the materials of strata have been transported to their present place, have caused them to be intermixed in

* The secondary strata are composed of extensive beds of sand and sandstone, mixed occasionally with pebbles, and alternating with deposits of clay, and marl, and limestone. The materials of most of these strata appear to have been derived from the detritus of primary and transition rocks; and the larger fragments, which are preserved in the form of pebbles, often indicate the sources from which these rounded fragments were supplied.

The transport of these materials from the site of older formations to their place in the secondary series, and their disposition in strata widely extended over the bottom of the early seas, seem to have resulted from forces, producing the destruction of more ancient lands, on a scale of magnitude unexampled among the actual phenomena of moving waters.

such manner, and in such proportions, as are in various degrees favourable to the growth of the different vegetable productions, which man requires for himself and the domestic animals he has collected around him.

The process is obvious whereby even solid rocks are converted into soil fit for the maintenance of vegetation, by simple exposure to atmospheric agency; the disintegration produced by the vicissitudes of heat and cold, moisture and dryness, reduces the surface of almost all strata to a comminuted state of soil, or mould, the fertility of which is usually in proportion to the compound nature of its ingredients.

The three principal materials of all strata are the earths of flint, clay, and lime; each of these, taken singly and in a state of purity, is comparatively barren: the admixture of a small proportion of clay gives tenacity and fertility to sand, and the further addition of calcareous earth produces a soil highly valuable to the agriculturist: and where the natural proportions are not adjusted in the most beneficial manner, the facilities afforded by the frequent juxta-position of lime, or marl, or gypsum, for the artificial improvement of those soils which are defective in these ingredients, add materially to the earth's capability of adaptation to the important office of producing food. Hence it happens that the great corn fields, and the greatest population of the world, are placed on strata of the secondary and tertiary formations; or on their detritus, composing still more compound, and consequently more fertile diluvial, and alluvial deposits.*

Another advantage in the disposition of stratified rocks consists in the fact that strata of limestone, sand, and sandstone which readily absorb water, alternate with beds of clay, or marl, which are impermeable to this most important fluid. All permeable strata receive rainwater at their surface, whence it descends until it is arrested by an impermeable subjacent bed of clay, causing it to accumulate throughout the lower region of each porous stratum, and to form extensive reservoirs, the overflowings of which on the sides of valleys constitute the ordinary supply of springs and rivers. These reservoirs are not only occasional crevices and caverns, but the entire space of all the small interstices

* It is no small proof of design in the arrangement of the materials that compose the surface of our earth, that whereas the primitive and granitic rocks are least calculated to afford a fertile soil, they are for the most part made to constitute the mountain districts of the world, which, from their elevation and irregularities, would otherwise be but ill adapted for human habitation; while the lower and more temperate regions are usually composed of derivative, or secondary strata, in which the compound nature of their ingredients qualifies them to be of the greatest utility to mankind, by their subserviency to the purposes of luxuriant vegetation.—Buckland's Inaugural Lecture, Oxford, 1820, p. 17.

of those lower parts of each permeable stratum,
which are beneath the level of the nearest flowing springs.   Hence if a well be sunk to the
water-bearing level of any stratum, it forms a
communication with a permanent subterranean
sheet of water, affording plentiful supplies to
the inhabitants of upland districts, which are
above the level of natural springs.

A further benefit which man derives from the
disposition of the mineral ingredients of the
secondary strata, results from the extensive diffusion of muriate of soda, or common salt, throughout certain portions of these strata, especially
those of the new red sandstone formation.   Had
not the beneficent providence of the Creator
laid up these stores of salt within the bowels
of the earth, the distance of inland countries
from the sea would have rendered this article of
prime and daily necessity, unattainable to a
large proportion of mankind : but, under the
existing dispensation, the presence of mineral
salt, in strata which are dispersed generally over
the interior of our continents and larger islands,
is a source of health, and daily enjoyment, to
the inhabitants of almost every region of the
earth.*   Muriate of soda is also among the most

* Although the most frequent position of rock salt, and of salt
springs, is in strata of the new red sandstone formation, which
has consequently been designated by some geologists as the sali-
ferous system, yet it is not exclusively confined to them.   The

abundant of the saline compounds formed by
sublimation in the craters of volcanos.

With respect to the state of animal life, during the deposition of the Secondary strata,
although the petrified remains of Zoophytes,
Crustacea, Testacea, and Fishes, show that the
seas in which these strata were formed, like
those which gave birth to the Transition series,
abounded with creatures referrible to the four
existing divisions of the animal kingdom, still
the condition of the globe seems not yet to have
been sufficiently advanced in tranquillity, to
admit of general occupation by warm-blooded
terrestrial Mammalia.

The only terrestrial Mammalia yet discovered
in any secondary stratum, are the small marsupial quadrupeds allied to the Opossum, which
occur in the oolite formation, at Stonesfield, near
Oxford.   The jaws of two species of this genus
are represented in Plate 2. A. B ; the double roots
of the molar teeth at once refer these jaws to
the class of Mammalia, and the form of their
crowns places them in the order of Marsupial
animals.   Two other small species have been
discovered by Cuvier, in the tertiary formations
of the basin of Paris, in the gypsum of Mont
Martre.

salt mines of Wieliezka and Sicily are in tertiary formations ;
those of Cardona in cretaceous; some of those in the Tyrol in
the oolites; and near Durham there are salt springs in the coal
formation.

The Marsupial Order comprehends a large number of existing genera, both herbivorous and carnivorous, which are now peculiar to North and South America, and to New Holland, with the adjacent islands. The kangaroo and opossum are its most familiar examples. The name Marsupialia is derived from the presence of a large external marsupium, or pouch, fixed on the abdomen, in which the fœtus is placed after a very short period of uterine gestation, and remains suspended to the nipple by its mouth, until sufficiently matured to come forth to the external air. The discovery of animals of this kind, both in the secondary and tertiary formations, shows that the Marsupial Order, so far from being of more recent introduction than other orders of mammalia, is in reality the first and most ancient condition, under which animals of this class appeared upon our planet: as far as we know, it was their only form during the secondary period; it was co-existent with many other orders in the early parts of the tertiary period; and its geographical distribution in the present creation, is limited to the regions we have above enumerated.*

---

* In a highly important physiological paper, in the Phil. Trans. London, 1834, part ii. p. 349, Mr. Owen has pointed out " the most irrefragible evidence of creative foresight, afforded by the existing Marsupialia, in the peculiar modifications both of the maternal and fœtal system, designed with especial reference to

The peculiar feature in the population of the whole series of secondary strata, was the prevalence of numerous and gigantic forms of Saurian reptiles. Many of these were exclusively marine; others amphibious; others were terrestrial, ranging in savannahs and jungles, clothed with a tropical vegetation, or basking on the margins of estuaries, lakes, and rivers. Even the air was tenanted by flying lizards, under the dragon

each other's peculiar condition." With respect to the final cause of these peculiarities, he conjectures that they have relation to an inferior condition of the brain and nervous system in the Marsupialia; and considers the more protracted period of viviparous utero-gestation in the higher orders of Mammalia to be connected with their fuller developement of the parts subservient to the sensorial functions; the more simple form, and inferior condition of the brain in Marsupialia, being attended with a lower degree of intelligence, and less perfect condition of the organs of voice.

As this inferior condition of living Marsupialia shows this order to hold an intermediate place between viviparous and oviparous animals, forming, as it were, a link between Mammalia and Reptiles; the analogies afforded by the occurrence of the more simple forms of other classes of animals in the earlier geological deposits, would lead us to expect also that the first forms of Mammalia would have been Marsupial.

In a recent letter to myself, Mr. Owen adds the following interesting particulars respecting the physiology of this remarkable class of animals. " Of the generality of the law, as regards the simple unconvoluted form of the cerebrum in the Marsupials, I have had additional confirmation from recent dissections of a *Dasyurus* and *Phalangista*. With an organization defective in that part which I believe to be essential to the docility of the horse, and sagacity of the dog, it is natural to suppose that the Marsupial series of warm-blooded quadrupeds would be insuf-

form of Pterodactyles. The earth was probably at that time too much covered with water, and those portions of land which had emerged above the surface, were too frequently agitated by earthquakes, inundations, and atmospheric irregularities, to be extensively occupied by any higher order of quadrupeds than reptiles.

As the history of these reptiles, and also that of the vegetable remains,* of the secondary for-

---

ficient for the great purposes of the Creator, when the earth was rendered fit for the habitation of man. They do, indeed, afford the wandering savages of Australia a partial supply of food; but it is more than doubtful that any of the species will be preserved by civilized man on the score of utility. The more valuable and tractable ruminants are already fast encroaching on the plains where the kangaroo was once the sole representative of the graminivorous Mammalia.

" It is interesting, however, to observe, that the Marsupials, including the Monotremes, form a very complete series, adapted to the assimilation of every form of organic matter; and, no doubt, with enough of instinctive precaution, to preserve themselves from extermination, when surrounded with enemies of no higher intellectual powers than the Reptilia. It would, indeed, be a strong support to the consideration of them as a distinct ovoviviparous sub-class of Mammals, if they should be found, as hitherto, to be the sole representatives of the highest class of Vertebrata, in the secondary strata."—R. Owen.

* The vegetable remains of the secondary strata differ from those of the transition period, and are very rarely accumulated into beds of valuable coal. The imperfect coal of the Cleveland Moorlands near Whitby, on the coast of Yorkshire, and that of Brora in the county of Sutherland, occurs in the lower region of the oolite formation; that of Bückeberg in Nassau, is in the upper region of the same formation, and is of superior quality.

mations, will be made a subject of distinct
inquiry, it will here suffice to state, that the
proofs of method and design in the adaptation of
these extinct forms of organization to the varied
circumstances and conditions of the earth's progressive stages of advancement, are similar to
those we trace in the structure of living animal
and vegetable bodies; in each case, we argue
that the existence of contrivances, adapted to
produce definite and useful ends, implies the
anterior existence and agency of creative intelligence.

Chapter IX.

Strata of the Tertiary Series.

The Tertiary Series introduces a system of new
phenomena, presenting formations in which the
remains of animal and vegetable life approach
gradually nearer to species of our own epoch.
The most striking feature of these formations
consists in the repeated alternations of marine
deposits, with those of fresh water (see Pl. 1,
sect. 25, 26, 27, 28).

We are indebted to Cuvier and Brogniart, for
the first detailed account of the nature and
relations of a very important portion of the

tertiary strata, in their inestimable history of the deposits above the chalk near Paris. For a short time, these were supposed to be peculiar to that neighbourhood ; further observation has discovered them to be parts of a great series of general formations, extending largely over the whole world, and affording evidences of, at least, four distinct periods, in their order of succession, indicated by changes in the nature of the organic remains that are imbedded in them.*

Throughout all these periods, there seems to have been a continually increasing provision for the diffusion of animal life, and we have certain evidence of the character and numbers of the

* In Vol. II. of his Principles of Geology, Mr. Lyell has given an interesting map, showing the extent of the surface of Europe, which has been covered by water since the commencement of the deposition of the tertiary strata.

M. Boué, also, has published an instructive map, representing the manner in which central Europe was once divided into a series of separate basins, each maintaining, for a long time, the condition of a fresh-water lake ; those which were subject to occasional irruptions of the sea, would, for a while, admit of the deposition of marine remains; the subsequent exclusion of the sea, and return to the condition of a fresh-water lake, would allow the same region to become the receptacle of the exuviæ of animals inhabiting fresh water.—Synoptische Darstellung der Erdrinde. Hanau, 1827. The same map, on a larger scale, appears in the second series of the Transactions of the Linnean Society of Normandy.

In the Annals of Philosophy, 1823, the Rev. W. D. Conybeare published an admirable memoir, illustrative of a similar geological map of Europe.

creatures that were permitted to enjoy it, in the multitude of shells and bones preserved in the strata that were deposited during each of the four epochs we are considering.

M. Deshayes and Mr. Lyell have recently proposed a fourfold division of the marine formations of the tertiary series, founded on the proportions which their fossil shells bear to marine shells of existing species. To these divisions Mr. Lyell has applied the terms *Eocene, Miocene, Older Pliocene,* and *Newer Pliocene;* and has most ably illustrated their history in the third volume of his Principles of Geology.

The term Eocene implies the commencement or *dawn* of the existing state of the animal creation; the strata of this series containing a very small proportion of shells referrible to living species. The Calcaire Grossier of Paris, and the London clay, are familiar examples of this older tertiary, or Eocene formation.

The term Miocene implies that a minority of the fossil shells, in formations of this period, are of recent species. To this era are referred the fossil shells of Bordeaux, Turin, and Vienna.

In formations of the Older, and Newer Pliocene, taken together, the majority of the shells belong to living species; the recent species in the newer, being much more abundant than in the older division.

To the Older Pliocene, belong the Sub-apenninè marine formations, and the English Crag ; and to the Newer Pliocene, the more recent marine deposits of Sicily, Ischia, and Tuscany.*

Alternating with these four great marine formations above the chalk, there intervenes a fourfold series of other strata, containing shells which show them to have been formed in fresh water, accompanied by the bones of many terrestrial and aquatic quadrupeds.

The greater number of shells, both in the fresh-water and marine formations of the tertiary series, are so nearly allied to existing genera, that we may conclude, the animals by which they were formed, to have discharged similar functions in the economy of nature, and to have been endowed with the same capacities of enjoyment as the cognate mollusks of living species. As the examination of these shells would disclose nearly the same arrangements and adaptations that prevail in living species, it will be more important to investigate the extinct

---

* The total number of known fossil shells in the tertiary series is 3,036. Of these 1,238 are found in the Eocene; 1,021 in the Miocene; and 777 in the Older, and Newer Pliocene divisions.

The numerical proportions of recent to extinct species may be thus expressed.—In the

Newer Pliocene period....	90 to 95	
Older Pliocene period	35 to 50	Per cent. are of
Miocene period	18	*recent* species.
Eocene period...........	3½	

—Lyell's Geology, 4. Ed. vol. iii. p. 308.

genera of the higher orders of animals, which seem to have been constructed with a view to the temporary occupation of the earth, whilst the tertiary strata were in process of formation. Our globe was no longer tenanted by those gigantic reptiles, which had been its occupants during the secondary period; neither was it yet fit to receive the numerous tribes of terrestrial mammalia that are its actual inhabitants. A large proportion of the lands which had been raised above the sea, being covered with fresh water, was best adapted for the abode of fluviatile and lacustrine quadrupeds.

Our knowledge of these quadrupeds is derived solely from their fossil remains; and as these are found chiefly (but not exclusively)* in the fresh-water formations of the tertiary series, it is to them principally that our present attention will be directed.

---

* The remains of Palæotherium occur, though very rarely, in the Calcaire Grossier of Paris. The bones of other terrestrial mammalia, occur occasionally in the Miocene and Pliocene marine formations, e. g. in Touraine and in the Sub-apennines. These are derived from carcases which, during these respective periods, were drifted into estuaries and seas.

No remains of mammalia have yet been found in the Plastic clay formation next above the chalk; the admixture of fresh-water and marine shells in this formation seems to indicate that it was deposited in an estuary. Beds of fresh-water shells are interposed more than once between the marine strata of the Calcaire Grossier, which are placed next above the plastic clay.

Mammalia of the Eocene Period.

In the first great fresh-water formation of the Eocene period, nearly fifty extinct species of mammalia have been discovered by Cuvier; the greater number of these belong to the following extinct genera, in the order Pachydermata,* viz. Palæotherium, Anoplotherium, Lophiodon, Anthracotherium, Cheropotamus, Adapis (see Plates 3 and 4).†

* Cuvier's order Pachydermata, i. e. *animals having thick skins*, includes three subdivisions of Herbivora, of which the Elephant, Rhinoceros, and Horse are respectively examples.

†   *Palæotherium.*

The place of the genus Palæotherium (see Plates 3 and 4) is intermediate between the rhinoceros, the horse, and tapir. Eleven or twelve species have already been discovered; some as large as a rhinoceros, others varying from the size of a horse to that of a hog. The bones of the nose show that, like the tapir, they had a short fleshy trunk. These animals probably lived and died upon the margins of the then existing lakes and rivers, and their dead carcases may have been drifted to the bottom in seasons of flood. Some perhaps retired into the water to die.

Anoplotherium.

Five species of Anothoplerium (see Plates 3, 4) have been found in the gypsum of the neighbourhood of Paris. The largest (A. Commune) being of the size of a dwarf ass, with a thick tail, equal in length to its body, and resembling that of an otter; its probable use was to assist the animal in swimming. Another (A. Medium) was of a size and form more nearly approaching the light and graceful character of the Gazelle; a third species was nearly of the size of a Hare.

The posterior molar teeth in the genus Anoplotherium resemble those of the rhinoceros; their feet are terminated by two large toes, like the ruminating animals, whilst the composition of their

G.   G

The nearest approach among living animals to the form of these extinct aquatic quadrupeds, is found in the Tapirs that inhabit the warm regions of South America, Malacca, and Sumatra, and in the Daman of Africa.

It is not easy to find a more eloquent and striking acknowledgment of the regularity and constancy of the systematic contrivances that pervade the animal remains of the fossil world, than is contained in Cuvier's Introduction to his account of the bones discovered in the gypsum

tarsus is like that of the camel. The place of this genus stands, in one respect, between the rhinoceros and the horse; and in another, between the hippopotamus, the hog, and the camel.

Lophiodon.

The Lophiodon is another lost genus, allied most nearly to the tapir and rhinoceros, and, in some respects, to the hippopotamus, and connected closely with the Palæotherium and Anoplotherium. Fifteen species of Lophiodon have been ascertained.

Anthracotherium.

The genus Anthracotherium was so called from its having been first discovered in the Tertiary coal, or Lignite of Cadibona in Liguria: it presents seven species, some of them approximating to the size and character of the hog; others approaching nearly to that of a hippopotamus.

Cheropotamus.

The Cheropotamus was an animal most nearly allied to the hogs; in some respects approaching the Babiroussa, and forming a link between the Anoplotherium and the Peccary.

Adapis.

The last of the extinct Pachydermata found in the gypsum quarries of Montmartre, is the Adapis. The form of this creature most nearly resembled that of a hedgehog, but it was three times the size of that animal: it seems to have formed a link connecting the Pachydermata with the Insectivorous Carnivora.

quarries of the neighbourhood of Paris. It affords, to persons unacquainted with the modern method of conducting physical researches, an example of the kind of evidence on which we found our conclusions, as to the form, character, and habits of extinct creatures, that are known only through the medium of their fossil remains. After stating by what slow degrees the cabinets of Paris had been filled with innumerable fragments of bones of unknown animals, from the gypsum quarries of Mont Martre, Cuvier thus records the manner in which he applied himself to the task of reconstructing their skeletons. Having gradually ascertained that there were numerous species, belonging to many genera, " I at length found myself," says he, " as if placed in a charnel house, surrounded by mutilated fragments of many hundred skeletons, of more than twenty kinds of animals, piled confusedly around me : the task assigned me was, to restore them all to their original position. At the voice of comparative anatomy, every bone, and fragment of a bone, resumed its place. I cannot find words to express the pleasure I experienced in seeing, as I discovered one character, how all the consequences, which I predicted from it, were successively confirmed ; the feet were found in accordance with the characters announced by the teeth ; the teeth in harmony with those indicated beforehand by the feet ; the bones

of the legs and thighs, and every connecting
portion of the extremities, were found set together
precisely as I had arranged them, before my
conjectures were verified by the discovery of the
parts entire : in short, each species was, as it
were, reconstructed from a single one of its
component elements." (Cuvier's Ossemens Fossiles, 1812, tom. iii. Introduction, p. 3, 4.)

Thus, by placing before his readers the
progress of his discovery, and restorations of
unknown species and genera, in the same
irregular succession in which they occurred to
him, he derives from this disorder the strongest
demonstration of the accuracy of the principles
which formed his guide throughout the whole
enquiry; the last found fragments confirming
the conclusions he had drawn from those first
brought to light, and his retrograde steps being
as nothing, in comparison with his predictions
which were verified.

Discoveries thus conducted, demonstrate the
constancy of the laws of co-existence that have
ever pervaded all animated nature, and place
these extinct genera in close connexion with the
living orders of Mammalia.   ·

We may estimate the number of the animals
collected in the gypsum of Mont Martre, from
the fact, stated by Cuvier, that scarcely a block
is taken from these quarries which does not
disclose some fragment of a fossil skeleton.
Millions of such bones, he adds, must have been

destroyed, before attention was directed to the subject.

The subjoined list of fossil animals found in the gypsum quarries of the neighbourhood of Paris, affords important information as to the population of this first lacustrine portion of the tertiary series.* (See Pl. 1. Figs. 73 to 96.)

---

* LIST OF VERTEBRAL ANIMALS FOUND IN THE GYPSUM
OF THE BASIN OF PARIS.

Pachydermata — Palæotherium .... Anoplotherium .... Cheropotamus........ Adapis............. — Extinct species, of extinct genera.

Carnivora ...
- Bat.
- Canis.. — Large Wolf, differing from any existing species. Fox.
- Coatis (Nasua, Storr), large Coati, now native of the warm parts of America.
- Racoon (Procyon, Storr), North America.
- Genette (Genetta, Cuv., Viverra Genetta, Linn.), now extending from South of Europe to Cape of Good Hope.

Marsupialia.. — Opossum, small (Didelphis, Linn.), allied to the Opossum of North and South America.

Rodentia .... — Dormouse (Myoxus, Gm.), two small species. Squirrel (Sciurus).

Birds .... — Birds, nine or ten species, referrible to the following genera : Buzzard, Owl, Quail, Woodcock, Sea-Lark (Tringa), Curlew, and Pelican.

Reptiles .... — Fresh-water Tortoises, Trionyx, Emys. Crocodile.

Fishes ...... Seven extinct species, of extinct Genera. *Agass.*

Extinct species belonging to existing genera.

Besides the many extinct species, and extinct genera of Mammalia that are enumerated in this list, the occurrence of nine or ten extinct species of fossil Birds in the Eocene period of the tertiary series, forms a striking phenomenon in the history of organic remains.*

In this small number of species, we have seven genera; and these afford examples of four, out of the six great Orders into which the existing Class of Birds is divided, viz. Accipitres, Gallinaceæ, Grallæ, and Palmipedes. Even the eggs of aquatic birds have been preserved in the lacustrine formations of Cournon, in Auvergne.†

* The only remains of Birds yet noticed in strata of the Secondary series are the bones of some Wader, larger than a common Heron, found by Mr. Mantell in the fresh-water formation of Tilgate Forest. The bones at Stonesfield, once supposed to be derived from Birds, are now referred to Pterodactyles. A discovery has recently been made in America by Professor Hitchcock, of the footsteps of Birds in the New Red sandstone of the valley of the Connecticut, which he refers to at least seven species, all apparently Waders, having very long legs, and of various dimensions from the size of a Snipe, to twice the size of an Ostrich. (See Pl. 26ᵃ. 26ᵇ.)

† In the same Eocene formation with these eggs, there occur also the remains of two species of Anoplotherium, a Lophidon, an Anthracrotherium, a Hippopotamus, a ruminating animal, a Dog, a Martin, a Lagomys, a Rat, one or two Tortoises, a Crocodile, a Serpent or Lizard, and three or four species of Birds. These remains are dispersed singly, as if the animals from which they were derived had decomposed slowly and at different intervals, and thus fragments of their bodies had been lodged irregularly in various parts of the bottom of the ancient lake: these bones are sometimes broken, but never rolled.

It appears that the animal kingdom was thus early established, on the same general principles that now prevail ; not only did the four present Classes of Vertebrata exist ; and among Mammalia, the Orders Pachydermata, Carnivora, Rodentia, and Marsupialia ; but many of the genera also, into which living families are distributed, were associated together in the same system of adaptations and relations, which they hold to each other in the actual creation. The Pachydermata and Rodentia were kept in check by the Carnivora—the Gallinaceous birds were controlled by the Accipitres.

" Le Règne Animal, à ces époques reculées, était composé d'après les mêmes lois ; il comprenoit les mêmes classes, les mêmes familles que de nos jours ; et en effet, parmi les divers systèmes sur l'origine des êtres organisés, il n'en est pas de moins vraisemblable que celui qui en fait naitre successivement les différens genres par des développemens ou des métamorphoses graduelles." (Cuvier, Oss. Foss. t. 3, p. 297.)

This numerical preponderance of Pachydermata, among the earliest fossil Mammalia, beyond the proportion they bear among existing quadrupeds, is a remarkable fact, much insisted on by Cuvier; because it supplies, from the relics of a former world, many intermediate forms which do not occur in the present distribution of that important Order. As the living genera of Pachydermata are more widely separated from one another, than those of any other Order of Mammalia, it is important to fill these vacant intervals with the fossil genera of a former state of the earth; thus supplying links that appeared deficient in the grand continuous chain which connects all past and present forms of organic life, as parts of one great system of Creation.

As the bones of all these animals found in the earliest series of the tertiary deposits are accompanied by the remains of reptiles, such as now inhabit the fresh waters of warm countries, e. g. the Crocodile, Emys, and Trionyx (see Pl. 1, Figs. 80, 81, 82), and also by the leaves and prostrate trunks of palm trees (Pl. 1, Figs. 66, 67, 68, and Pl. 56), we cannot but infer that the temperature of France was much higher than it is at present, at the time when it was occupied by these plants and reptiles, and by Mammalia allied to families which are natives of some of the warmest latitudes of the present earth, e. g. the Tapir, Rhinoceros, and Hippopotamus.

The frequent intrusion of volcanic rocks is a remarkable accompaniment of the tertiary strata of the Eocene period, in various parts of Europe; and changes of level, resulting from volcanic agency, may partially explain the fact, that portions of the same districts became alternately the receptacles of fresh and salt water.

The fresh-water calcareous deposits of this period are also highly important, in relation to the general history of the origin of limestone, from their affording strong evidence of the sources whence carbonate of lime has been derived.*

* We see that thermal springs, in volcanic districts, issue from the earth, so highly charged with carbonate of lime, as to overspread large tracts of country with beds of calcareous tufa, or travertino. The waters that flow from the Lago di Tartaro, near Rome, and the hot springs of San Filippo, on the borders of Tuscany, are well known examples of this phenomenon. These existing operations afford a nearly certain explanation of the origin of extensive beds of limestone in fresh-water lakes of the tertiary period, where we know them to have been formed during seasons of intense volcanic activity. They seem also to indicate the probable agency of thermal waters in the formation of still larger calcareous deposits at the bottom of the sea, during preceding periods of the secondary and transition series.

It is a difficult problem to account for the source of the enormous masses of carbonate of lime that compose nearly one-eighth part of the superficial crust of the globe. Some have referred it entirely to the secretions of marine animals; an origin to which we must obviously assign those portions of calcareous strata which are composed of comminuted shells and corallines: but, until it can be shown that these animals have the power of forming lime from other elements, we must suppose that they derived it from the sea, either directly, or through the medium of its plants. In either case, it remains to find the source whence the sea obtained, not only these supplies of carbonate of lime for its animal inhabitants, but also the still larger quantities of the same substance, that have been precipitated in the form of calcareous strata.

We cannot suppose it to have resulted, like sands and clays, from the mechanical detritus of rocks of the granitic series, because the quantity of lime these rocks contain, bears no proportion to its large amount among the derivative rocks. The only remaining hypothesis seems to be, that lime was continually introduced to lakes and seas, by water that had percolated rocks through which calcareous earth was disseminated.

Mammalia of the Miocene Period.

The second, or Miocene System of Tertiary Deposits, contains an admixture of the extinct genera of lacustrine mammalia, of the first or Eocene series, with the earliest forms of genera which exist at the present time. This admixture was first noticed by M. Desnoyers, in the marine formations of the Faluns of Touraine.*

Although carbonate of lime occurs not in distinct masses among rocks of igneous origin, it forms an ingredient of lava and basalt, and of various kinds of trap rocks. The calcareous matter thus dispersed through the substance of these volcanic rocks, seems to afford a magazine from which percolating water, charged with carbonic acid gas, may, in the lapse of ages, have derived sufficient carbonate of lime to form all the existing strata of limestone, by successive precipitates at the bottom of ancient lakes and seas. Mr. De la Beche states the quantity of lime in granite composed of two-fifths quartz, two-fifths felspar, and one-fifth mica, to be 0.37 ; and in greenstone, composed of equal parts of felspar and hornblende, to be 7.29. (Geol. Researches, p. 379.)—The compact lava of Calabria contains 10. of carbonate of lime, and the basalt of Saxony 9.5.

We may, in like manner, refer the origin of those large quantities of silex, which constitute the chert and flint beds of stratified formations, to the waters of hot springs, holding siliceous earth in solution, and depositing it on exposure to reduced degrees of temperature and pressure, as silex is deposited by the hot waters that issue from the geysers of Iceland.

* Here, the remains of Palæotherium, Anthracotherium, and Lophiodon, which formed the prevailing genera in the Eocene period, are found mixed with bones of the Tapir, Mastodon, Rhinoceros, Hippopotamus, and Horse : these bones are fractured and rolled, and sometimes covered with flustra, and must have been derived from carcases drifted into an estuary, or sea. *Annales des Sciences Naturelles. Février,* 1828.

Similar admixtures have been found in Bavaria,* and near Darmstadt.† Many of these animals also indicate a lacustrine, or swampy condition of the regions they inhabited : one of them, the Dinotherium giganteum (gigantic

* Count Munster and Mr. Murchison have discovered, at Georgensgemünd, in Bavaria, the bones of Palæotherium, Anoplotherium, and Anthracotherium, mixed with those of Mastodon, Rhinoceros, Hippopotamus, Horse, Ox, Bear, Fox, &c. ; and several species of land shells.

A very interesting detailed description of the remains found at this place has been published by Hermann von Meyer, Frankfurt, 1834, 4to. with 14 plates.

† We learn from the excellent publication of Professor Kaup, of Darmstadt, that at Epplesheim, near Altzey, about twelve leagues south of Mayence, remains of the following animals have been found, in strata of sand, referrible to the second or Miocene period of the tertiary formations. These are preserved in the museum at Darmstadt.

	Number of Species.	
Dinotherium	2	.... { Gigantic Herbivorous Animals fifteen and eighteen feet long.
Tapirus	2	.... { Larger than living species.
Chalicotherium	2	.... Allied to Tapirs.
Rhinoceros	2	
Tetracaulodon	1	. . . Allied to Mastodon.
Hippotherium	1	.... Allied to the Horse
Sus	3	.... Hog.
Felis	4	.... { Large Cats, some as large as a Lion.
Machairodus	1	.... { Allied to Bear. Ursus Cultridens.
Gulo	1	.... Glutton.
Agnotherium........	1	.... { Allied to Dog, large as a Lion.

See Description d'Ossemens Fossiles, par Kaup. Darmst. 1832.

Tapir of Cuvier), is calculated to have been eighteen feet in length, and was much the largest of all terrestrial Mammalia yet discovered, exceeding even the largest fossil elephant.

The Dinotherium will be described in a subsequent chapter.

Mammalia of the Pliocene Periods.

The third, and fourth, or Pliocene divisions of the tertiary fresh-water deposits, contain no more traces of the extinct lacustrine genera of the Palæotherian family, but abound in extinct species of existing genera of Pachydermata, e. g. Elephant, Rhinoceros, Hippopotamus, and Horse, together with the extinct genus Mastodon. With these also occur the first abundant traces of Ruminantia, e. g. Oxen and Deer. The number of Rodentia becomes also enlarged ; and the Carnivora assume a numerical importance commensurate with the increased numbers of terrestrial herbivora.

The seas, also, of the Miocene and Pliocene periods, were inhabited by marine Mammalia, consisting of Whales, Dolphins, Seals, Walrus, and the Lamantin, or Manati, whose existing species are chiefly found near the coasts and mouths of rivers in the torrid zone (see Pl. 1, Figs. 97 to 101). The presence of the Lamantin adds another argument to those which arise from the tropical character of many other animals, even of the latest tertiary strata, in favour of the opinion, that the climate of Europe maintained a high, though probably a gradually decreasing temperature, even to the latest period of the tertiary formations.

We have many sources of evidence whereby the history of the Pliocene periods is illustrated : First, we have the remains of terrestrial animals, drifted into estuaries or seas, and preserved together with marine shells ; such are the Sub-apennine marine formations, containing the remains of Elephant, Rhinoceros, &c. and the Crag of Norfolk.*

Secondly, we have similar remains of terrestrial quadrupeds, mixed, with fresh-water shells, in strata formed during the same epoch, at the bottom of fresh-water lakes and ponds ; such as those which occur in the Val D'Arno, and in the small lacustrine deposit at North Cliff, near Market Weighton, in Yorkshire. (See Phil. Mag. 1829, vol. vi. p. 225.)

Thirdly, we have remains of the same animals

---

* In the museum at Milan, I have seen a large part of the skeleton of a Rhinoceros, from the Sub-apennine formation, having oyster shells attached to many of its bones, in such a manner as to show that the skeleton must have remained undisturbed for a considerable time at the bottom of the sea. Cuvier also states that in the museum at Turin there is the head of an elephant, to which shells of the same kind similarly attached, and fitted to the form of the bones.

in caverns and fissures of rocks, which formed parts of the dry land during the more recent portions of the same period. Such are the bones collected by Hyænas, in the caves of Kirkdale, Kent's Hole, Lunel, &c.: and the bones of Bears in caverns of the limestone rocks of central Germany, and the Grotte d'Osselles, near Besançon. Such also are the bones of the osseous breccia, found in fissures of limestone rocks on the northern shores of the Mediterranean, and in similar fissures of limestone at Plymouth, and in the Mendip Hills in Somerset. These are derived chiefly from herbivora which fell into the fissures before they were partially filled with the detritus of a violent inundation.

Fourthly, we have the same remains contained in deposits of diluvial detritus, dispersed over the surface of formations of all ages.

As I have elsewhere (Reliquiæ Diluvianæ)*

* The evidence which I have collected in my Reliquiæ Diluvianæ, 1823, shows, that one of the last great physical events that have affected the surface of our globe, was a violent inundation, which overwhelmed great part of the northern hemisphere, and that this event was followed by the sudden disappearance of a large number of the species of terrestrial quadrupeds, which had inhabited these regions in the period immediately preceding it. I also ventured to apply the name *Diluvium* to the superficial beds of gravel, clay, and sand, which appear to have been produced by this great irruption of water.

The description of the facts that form the evidence presented in this volume, is kept distinct from the question of the identity of the event attested by them, with any deluge recorded in

entered into the evidences illustrating the state of animal life, during the period immediately preceding the formation of this diluvium, I must refer to that work for details respecting the nature and habits of the then existing population of the earth. It appears that at this epoch, the whole surface of Europe was densely peopled by various orders of Mammalia; that the numbers of the herbivora were maintained in due proportion by the controlling influence of carnivora; and that the individuals of every species were constructed in a manner fitting

history. Discoveries which have been made, since the publication of this work, show that many of the animals therein described, existed during more than one geological period preceding the catastrophe by which they were extirpated. Hence it seems more probable, that the event in question, was the last of the many geological revolutions that have been produced by violent irruptions of water, rather than the comparatively tranquil inundation described in the Inspired Narrative.

It has been justly argued, against the attempt to identify these two great historical and natural phenomena, that as the rise and fall of the waters of the Mosaic deluge are described to have been gradual, and of short duration, they would have produced comparatively little change on the surface of the country they overflowed. The large preponderance of extinct species among the animals we find in caves, and in superficial deposits of diluvium, and the non-discovery of human bones along with them, afford other strong reason for referring these species to a period anterior to the creation of man. This important point, however, cannot be considered as completely settled, till more detailed investigations of the newest members of the Pliocene, and of the diluvial and alluvial formations shall have taken place.

each to its own enjoyment of the pleasures of existence, and placing it in due and useful relations to the animal and vegetable kingdoms by which it was surrounded.

Every comparative anatomist is familiar with the beautiful examples of mechanical contrivance and compensations, which adapt each existing species of herbivora and carnivora to its own peculiar place and state of life. Such contrivances began not with living species : the geologist demonstrates their prior existence in the extinct forms of the same genera which he discovers beneath the surface of the earth, and he claims for the Author of these fossil forms under which the first types of such mechanisms were embodied, the same high attributes of Wisdom and Goodness, the demonstration of which exalts and sanctifies the labours of science, in her investigation of the organizations of the living world.

Chapter X.

Relations of the Earth and its Inhabitants to Man.

From the statements which have been made in the preceding chapters, it appears that five principal causes have been instrumental in producing the actual condition of the surface of our globe. First, The passage of the unstratified crystalline rocks, from a fluid to a solid state.—Secondly, The deposition of stratified rocks at the bottom of the ancient seas.—Thirdly, The elevation both of stratified and unstratified rocks from beneath the sea, at successive intervals, to form continents and islands.—Fourthly, Violent inundations; and the decomposing Power of atmospheric agents; producing partial destruction of these lands, and forming, from their detritus, extensive beds of gravel, sand, and clay.—Fifthly, Volcanic eruptions.

We shall form a better estimate of the utility of the complex disposition of the materials of the earth, which has resulted from the operations of all these mighty conflicting forces, if we consider the inconveniences that might have attended other arrangements, more simple than

G.   H

those which actually exist. Had the earth's surface presented only one unvaried mass of granite or lava; or, had its nucleus been surrounded by entire concentric coverings of stratified rocks, like the coats of an onion, a single stratum only would have been accessible to its inhabitants; and the varied intermixtures of limestone, clay, and sandstone, which, under the actual disposition, are so advantageous to the fertility, beauty, and habitability, of the globe, would have had no place.

Again, the inestimably precious treasures of mineral salt and coal, and of metallic ores, confined as these latter chiefly are, to the older series of formations, would, under the supposed more simple arrangement of the strata, have been wholly inaccessible; and we should have been destitute of all these essential elements of industry and civilization. Under the existing disposition, all the various combinations of strata with their valuable contents, whether produced by the agency of subterranean fire, or by mechanical, or chemical deposition beneath the water, have been raised above the sea, to form the mountains and the plains of the present earth; and have still further been laid open to our reach, by the exposure of each stratum, along the sides of valleys.

With a view to human uses, the production of a soil fitted for agriculture, and the general dispersion of metals, more especially of that most important metal iron, were almost essential conditions of the earth's habitability by civilized man.

I would in this, as in all other cases, be unwilling to press the theory of relation to the human race, so far as to contend that all the great geological phenomena we have been considering were conducted *solely* and *exclusively* with a view to the benefit of man. We may rather count the advantages he derives from them as incidental and residuary consequences; which, although they may not have formed the exclusive object of creation, were all foreseen and comprehended in the plans of the Great Architect of that Globe, which, in his appointed time, was destined to become the scene of human habitation.*

---

* " It is true that by applying ourselves to the study of nature, we daily find more and more uses in things that at first appeared useless. But some things are of such a kind as not to admit of being applied to the benefit of man, and others too noble for us to claim the sole use of them. Man has no farther concern with this earth than a few fathoms under his feet: was then the whole solid globe made only for a foundation to support the slender shell he treads upon? Do the magnetic effluvia course incessantly over land and sea, only to turn here and there a mariner's compass? Are those immense bodies, the fixed stars, hung up for nothing but to twinkle in our eyes by night, or to find employment for a few astronomers? Surely he must have an overweening conceit of man's importance, who can imagine this stupendous frame of the universe made for him alone.

With respect to the animal kingdom, we
acknowledge with gratitude, that among the
higher classes, there is a certain number of
living species, which are indispensable to the
supply of human food and raiment, and to the
aid of civilized man in his various labours and
occupations; and that these are endowed with
dispositions and faculties which adapt them in
a peculiar degree for domestication :* but their
number bears an extremely small proportion to
the total amount of existing species; and with
regard to the lower classes of animals, there are
but very few among their almost countless multitudes, that minister either to the wants or luxuries

Nevertheless, we may so far acknowledge all things made for
man as that his uses are regarded conjointly with those of other
creatures, and that he has an interest in every thing reaching his
notice, and contributing either to the support of his body, the
improvement or entertainment of his mind. The satellites that
turn the night of Jupiter into day, assist him in ascertaining the
longitude, and measuring the velocity of light : the mighty sun,
that like a giant holds the planets and comets in their orbits,
enlightens him with its splendour, and cherishes him with its
warmth : the distant stars, whose attraction probably confines
other planets within their vortices, direct his course over the
boundless sea, and the inhospitable desert."—Tucker's Light of
Nature, book iii. chap. ix. p. 9.

See an excellent note on prospective provisions, to afford materials for human arts, and having reference to the future discoveries of human science, in Rev. W. D. Conybeare's Inaugural
Address to Bristol College, 1831.

* See Lyell's Principles of Geology, 3rd edit. vol. ii. book 3,
c. 3.

of the human race. Even could it be proved that all existing species are serviceable to man, no such inference could be drawn with respect to those numerous extinct animals which Geology shows to have ceased to live, long before our race appeared upon the earth. It is surely more consistent with sound philosophy, and with all the information that is vouchsafed to us respecting the attributes of the Deity, to consider each animal as having been created first for its own sake, to receive its portion of that enjoyment which the Universal Parent is pleased to impart to each creature that has life; and secondly, to bear its share in the maintenance of the general system of co-ordinate relations, whereby all families of living beings are reciprocally subservient to the use and benefit of one another. Under this head only can we include their relations to man; forming, as he does, but a small, although it be the most noble and exalted part, of that vast system of universal life, with which it hath pleased the Creator to animate the surface of the globe.

" More than three-fifths of the earth's surface," says Mr. Bakewell, " are covered by the ocean; and if from the remaining part we deduct the space occupied by polar ice and eternal snow, by sandy deserts, sterile mountains, marshes, rivers and lakes, the habitable portion will scarcely exceed one-fifth of the whole of the globe.

Nor have we reason to believe that at any former period the dominion of man over the earth was more extensive than at present. The remaining four-fifths of our globe, though untenanted by mankind, are for the most part abundantly stocked with animated beings, that exult in the pleasure of existence, independent of human control, and no way subservient to the necessities or caprices of man. Such is, and has been for several thousand years, the actual condition of our planet; nor is the consideration foreign to our subject, for hence we may feel less reluctance in admitting the prolonged ages or days of creation, when numerous tribes of the lower orders of aquatic animals lived and flourished, and left their remains imbedded in the strata that compose the outer crust of our planet." Bakewell's Introduction to Geology, 4th edit. p. 6.

Chapter XI.

Supposed Cases of Fossil Human Bones.

BEFORE we enter on the consideration of the fossil remains of other animals, it may be right to enquire whether any traces of the human species have yet been found in the strata of the earth.

The only evidence that has been yet collected upon this subject is negative; but as far as this extends, no conclusion is more fully established, than the important fact of the total absence of any vestiges of the human species throughout the entire series of geological formations.* Had the case been otherwise, there would indeed have been great difficulty in reconciling the early and extended periods which have been assigned to the extinct races of animals with our received chronology. On the other hand, the fact of no human remains having as yet been found in conjunction with those of extinct animals, may be alleged in confirmation of the hypothesis that these animals lived and died before the creation of man.

* See Lyell's Principles of Geology, vol. i. pp. 153 and 159, first edit. 1830.

The occasional discovery of human bones and works of art in any stratum, within a few feet of the surface, affords no certain evidence of such remains being coeval with the matrix in which they are deposited. The universal practice of interring the dead, and frequent custom of placing various instruments and utensils in the ground with them, offer a ready explanation of the presence of bones of men in situations accessible for the purposes of burial.

The most remarkable and only recorded case of human skeletons imbedded in a solid limestone rock, is that on the shore of Guadaloupe.*

* One of these skeletons is preserved in the British Museum, and has been described by Mr. König, in the Phil. Trans. for 1814, vol. civ. p. 101. According to General Ernouf, (Lin. Trans. 1818, vol. xii. p. 53), the rock in which the human bones occur at Guadaloupe, is composed of consolidated sand, and contains also shells, of species now inhabiting the adjacent sea and land, together with fragments of pottery, arrows, and hatchets of stone. The greater number of the bones are dispersed. One entire skeleton was extended in the usual position of burial; another, which was in a softer sandstone, seemed to have been buried in the sitting position customary among the Caribs. The bodies thus differently interred, may have belonged to two different tribes. General Ernouf also explains the occurrence of the scattered bones, by reference to a tradition of a battle and massacre on this spot, of a tribe of Gallibis by the Caribs, about the year 1710. These scattered bones of the massacred Gallibis were probably covered, by the action of the sea, with sand, which soon after became converted to solid stone.

On the west coast of Ireland, near Killery Harbour, a sand bank, which is surrounded by the sea at high water, is at this time employed by the natives as a place of interment.

There is, however, no reason to consider these bones to be of high antiquity, as the rock in which they occur is of very recent formation, and is composed of agglutinated fragments of shells and corals which inhabit the adjacent water. Such kind of stone is frequently formed in a few years from sand-banks composed of similar materials, on the shores of tropical seas.

Frequent discoveries have also been made of human bones, and rude works of art, in natural caverns, sometimes inclosed in stalactite, at other times in beds of earthy materials, which are interspersed with bones of extinct species of quadrupeds. These cases may likewise be explained by the common practice of mankind in all ages, to bury their dead in such convenient repositories. The accidental circumstance that many caverns contained the bones of extinct species of other animals, dispersed through the same soil in which human bodies may, at any subsequent period have been buried, affords no proof of the time when these remains of men were introduced.

Many of these caverns have been inhabited by savage tribes, who, for convenience of occupation, have repeatedly disturbed portions of soil in which their predecessors may have been buried. Such disturbances will explain the occasional admixture of fragments of human skeletons, and the bones of modern quadrupeds, with those of extinct species, introduced at more early periods, and by natural causes.

Several accounts have been published within the last few years of human remains discovered in the caverns of France, and the province of Liege, which are described as being of the same antiquity with the bones of Hyænas, and other extinct quadrupeds, that accompany them. Most of these may probably admit of explanation by reference to the causes just enumerated. In the case of caverns which form the channels of subterranean rivers, or which are subject to occasional inundations, another cause of the admixture of human bones, with the remains of animals of more ancient date, may be found in the movements occasioned by running water.

Chapter XII.

General History of Fossil Organic Remains.

As "the variety and formation of God's creatures in the animal, vegetable, and mineral kingdoms" are specially marked out by the founder of this Treatise, as the subjects from which he desires that proofs should be sought of the power, wisdom, and goodness of the Creator; I shall enter at greater length into the Evidences of this

kind, afforded by fossil organic remains, than I
might have done, without such specific directions
respecting the source from which my arguments
are to be derived. I know not how I can better
fulfil the object thus proposed, than by attempting to shew that the extinct species of Animals
and Vegetables which have, in former Periods,
occupied our Planet, afford in their fossil remains, the same evidences of contrivance and
design that have been shewn by Ray, Derham,
and Paley, to pervade the structure of existing
Genera and species of organized Beings.

From the high preservation in which we find
the remains of animals and vegetables of each
geological formation, and the exquisite mechanism which appears in many fossil fragments of
their organization, we may collect an infinity
of arguments, to show that the creatures from
which all these are derived were constructed
with a view to the varying conditions of the
surface of the Earth, and to its gradually increasing capabilities of sustaining more complex
forms of organic life, advancing through successive stages of perfection.*

* When we speak of different forms of animal life, as possessing various degrees of *perfection*, we do not impute to any
creature the presence of absolute *imperfection*, we mean only,
that animals of more simple structure discharge a lower office
in the gradually descending scale of animated beings. All perfection has relation to the object proposed to be attained by
each form of organization that occurs in nature, and nothing
can be called imperfect which fully accomplishes the end pro-

Few facts are more remarkable in the history of the progress of human discovery, than that it should have been reserved almost entirely for the researches of the present generation, to arrive at any certain knowledge of the existence of the numerous extinct races of animals, which occupied the surface of our planet, in ages preceding the creation of man. The rapid progress, which during the last half century, has been made in the physical sciences, enables us now to enter into the history of Fossil Organic Remains, in a manner which, till within a very few years, would have been quite impracticable; during these years the anatomy of extinct species of Quadrupeds has been most extensively investi-

posed: thus a Polype, or an Oyster, are as perfectly adapted to their functions at the bottom of the sea, as the wings of the Eagle are perfect, as organs of rapid passage through the air, and the feet of the stag perfect, in regard to their functions of effecting swift locomotion upon the land.

Unusual deviations from ordinary structure appear monstrosities only, until considered with reference to their peculiar use, but are proved to be instruments of perfect contrivance, when we understand the nature of the service to which they are applied: thus; the beak of the Cross Bill (Loxia curvirostra, Linn.) would be an awkward instrument if applied to the ordinary service of the beaks of the Passerine Order, to which this bird belongs; but viewed in relation to its peculiar function of extracting seeds from between the indurated scales of Fir cones, it is at once seen to be an instrument of perfect adaptation to its intended work.

The *Perfection* of an organized Body is usually considered to be in proportion to the Variety and compound Nature of its parts, as the *imperfection* is usually considered to be in the Ratio of its Simplicity.

gated, and the greatest of comparative anatomists has devoted much of his time and talent to illustrate their organization. Similar inquiries have been carried on also by a host of other enlightened and laborious individuals, conducting independent researches in various countries, since the commencement of the present century; hence our knowledge of the osteology of a large number of extinct genera and species, now rests on nearly the same foundation, and is established with scarcely less certainty, than the anatomical details of those creatures that present their living bodies to our examination.

We can hardly imagine any stronger proof of the Unity of Design and Harmony of Organizations that have ever pervaded all animated nature, than we find in the fact established by Cuvier, that from the character of a single limb, and even of a single tooth or bone, the form and proportions of the other bones, and condition of the entire Animal may be inferred. This law prevails, no less universally, throughout the existing kingdoms of animated nature, than in those various races of extinct creatures that have preceded the present tenants of our planet; hence not only the framework of the fossil skeleton of an extinct animal, but also the character of the muscles, by which each bone was moved, the external form and figure of the body, the food, and habits, and haunts, and mode of life of creatures that ceased to exist before the creation of the human race, can with a high degree of probability be ascertained.

Concurrent with this rapid extension of our knowledge of the comparative anatomy of extinct families of the ancient inhabitants of the earth, has been the attention paid to fossil Conchology; a subject of vast importance in investigating the records of the changes that have occurred upon the surface of our globe.

Still more recently, the study of botanists has been directed to the History of fossil vegetables; and although, from the late hour at which this subject has been taken up, our knowledge of fossil plants is much in arrear of the progress made in Anatomy and Conchology, we have already a mass of most important evidence, showing the occurrence of a series of changes in vegetable life, coextensive and contemporaneous with those that have pervaded both the higher and lower orders of the animal kingdom.

The study of Organic Remains indeed, forms the peculiar feature and basis of modern Geology, and is the main cause of the progress this science has made, since the commencement of the present century. We find certain families of Organic Remains pervading strata of every age, under nearly the same generic forms which they present among existing organizations.*

* e. g. The Nautilus, Echinus, Terebratula, and various forms of Corals; and among Plants, the Ferns, Lycopodiaceæ, and Palms.

Other families, both of animals and vegetables, are limited to particular formations, there being certain points where entire groups ceased to exist, and were replaced by others of a different character. The changes of genera and species are still more frequent; hence, it has been well observed, that to attempt an investigation of the structure and revolutions of the earth, without applying minute attention to the evidences afforded by organic remains, would be no less absurd than to undertake to write the history of any ancient people, without reference to the documents afforded by their medals and inscriptions, their monuments, and the ruins of their cities and temples. The study of Zoology and Botany has therefore become as indispensable to the progress of Geology, as a knowledge of Mineralogy. Indeed the mineral character of the inorganic matter of which the Earth's strata are composed, presents so similar a succession of beds of sandstone, clay, and limestone, repeated irregularly, not only in different, but even in the same formations,* that similarity of mineral composition is but an uncertain proof of contemporaneous origin, while the surest test of

---

* The same formation which in England constitutes the argillaceous deposits of the London Clay, presents at Paris the sand and freestone of the Calcaire Grossier; whilst the resemblance of their Organic remains, proves the period of their deposition to have been the same, notwithstanding the difference in the character of their mineral ingredients.

identity of time is afforded by the correspondence of the organic remains: in fact without
these, the proofs of the lapse of such long periods
as Geology shews to have been occupied in the
formation of the strata of the Earth, would have
been comparatively few and indecisive.

The secrets of Nature, that are revealed to us,
by the history of fossil Organic Remains, form
perhaps the most striking results at which we
arrive from the study of Geology. It must
appear almost incredible to those who have not
minutely attended to natural phenomena, that
the microscopic examination of a mass of rude
and lifeless limestone should often disclose the
curious fact, that large proportions of its substance have once formed parts of living bodies.
It is surprising to consider that the walls of our
houses are sometimes composed of little else
than comminuted shells, that were once the
domicile of other animals, at the bottom of
ancient seas and lakes.

It is marvellous that mankind should have
gone on for so many centuries in ignorance of
the fact, which is now so fully demonstrated,
that no small part of the present surface of
the earth is derived from the remains of animals, that constituted the population of ancient
seas. Many extensive plains and massive
mountains form, as it were, the great charnel-
houses of preceding generations, in which the
petrified exuviæ of extinct races of animals and

vegetables are piled into stupendous monuments of the operations of life and death, during almost immeasurable periods of past time. " At the sight of a spectacle," says Cuvier,* " so imposing, so terrible as that of the wreck of animal life, forming almost the entire soil on which we tread, it is difficult to restrain the imagination from hazarding some conjectures as to the causes by which such great effects have been produced."

The deeper we descend into the strata of the Earth, the higher do we ascend into the archæological history of past ages of creation. We find successive stages marked by varying forms of animal and vegetable life, and these generally differ more and more widely from existing species, as we go further downwards into the receptacles of the wreck of more ancient creations.

When we discover a constant and regular assemblage of organic Remains, commencing with one series of strata, and ending with another, which contains a different assemblage, we have herein the surest grounds whereon to establish those Divisions which are called geological formations, and we find many such Divisions succeeding one another, when we investigate the mineral deposits on the surface of the Earth.

* Cuvier rapport sur le progrès des sciences naturelles, p. 179.

G.   I

The study of these Remains presents to the
Zoologist a large amount of extinct species and
genera, bearing important relations to existing
forms of animals and vegetables, and often supplying links that had hitherto appeared deficient,
in the great chain whereby all animated beings
are held together in a series of near and gradual
connexions.

This discovery, amid the relics of past creations, of links that seemed wanting in the present
system of organic nature, affords to natural
Theology an important argument, in proving
the unity and universal agency of a common
great first cause; since every individual in such
an uniform and closely connected series, is thus
shewn to be an integral part of one grand
original design.

The non-discovery of such links indeed, would
form but a negative and feeble argument against
the common origin of organic beings, widely
separated from one another; because, for aught
we know, the existence of intervals may have
formed part of the original design of a common
creator; and because such apparent voids may
perhaps exist only in our own imperfect knowledge; but the presence of such links throughout
all past and present modifications of being,
shews an unity of design which proves the unity
of the intelligence in which it originated.

It is indeed true that animals and vegetables

of the lower classes prevailed *chiefly* at the commencement of organic life, but they did not prevail *exclusively*; we find in rocks of the transition formation, not only remains of radiated and articulated animals and mollusks, such as Corals, Trilobites, and Nautili; but we see the vertebrata also represented by the Class of Fishes. Reptiles have been found in some of the earliest strata of the secondary formations.* In the footsteps on the New Red sandstone, we have probably the first traces of Birds and Marsupialia. (See Pl. 26ª. and 26'.) The bones of Birds occur in the Wealden formation of Tilgate forest, and those of Marsupialia in the Oolite at Stonesfield. (See Pl. 2. Figs. A. B.) In the midway regions of the secondary strata, are the earliest remains yet discovered of Cetacea.† In the tertiary formations, we find both Birds, Cetacea, and terrestrial Mammalia, some referrible to existing genera, and all to existing orders. See Pl. 1, fig. 73—101.

Thus it appears, that the more perfect forms of animals become gradually more abundant, as we advance from the older into the newer series of depositions: whilst the more simple orders, though often changed in genus and species, and

* E. g. In the Magnesian Conglomerate of Durdham Down near Bristol, and in the bituminous marl slate, (Kupferschiefer) of Mansfeld in the Hartz.

† There is, in the Oxford Museum, an ulna from the Great Oolite of Enstone near Woodstock, Oxon, which was examined by Cuvier, and pronounced to be cetaceous; and also a portion of a very large rib, apparently of a whale, from the same locality.

sometimes losing whole families, which are replaced by new ones, have pervaded the entire
range of fossiliferous formations.

The most prolific source of organic remains
has been the accumulation of the shelly coverings
of animals which occupied the bottom of the sea
during a long series of consecutive generations.
A large proportion of the entire substance of
many strata is composed of myriads of these
shells reduced to a comminuted state by the
long continued movements of water. In other
strata, the presence of countless multitudes of
unbroken corallines, and of fragile shells, having
their most delicate spines, still attached and undisturbed, shows that the animals which formed
them, lived and died upon or near the spot
where these remains are found.

Strata thus loaded with the exuviæ of innumerable generations of organic beings, afford
strong proof of the lapse of long periods of time,
wherein the animals from which they have been
derived lived and multiplied and died, at the
bottom of seas which once occupied the site of
our present continents and islands. Repeated
changes in species, both of animals and vegetables, in succeeding members of different formations, give further evidence, not only of the
lapse of time, but also of important changes in
the physical condition and climate of the ancient
earth.

Besides these more obvious remains of Testacea and of larger animals, minute examination discloses occasionally prodigious accumulations of miscrocopic shells that surprise us no less by their abundance than their extreme minuteness ; the mode in which they are sometimes crowded together, may be estimated from the fact that Soldani collected from less than an ounce and a half of stone found in the hills of Casciana, in Tuscany, 10,454 microscopic chambered shells. The rest of the stone was composed of fragments of shells, of minute spines of Echini, and of a sparry calcareous matter.

Of several species of these shells, four or five hundred weigh but a single grain ; of one species he calculates that a thousand individuals would scarcely weigh one grain. (Saggio Orittografico, 1780, pag. 103, Tab. III. fig. 22, H. 1.) He further states that some idea of their diminutive size may be formed from the circumstance that immense numbers of them pass through a paper in which holes have been pricked with a needle of the smallest size.

Our mental, like our visual faculties, begin rapidly to fail us when we attempt to comprehend the infinity of littleness towards which we are thus conducted, on approaching the smaller extremes of creation.

Similar accumulations of microscopic shells have been observed also in various sedimentary deposits of freshwater formation. A striking example of this kind is found in the abundant diffusion of the remains of a microscopic crustaceous animal of the genus Cypris. Animals of this genus are enclosed within two flat valves, like those of a bivalve shell, and now inhabit the waters of lakes and marshes. Certain clay beds of the Wealden formation below the chalk, are so abundantly charged with microscopic shells of the Cypris Faba, that the surfaces of many laminæ into which this clay is easily divided, are often entirely covered with them as with small seeds. The same shells occur also in the Hastings sand and sandstone, in the Sussex marble, and in the Purbeck limestone, all of which were deposited during the same geological epoch in an ancient lake or estuary, wherein strata of this formation have been accumulated to the thickness of nearly 1000 feet. (See Dr. Fitton's Geol. sketch of Hastings, 1833, p. 68.)

We have similar evidence of the long duration of time, in another series of Lacustrine formations, more recent than the chalk, viz. in the great freshwater deposits of the tertiary period in central France; here the district of Auvergne presents an area of twenty miles in width, and eighty miles in length, within which strata of gravel, sand, clay, and limestone have been accumulated by the operations of fresh water, to the thickness of at least seven hundred feet. Mr. Lyell, in his Principles of Geology, 3rd ed. vol. iv. p. 98, states that the foliated character of

many of the marly beds of this formation is due
to the presence of countless myriads of similar
exuviæ of the Cypris which give rise to divisions
in the marl as thin as paper. Taking this fact
in conjunction with the habit of these animals to
moult and change their skin annually, together
with their shell, he justly observes that a more
convincing proof of the tranquillity of the waters,
and of the slow and gradual process by which
the lake was filled up with fine mud cannot be
desired.

Another proof of the length of time that must
have elapsed during the deposition of these
tertiary freshwater formations in Auvergne, is
afforded near Cleremont by the occurrence of
beds of limestone several feet in thickness,
almost wholly made up of the Indusiæ, or
Caddis-like coverings, resembling the cases that
enclose the larvæ of our common May-fly.

Mr. Lyell states that a single individual of
these Indusiæ is often surrounded by no less
than a hundred minute shells of a small spiral
univalve, (Paludina), fixed to the outside of this
tubular case of a larva of the genus Phryganea.
See Lyell's Principles of Geology, 3rd edit. vol.
iv. p. 100. It is difficult to conceive how strata
like these, extended over large tracts of country,
and laid one above another, with beds of marl
and clay between them, should have contained
the coverings of such multitudes of aquatic
animals, by any other process than that of

gradual accumulation during a long series of years.

In the case of deposits formed in estuaries, the admixture and alternation of the remains of fluviatile and lacustrine shells with marine Exuviæ, indicate conditions analogous to those under which we observe the inhabitants both of the sea and rivers existing together in brackish water near the Deltas of the Nile,* and other great rivers. Thus, we find a stratum of oyster shells, that indicate the presence either of salt or brackish water, interposed between limestone strata filled with freshwater shells among the Purbeck formations; so also in the sands and clays of the Wealden formation of Tilgate forest, we have freshwater and lacustrine shells intermixed with remains of large terrestrial reptiles, e. g. Megalosaurus, Iguanodon, and Hylæosaurus; with these we find also the bones of the marine reptiles Plesiosaurus, and from this admixture we infer that the former were drifted from the land into an estuary which the Plesiosaurus also having entered from the sea, left its bones in this common receptacle of the animal and mineral exuviæ of some not far distant land.†

Another condition of organic remains is that

* See Madden's Travels in Egypt, vol. ii. p. 171-175.

† For the detailed history of the organic remains of the Wealden formation, see Mr. Mantell's highly instructive and accurate volumes on the geology of Sussex.

of which a well known example occurs in the oolitic slate of Stonesfield, near Oxford. At this place a single bed of calcareous and sandy slate not six feet thick, contains an admixture of terrestrial animals and plants with shells that are decidedly marine; the bones of Didelphys, Megalosaurus, and Pterodactyle are so mixed with Ammonites, Nautili, and Belemnites, and many other species of marine shells, that there can be little doubt that this formation was deposited at the bottom of a sea not far distant from some .ancient shore. We may account for the presence of remains of terrestrial animals in such a situation by supposing their carcases to have been floated from land at no great distance from their place of submarine interment.

A similar explanation may be given of the mixture of the bones of large terrestrial mammalia with marine shells, in the Miocene Tertiary formations of Touraine, and in the Crag of Norfolk.

Cases of Animals destroyed suddenly.

The cases hitherto examined, are examples of the processes of slow and gradual accumulations in which are preserved the remains of marine, lacustrine, and terrestrial animals that perished during extended periods of time, by natural death. It remains to state that other causes seem to have operated occasionally, and at distant intervals, to produce a rapid accumulation of certain strata, accompanied by the sudden destruction, not only of testacea, but also of the higher classes of the then existing inhabitants of the seas. We have analogous instances of sudden destruction operating locally at the present time, in the case of fishes that perish from an excessive admixture of mud with the water of the sea, during extraordinary tempests; and also from the sudden imparting of heat, and noxious gases, to water in immediate contact with the site of submarine volcanoes. A sudden irruption of salt water into lakes or estuaries, previously occupied by fresh water, or the sudden occupation of a portion of the sea, by an immense body of freshwater from a bursting lake, or unusual land flood, is often fatal to large numbers of the inhabitants of the waters thus respectively interchanged.*

The greater number of fossil fishes present no appearance of having perished by mechanical violence; they seem rather to have been destroyed by some noxious qualities imparted to the waters in which they moved; either by sudden change of temperature,† or an admix-

---

* See account of the effects of an irruption of the sea into the freshwater of the lake of Lowestoffe, on the coast of Suffolk. Edinburgh Philosophical Journal, No. 25, p. 372.

† M. Agassiz has observed that a sudden depression to the amount of 1.5° of the temperature of the water in the river Glat, which falls into the lake of Zurich, caused the immediate death of thousands of Barbel.

ture of carbonic acid, or sulphuretted hydrogen gas, or of bituminous or earthy matter in the form of mud.

The circumstances under which the fossil fishes are found at Molte Bolca seem to indicate that they perished suddenly on arriving at a part of the then existing seas, which was rendered noxious by the volcanic agency, of which the adjacent basaltic rocks afford abundant evidence. The skeletons of these fish lie parallel to the laminæ of the strata of the calcareous slate; they are always entire, and so closely packed on one another, that many individuals are often contained in a single block. The thousands of specimens which are dispersed over the cabinets of Europe, have nearly all been taken from one quarry. All these fishes must have died suddenly on this fatal spot, and have been speedily buried in the calcareous sediment then in the course of deposition. From the fact that certain individuals have even preserved traces of colour upon their skin, we are certain that they were entombed before decomposition of their soft parts had taken place.*

* The celebrated fish (Blochius longirostris) from this quarry, described as petrified in the act of swallowing another fish (Ithiolitologia Veronese, Tab. XII.) has been ascertained by M. Agassiz to be a deception, arising from the accidental juxtaposition of two fishes. The size of the head of the smaller fish supposed to be swallowed, is such as never could have entered the diminutive stomach of the putative glutton; moreover it does not enter within the margin of its jaws.

The fishes of Torre d'Orlando, in the Bay of Naples, near Castelamare, seem also to have perished suddenly. M. Agassiz finds that the countless individuals which occur there in Jurassic limestone, all belong to a single species of the genus Tetragonolepis. An entire shoal seems to have been destroyed at once, at a place where the waters were either contaminated with some noxious impregnation, or overcharged with heat.*

In the same manner also, we may imagine deposits from muddy water, mixed perhaps with noxious gases, to have formed by their sediments a succession of thick beds of marl and clay, such as those of the Lias formation ; and at the same time to have destroyed, not only the Testacea and lower orders of animals inhabiting the bottom, but also the higher orders of marine creatures within the regions thus invaded. Evidence of the fact of vast numbers of fishes and saurians having met with sudden death and immediate burial, is also afforded by the state of entire preservation in which the bodies of hundreds of them are often found in the Lias.

---

* The proximity of this rock to the Vesuvian chain of volcanic eruptions, offers a cause sufficient to have imparted either of these destructive powers to the waters of a limited space in the bay of Naples, at a period preceding those intense volcanic actions which prevailed in this district during the deposition of the Tertiary strata, and which are still going on there.

It sometimes happens that scarcely a single bone, or scale, has been removed from the place it occupied during life ; this condition could not possibly have been retained, had the uncovered bodies of these animals been left, even for a few hours, exposed to putrefaction, and to the attacks of fishes and other smaller animals at the bottom of the sea.*

Another celebrated deposit of fossil fishes is that of the cupriferous slate surrounding the Hartz. Many of the fishes of this slate at Mansfeldt, Eiseleben, &c. have a distorted attitude, which has often been assigned to writhing in the agonies of death. The true origin of this condition, is the unequal contraction of the muscular fibres, which causes fish and other animals to become stiff, during a short interval between death and the flaccid state preceding decomposition. As these fossil fishes maintain the altitude of the rigid stage immediately succeeding death, it follows that they were buried before putrefaction had commenced, and apparently in the same bituminous mud, the influx of which had caused their destruction. The dissemination of Copper and Bitumen through the slate

---

* Although it appears from the preservation of these animals, that certain parts of the Lias were deposited rapidly, there are also proofs of the lapse of much time during the deposition of other parts of this formation. See Notes in future Chapters on Coprolites and fossil Loligo.

that contains so many perfect fishes around the Hartz, seems to offer two other causes, either of which may have produced their sudden death.*

From what has been said respecting the general history of fossil organic Remains, it appears that not only the relics of aquatic, but also those of terrestrial animals and plants, are found almost exclusively in strata that have been accumulated by the action of water. This circumstance is readily explained, when we consider that the bones of all dead creatures that may be left uncovered upon dry land, are in a few years entirely destroyed by various animals, and the decomposing influence of the atmosphere. If we except the few bones that may have been collected in caves, or buried under land slips, or the products of volcanic Eruptions, or in sand drifted by the winds,† it is only in strata formed

* Under the turbulent conditions of our planet, whilst stratification was in progress, the activity of volcanic agents, then frequent and intense, was probably attended also with atmospheric disturbances affecting both the air and water, and producing the same fatality among the then existing Tribes of fishes, that is now observed to result from sudden and violent changes in the electric condition of the atmosphere. M. Agassiz has observed that rapid changes in the degree of atmospheric pressure upon the water, affect the air within the swimming bladders of fishes, sometimes causing them to be distended to a fatal degree, and even to burst. Multitudes of dead fishes, that have thus perished during tempests, are often seen floating on the surface, and cast on the shores of the lakes of Switzerland.

† Captain Lyon states, that in the deserts of Africa, the bodies of camels are often desiccated by the heat and dryness

by water that any remains of land animals can have been preserved.

We continually see the carcases of such animals drifted by rivers in their seasons of flood, into lakes, estuaries, and seas; and although it may at first seem strange to find terrestrial remains, imbedded in strata formed at the bottom of the water, the difficulty vanishes on recollection that the materials of stratified rocks are derived in great part from the Detritus of more

of the atmosphere, and become the nucleus of a sand hill; which the wind accumulates around them. Beneath this sand they remain interred like the stumps of palm trees, and the buildings of ancient Egypt.

In a recent paper on the geology of the Bermudas (Proceedings of Geol. Soc. Lond. Ap. 9, 1834), Lieutenant Nelson describes these islands as composed of calcareous sand and limestone, derived from comminuted shells and corals; he considers great part of the materials of these strata to have been drifted up from the shore by the action of the wind. The surface in many parts is composed of loose sand, disposed in all the irregular forms of drifted snow, and presents a surface covered with undulations like those produced by the ripple of water upon sand on the sea shore. Recent shells occur both in the loose sand and solid limestone, and also roots of the Palmetto now growing in the island. The N. W. coast of Cornwall affords examples of similar invasions of many thousand acres of land by Deluges of sand drifted from the sea shore, at the villages of Bude, and Perran Zabulo; the latter village has been twice destroyed, and buried under sand, drifted inland during extraordinary tempests, at distant intervals of time. See Trans. of Geol. Soc. of Cornwall, vol. ii. p. 140. and vol. iii. p. 12. See also De la Beche's Geological Manual, 3rd edit. p. 84, and Jameson's Translation of Cuvier's Theory of the Earth, 5th ed. Note G.

ancient lands. As the forces of rains, torrents, and inundations have conveyed this detritus into lakes, estuaries, and seas, it is probable that many carcases of terrestrial and amphibious animals, should also have been drifted to great distances by currents which swept such enormous quantities of abraded matter from the lands; and accordingly we find, that strata of aqueous formation have become the common repository not only of the Remains of aquatic, but also of terrestrial animals and vegetables.

The study of these Remains will form our most interesting and instructive subject of inquiry, since it is in them that we shall find the great master key whereby we may unlock the secret history of the earth. They are documents which contain the evidences of revolutions and catastrophes, long antecedent to the creation of the human race; they open the book of nature, and swell the volumes of science, with the Records of many successive series of animal and vegetable generations, of which the Creation and Extinction would have been equally unknown to us, but for recent discoveries in the science of Geology.

Chapter XIII.

Aggregate of Animal Enjoyment increased, and that of Pain diminished, by the existence of Carnivorous Races.

BEFORE we proceed to consider the evidences of design, discoverable in the structure of the extinct carnivorous races, which inhabited our planet during former periods of its history; we may briefly examine the nature of that universal dispensation, whereby a system of perpetual destruction, followed by continual renovation, has at all times tended to increase the aggregate of animal enjoyment, over the entire surface of the terraqueous globe.

Some of the most important provisions which will be presented to us in the anatomy of these ancient animals, are found in the organs with which they were furnished for the purpose of capturing and killing their prey; and as contrivances exhibited in instruments formed expressly for destruction, may at first sight, seem inconsistent with the dispensations of a creation founded in benevolence, and tending to produce the greatest amount of enjoyment to

G.　　　　　　　K

the greatest number of individuals; it may be proper to premise a few words upon this subject, before we enter on the history of that large portion of the animals of a former world, whose office was to effect the destruction of life.

The law of universal mortality being the established condition, on which it has pleased the Creator to give being to every creature upon earth, it is a dispensation of kindness to make the end of life to each individual as easy as possible. The most easy death is, proverbially, that which is the least expected; and though, for moral reasons peculiar to our own species, we deprecate the *sudden* termination of our mortal life; yet, in the case of every inferior animal, such a termination of existence is obviously the most desirable. The pains of sickness, and decrepitude of age, are the usual precursors of death, resulting from gradual decay: these, in the human race alone, are susceptible of alleviation from internal sources of hope and consolation; and give exercise to some of the highest charities, and most tender sympathies of humanity. But, throughout the whole creation of inferior animals, no such sympathies exist; there is no affection or regard for the feeble and aged; no alleviating care to relieve the sick; and the extension of life through lingering stages of decay and of old age, would to each individual be a scene of protracted misery. Under such a system, the natural world would present a mass of daily suffering, bearing a large proportion to the total amount of animal enjoyment. By the existing dispensations of sudden destruction and rapid succession, the feeble and disabled are speedily relieved from suffering, and the world is at all times crowded with myriads of sentient and happy beings; and though to many individuals their allotted share of life be often short, it is usually a period of uninterrupted gratification; whilst the momentary pain of sudden and unexpected death is an evil infinitely small, in comparison with the enjoyments of which it is the termination.

The inhabitants of the earth have ever been divided into two great classes, the one herbivorous, the other carnivorous; and though the existence of the latter may, at first sight, seem calculated to increase the amount of animal pain; yet, when considered in its full extent, it will be found materially to diminish it.

To the mind which looks not to general results in the economy of Nature, the earth may seem to present a scene of perpetual warfare, and incessant carnage: but the more enlarged view, while it regards individuals in their conjoint relations to the general benefit of their own species, and that of other species with which they are associated in the great family of Nature, resolves each apparent case of individual evil, into an example of subserviency to universal good.

Under the existing system, not only is the aggregate amount of animal enjoyment much increased, by adding to the stock of life all the races which are carnivorous, but these are also highly beneficial even to the herbivorous races, that are subject to their dominion.

Besides the desirable relief of speedy death on the approach of debility or age, the carnivora confer a further benefit on the species which form their prey, as they control their excessive increase, by the destruction of many individuals in youth and health. Without this salutary check, each species would soon multiply to an extent, exceeding in a fatal degree their supply of food, and the whole class of herbivora would ever be so nearly on the verge of starvation, that multitudes would daily be consigned to lingering and painful death by famine. All these evils are superseded by the establishment of a controlling Power in the carnivora; by their agency the numbers of each species are maintained in due proportion to one another—the sick, the lame, the aged, and the supernumeraries, are consigned to speedy death; and while each suffering individual is soon relieved from pain, it contributes its enfeebled carcase to the support of its carnivorous benefactor, and leaves more room for the comfortable existence of the healthy survivors of its own species.

The same "police of Nature," which is thus beneficial to the great family of the inhabitants of the land, is established with equal advantage among the tenants of the sea. Of these also, there is one large division that lives on vegetables, and supplies the basis of food to the other division that is carnivorous. Here again we see, that in the absence of carnivora, the uncontrolled herbivora would multiply indefinitely, until the lack of food brought them also to the verge of starvation; and the sea would be crowded with creatures under the endurance of universal pain from hunger, while death by famine would be the termination of ill fed and miserable lives.

The appointment of death by the agency of carnivora, as the ordinary termination of animal existence, appears therefore in its main results to be a dispensation of benevolence; it deducts much from the aggregate amount of the pain of universal death; it abridges, and almost annihilates, throughout the brute creation, the misery of disease, and accidental injuries, and lingering decay; and imposes such salutary restraint upon excessive increase of numbers, that the supply of food maintains perpetually a due ratio to the demand. The result is, that the surface of the land and depths of the

waters are ever crowded with myriads of animated beings, the pleasures of whose life are
co-extensive with its duration; and which,
throughout the little day of existence that is
allotted to them, fulfil with joy the functions
for which they were created.   Life to each individual is a scene of continued feasting, in a
region of plenty; and when unexpected death
arrests its course, it repays with small interest
the large debt, which it has contracted to the
common fund of animal nutrition, from whence
the materials of its body have been derived.
Thus the great drama of universal life is perpetually sustained; and though the individual
actors undergo continual change, the same parts
are ever filled by another and another generation; renewing the face of the earth, and the
bosom of the deep, with endless successions of
life and happiness.

Chapter XIV.

Proofs of Design in the Structure of Fossil Vertebrated Animals.

SECTION I.

FOSSIL MAMMALIA.—DINOTHERIUM.

ENOUGH has, I trust, been stated in the preceding chapter, to show the paramount importance of appealing to organic remains, in illustration of that branch of physico-theology with which we are at present occupied.

The structure of the greater number, even of the earliest fossil Mammalia, differs in so few essential points from that of the living representatives of their respective Orders, that I forbear to enter on details which would indeed abound with evidences of creative design, but would offer little that is not equally discoverable in the anatomy of existing species. I shall, therefore, limit my observations to two extinct genera, which are perhaps the most remarkable of all fossil Mammalia, for size and unexampled peculiarities of anatomical construction; the first of these, the Dinotherium, having been the largest of terrestrial Mammalia; and the second, the Megatherium, presenting greater deviations from ordinary animal forms, than occur in any other species, either of recent or fossil quadrupeds.

It has been already stated, in our account of the Mammalia of the Miocene period of the tertiary series, that the most abundant remains of the Dinotherium are found at Epplesheim, in the province of Hesse Darmstadt, and are described, in a work now in process of publication, by Professor Kaup. Fragments of the same genus are mentioned by Cuvier, as occurring in several parts of France, and in Bavaria and Austria.

The form of the molar teeth of the Dinotherium (Pl. 2, C. Fig. 3), so nearly resembles that of the Tapirs, that Cuvier at first referred them to a gigantic species of this genus. Professor Kaup has since placed this animal in the new genus Dinotherium, holding an intermediate place between the Tapir and the Mastodon, and supplying another important extinct link in the great family of Pachydermata. The largest species of this genus, D. Giganteum, is calculated, both by Cuvier and Kaup, to have attained the extraordinary length of eighteen feet. The most remarkable bone of the body yet found is the shoulder-blade, the form of which more nearly resembles that of a Mole than of any other animal, and seems to indicate a peculiar adaptation of the fore leg to the purposes of digging, an indication which is corroborated by the remarkable structure of the lower jaw.

The lower jaws of two species of Dinotherium, figured in Plate 2. C. Figs. 1. 2. exhibit peculiarities in the disposition of the tusks, such as are found in no other living or fossil animal.

The form of the molar teeth, Pl. 2. C. Fig. 3, approaches, as we have stated, most nearly to that of the molar teeth in Tapirs ; but a remarkable deviation from the character of Tapirs, as well as of every other quadruped, consists in the presence of two enormous tusks, placed at the anterior extremity of the lower jaw, and curved downwards, like the tusks in the upper jaw of the Walrus. (Pl. 2. C. 1. 2.)

I shall confine my present remarks to this peculiarity in the position of the tusks, and endeavour to show how far these organs illustrate the habits of the extinct animals in which they are found. It is mechanically impossible that a lower jaw, nearly four feet long, loaded with such heavy tusks at its extremity, could have been otherwise than cumbrous and inconvenient to a quadruped living on dry land. No such disadvantage would have attended this structure in a large animal destined to live in water ; and the aquatic habits of the family of Tapirs, to which the Dinotherium was most nearly allied, render it probable that, like them, it was an inhabitant of fresh-water lakes and

rivers. To an animal of such habits, the weight
of the tusks sustained in water would have been
no source of inconvenience; and, if we suppose
them to have been employed, as instruments for
raking and grubbing up by the roots large
aquatic vegetables from the bottom, they would,
under such service, combine the mechanical
powers of the pick-axe with those of the horse-
harrow of modern husbandry. The weight of
the head, placed above these downward tusks,
would add to their efficiency for the service here
supposed, as the power of the harrow is increased
by being loaded with weights.

The tusks of the Dinotherium may also have
been applied with mechanical advantage to hook
on the head of the animal to the bank, with the
nostrils sustained above the water, so as to
breathe securely during sleep, whilst the body
remained floating, at perfect ease, beneath
the surface : the animal might thus repose,
moored to the margin of a lake or river,
without the slightest muscular exertion, the
weight of the head and body tending to fix and
keep the tusks fast anchored in the substance of
the bank; as the weight of the body of a sleep-
ing bird keeps the claws clasped firmly around
its perch. These tusks might have been further
used, like those in the upper jaw of the Walrus, to assist in dragging the body out of the
water; and also as formidable instruments of
defence.

The structure of the scapula, already noticed, seems to show that the fore leg was adapted to co-operate with the tusks and teeth, in digging and separating large vegetables from the bottom. The great length attributed to the body, would have been no way inconvenient to an animal living in the water, but attended with much mechanical disadvantage to so weighty a quadruped upon land. In all these characters of a gigantic, herbivorous, aquatic quadruped, we recognize adaptations to the lacustrine condition of the earth, during that portion of the tertiary periods, to which the existence of these seemingly anomalous creatures appears to have been limited.

SECTION II.

MEGATHERIUM.

As it will be quite impossible, in the present Treatise, to give particular descriptions of the structure, even of a few of the fossil Mammalia, which have been, as it were, restored again to life by the genius and industry of Cuvier; I shall endeavour to illustrate, by the details of a single species, the method of analytical investigation, that has been applied by that great philosopher to the anatomy both of fossil and recent animals.

The result of his researches, as recorded in the Ossemens Fossiles, has been to show that all fossil quadrupeds, however differing in generic, or specific details, are uniformly constructed on the same general plan, and systematic basis of organization as living species; and that throughout the various adaptations of a common type to peculiar functions, under different conditions of the earth, there prevails such universal conformity of design, that we cannot rise from the perusal of these inestimable volumes, without a strong conviction of the agency of one vast and mighty Intelligence, ever directing the entire fabric, both of past and present systems of creation.

Nothing can exceed the accuracy of the severe and logical demonstrations, that fill these volumes with proofs of wise design, in the constant relation of the parts of animals to one another, and to the general functions of the whole body. Nothing can surpass the perfection of his reasoning, in pointing out the beautiful contrivances, which are provided in almost endless variety, to fit every living creature to its own peculiar state and mode of life. His illustration of the curious conditions, and concurrent compensations that are found in the living Elephants, apply equally to the extinct fossil species of the same genus; and similar exemplifications may be extended from the living to the extinct species of other genera, e. g. Rhinoceros, Hippopotamus, Horse, Ox, Deer, Tiger, Hyæna, Wolf, &c. that are usually associated with the Elephant in a fossil state.

The animal I shall select for my present purpose is that most extraordinary fossil creature, the Megatherium, (see Pl. 5), an animal, in some parts of its organization, nearly allied to the Sloth, and, like the Sloth, presenting an apparent monstrosity of external form, accompanied by many strange peculiarities of internal structure, which have hitherto been but little understood.

The Sloths have afforded a remarkable exception to the conclusions which naturalists have usually drawn, from their study of the organic structure and mechanism of other animals. The adaptation of each part of the body of the Elephant, to produce extraordinary strength, and of every member of the Deer and Antelope to give agility and speed, are too obvious to have escaped the attention of any scientific observer; but, it has been the constant practice of naturalists, to follow Buffon in misrepresenting the Sloths, as the most imperfectly constructed among all the members of the animal kingdom, as creatures incapable of enjoyment, and formed only for misery.

The Sloth does, indeed, afford the greatest deviations from the ordinary structure of living quadrupeds; and these have been erroneously considered as imperfections in its organization, without any compensating advantage. I have elsewhere * attempted to show that these ano-

* Linnean Transactions, Vol. XVII. Part 1.

malous conditions are so far from being defects, or sources of inconvenience in the Sloth, that they afford striking illustrations of the varied contrivances, whereby the structure of every creature is harmoniously adapted to the state in which it was destined to live. The peculiarities of the Sloth, that render its movements so awkward on the earth, are fitted with much advantage to its destined office of living entirely upon trees, and feeding upon their leaves: so also, if we consider the Megatherium with a view to its province of digging and feeding upon roots, we shall, in this habit, discover the explanation of its unusual structure, and apparently incongruous proportions; and find, in every organ, a relation of obvious convenience, and of adaptation to the office it had to discharge.*

It will be my present object to enter into such a minute investigation of some of the more remarkable parts of this animal, viewing them with a constant reference to a peculiar mode of life, as may lead to the recognition of a system of well

* The remains of the Megatherium have been found chiefly in the southern regions of America, and most abundantly in Paraguay; it appears also to have extended on the north of the equator as far as the United States. We have, for some time, possessed detailed descriptions of this animal by Cuvier, Oss. Foss. vol. 5. and a series of large engravings, by Pander and D'Alton, taken from a nearly perfect skeleton, sent in 1789 from Buenos Ayres to Madrid. Dr. Mitchell and Mr. Cooper have described, in the Annals of the Lyceum of Nat. Hist. of New York, May, 1824, some teeth and bones found in the marshes of

connected contrivances, in the mechanism of a creature apparently the most monstrous, and seeming to present the most ill-assorted proportions, that occur throughout the entire range of the animal kingdom.

We have here before us a gigantic quadruped, (see Pl. 5, Fig. 1), which at first sight appears not only ill-proportioned as a whole, but whose members also seem incongruous, and clumsy, if considered with a view to the functions and corresponding limbs of ordinary quadrupeds : let us only examine them with the aid of that clue, which is our best and essential guide in every investigation of the mechanism of the animal frame ; let us first infer from the total composition and capabilities of the machinery, what was the general nature of the work it was destined to perform ; and from the character of the most important parts, namely, the feet and teeth, make ourselves acquainted with the food these organs were adapted to procure and masticate; and we shall find every other member of the body act-

---

the Isle of Skiddaway, on the coast of Georgia, which correspond with the skeleton at Madrid.   Cuvier, Vol. V. part 2, p. 519.— In the year 1832, many parts of another skeleton were brought to England by Woodbine Parish, esq. from the bed of the river Salado, near Buenos Ayres : these are placed in the museum of the Royal College of Surgeons in London, and will be described in the Trans. Geol. Soc. Lond. Vol. III. N. S. Part 3, by my friend Mr. Clift, a gentleman from whose great anatomical knowledge, I have derived most important aid, in my investigation of this animal.

ing in harmonious subordination to this chief purpose in the animal economy.

In the case of ordinary animals, the passage from one form to another is so gradual, and the functions of one species receive such ample and obvious illustrations from those of the species adjacent to it, that we are rarely at a loss, to see the final cause of almost every arrangement that is presented to the anatomist. This is more especially the case with respect to the skeleton, which forms the foundation of all the other mechanisms within the body, and is of the highest importance in the history of fossil animals, of which we rarely find any other remains besides the bones, and teeth, and the scaly or osseous integuments. I select the Megatherium, because it affords an example of most extraordinary deviations, and of egregious apparent monstrosity; viz. the case of a gigantic animal exceeding the largest Rhinoceros in bulk, and to which the nearest approximations that occur in the living world, are found in the not less anomalous genera of Sloth, Armadillo, and Chlamyphorus; the former adapted to the peculiar habit of residing upon trees; the two latter constructed with unusual adaptations to the habit of burrowing in search of their food and shelter in sand; and all limited in their geographical distribution, nearly to the same regions of America that were once the residence of the Megatherium.

I shall not here enter on the unsettled questions as to the precise age of the deposits in which the Megatherium is found, or the causes by which it has been extirpated; my object is to show that the apparent incongruities of all its parts, are in reality systems of wise and well contrived adaptation to a peculiar mode of life. I proceed therefore to consider, in the order in which they are described by Cuvier, the most important organs of the Megatherium, beginning with the head, and from thence advancing to the trunk and extremities.

Head.

The bones of the head (Pl. 5, Fig. 1. a.) most nearly resemble those of a Sloth. The long and broad bone, (b,) descending the cheek from the zygomatic arch, connects it more nearly with the Ai than with any other animal: this extraordinary bone must have been auxiliary to the power of muscles, acting with more than usual advantage, in giving motion to the lower jaw (d).

The anterior part of the muzzle (c) is so strong and substantial, and so perforated with holes for the passage of nerves and vessels, that we may be sure it supported some organ of considerable size: a long trunk was needless to an animal possessing so long a neck; the organ was probably a snout, something like that of

G. L

the Tapir, sufficiently elongated to gather up roots from the ground. The septum of the nostrils also being strong and bony, gives further indication of the presence of a powerful organ appended to the nose; such an apparatus would have afforded compensation for the absence of incisor teeth and tusks. Having no incisors, the Megatherium could not have lived on grass. The structure of the molar teeth (Pl. 5, Fig. 6—11, and Pl. 6, No. 1), shows that it was not carnivorous.

The composition of a single molar tooth resembles that of one, of the many denticules, that are united in the compound molar of the Elephant; and affords an admirable exemplification of the method employed by Nature, whereby three substances, of unequal density, viz. *ivory, enamel*, and *crusta petrosa*, or *cœmentum*, are united in the construction of the teeth of graminivorous animals. The teeth are about seven inches long, and nearly of a prismatic form (Pl. 5, Fig. 7. 8). The grinding surfaces (Pl. 5. Fig. 9. a. b. c. and Pl. 6, Z. a. b. c.) exhibit a peculiar and beautiful contrivance for maintaining two cutting wedge-shaped salient edges, in good working condition during the whole existence of the tooth; being, as I before stated, a modification of the contrivance employed in the molars of the Elephant, and other herbivora. The same principle is applied by tool-makers for the purpose of maintaining a sharp edge in axes, scythes, bill-hooks, &c. An axe, or bill-hook, is not made entirely of steel, but of one thin plate of steel, inserted between two plates of softer iron, and so enclosed that the steel projects beyond the iron, along the entire line of the cutting edge of the instrument. A double advantage results from this contrivance; first, the instrument is less liable to fracture than if it were entirely made of the more brittle material of steel; and secondly, the cutting edge is more easily kept sharp by grinding down a portion of exterior soft iron, than if the entire mass were of hard steel. By a similar contrivance, two cutting edges are produced on the crown of the molar teeth of the Megatherium. (See Pl. 6, W. X. Y. Z. and Pl. 5, Figs. 6—10.) *

---

* The outside of the tooth, like that of an axe, is made of a comparatively soft material, viz. the crusta petrosa, (a a), inclosing a plate of enamel, (b b), which is the hardest substance, or steel of the tooth. This enamel passes twice across the grinding surface, (z), and forms the cutting edges of two parallel wedges, Y. b. b.: a longitudinal section of these wedges is seen, Pl. 6. v. w. x. y. Within the enamel, (b b), is a central mass of ivory, (c), which, like the external crust, (a) is softer than the enamel. A tooth, thus constructed of materials of unequal density, would have its softer parts, (a c), worn down more readily than the harder plates of enamel, (b b).

We find a further nicety of mechanical contrivance, for producing and maintaining two transverse wedges upon the surface

Pl. 6, W. X. represents the manner in which each lower tooth was opposed to the tooth above it, so that the hard enamel of the one should come in contact only with the softer materials of the other; viz. the edges of the plates of enamel, (b) rubbing upon the ivory, (c); and the enamel, (b'), upon the crusta petrosa, (a), of the two teeth opposite to it. Hence the act of mastication formed and perpetually maintained a series of wedges, locking into each other like the alternate ridges on the rollers of a crushing-mill; and the mouth of the Megatherium became an engine of prodigious power, in which thirty-two such wedges formed the grinding surfaces of sixteen molar teeth; each from seven to nine inches long, and having the greater part of this length fixed firmly in a socket of great depth.

As the surfaces of these teeth must have worn away with much rapidity, a provision, unusual in molar teeth, and similar to that in the incisor teeth

of each tooth, in the relative adjustment of the thickness, of the lateral and transverse portions of the plate of enamel, which is interposed between the external crust, (a), and the central ivory, (c). Had this enamel been of uniform thickness all round the central ivory, the tooth would have worn down equally to a horizontal surface. In the crown of the tooth, Pl. 6. Z. the plate of enamel is seen to be thin on the two sides of the tooth, whilst the transverse portions of the same plate, (b. b.) are comparatively thick and strong. Hence the weaker lateral portions of thin enamel wear away more rapidly, than the thicker and stronger transverse portions, (b b), and do not prevent the excavation of the furrow across the surface of the ivory, c.

of the Beaver and other Rodentia,* supplied the loss that was continually going on at the crown, by the constant addition of new matter at the root, which for this purpose remained hollow, and filled with pulp during the whole life of the animal.†

It is scarcely possible to find any apparatus in the mechanism of dentition, which constitutes a more powerful engine for masticating roots, than was formed by these teeth of the Megatherium; accompanied also by a property, which is the perfection of all machinery, namely, that of maintaining itself perpetually in perfect order, by the act of performing its work.

Lower Jaw.

The lower jaw (Pl. 5, 1. d.) is very large and weighty in proportion to the rest of the head; the object of this size being to afford deep sockets

---

* The incisors of the Beaver, and other Rodentia, and tusks of the Hog and Hippopotamus, which require only an external cutting edge, and not a grinding surface, are constructed on the same principle as the cutting edge of a chisel or an adze; viz. a plate of hard enamel is applied to the outer surface only, of the ivory of these teeth, in the same manner as the outer cutting edge of the chisel and adze is faced with a plate of steel, welded against an inner plate of softer iron. A tooth thus constructed maintains its cutting edge of enamel continually sharp, by the act of working against the similarly constructed extremity of the tooth opposed to it.

† Pl. 5, Fig. 11, represents the section of the cavity containing this pulp.

for the continual growth and firm fixture of the long and vertical molar teeth ; the extraordinary and strong process (b) descending from the zygomatic arch in the Megatherium, as well as in the Sloths, seems intended to support the unusual weight of the lower jaw consequent upon the peculiar form of the molar teeth.

Bones of the Trunk.

The vertebræ of the neck, though strong, are small in comparison with those towards the opposite extremity of the body ; being duly proportioned to the size of a head, comparatively light, and without tusks. The dorsal portion of the vertebral column is of moderate size, but there is an enlargement of the vertebræ of the loins, corresponding with the extraordinary bulk of the pelvis and hind legs ; the summits of the spinous processes, (e,) are flattened like those in the Armadillo, as if by the pressure of a cuirass.

The sacral bone, (Pl. 5, Fig. 2, a), is united to the pelvis, (p), in a manner peculiar to itself, and calculated to produce extraordinary strength ; its processes indicate the existence of very powerful muscles for the movement of the tail. The tail was long, and composed of vertebræ of enormous magnitude, (Pl. 6, Fig. 2), the body of the largest being seven inches in diameter, and

the horizontal distance between the extremities
of the two transverse processes, being twenty
inches. If to this we add the thickness of the
muscles and tendons, and of the shelly integument, the diameter of the tail, at its largest end,
must have been at least two feet; and its circumference, supposing it to be nearly circular
like the tail of the Armadillo, about six feet.
These vast dimensions are not larger in proportion to the adjacent parts of the body, than
those of the tail of the Armadillo, and as this
animal applies its tail, to aid in supporting the
weight of its body and armour, it is probable
that the Megatherium made a similar use of
the same organ.* To the caudal vertebræ were
attached also large inferior spines, or additional
Chevron bones, which must have added to the
strength of the tail, in assisting to support the
body. The tail also probably served for a
formidable instrument of defence, as in the
Pangolins and Crocodiles. In 1822, Sellow saw
portions of armour that had covered a tail, found
near Monte Video.

The ribs are more substantial, and much
thicker, and shorter, than those of the Elephant

---

* The tail of the Elephant is remarkably light and slender,
with a tuft of coarse hair at its extremity, to brush off flies; that
of the Hippopotamus is a few inches only in length, and flattened
vertically, to act as a small rudder in swimming.

or Rhinoceros; and the upper convex surfaces of some of them exhibit a rugous and flattened condition of that part, on which the weight of a bony cuirass would most immediately have rested.

Anterior Extremity.

The *scapula* or shoulder blade, (Pl. 5, Fig. 1, f,) resembles that of no other family except the Sloths, and exhibits in the Acromion (g,) contrivances for strength, peculiar to itself and them, in its mode of articulation with the collar bone (h); it exhibits also unusual provisions for the support of the most powerful muscles for the movement of the arm.

The *clavicle* or collar bone (h) is strong, and curved nearly as in the human subject; the presence of this bone in the Megatherium, whilst it is wanting in the Elephant, Rhinoceros, and all the large ruminating animals, shews that the fore leg discharged some other office, than that of an organ of locomotion. This clavicle would give a steady and fixt position to the socket, or glenoid cavity of the scapula, admitting of rotatory motion in the fore leg, analogous to that of the human arm. There is in these circumstances a triple accommodation to the form and habits of the Megatherium; 1°. a free rotatory power of the arm was auxiliary to its office, as an instrument to be employed continually in digging food out of the ground ; 2°. this act of perpetual digging in search of stationary objects like roots, required but little locomotive power; 3°. the comparatively small support afforded to the weight of the body by the fore leg, was compensated by the extraordinary and colossal strength of the haunches and hind legs. In the Elephant, the great weight of the head and tusks require shortness of neck, and unusual enlargement and strength in the fore legs; hence, the anterior parts of this animal are much stronger and larger than its hinder parts. In the case of the Megatherium, the relative proportions are reversed; the head is comparatively small, the neck is long, and the anterior part of the body but slightly loaded in comparison with its abdominal and posterior regions. In the shoulder blade and collar bone there is great provision to give strength and motion to the fore legs ; but this motion is not progressive, nor is the strength calculated merely to support the weight of the body. The humerus, (k) articulates with the scapula by a round head, admitting of free motion in various directions, and is small at its upper and middle part, but at its lower end attains extraordinary breadth, in consequence of an enormous expansion of the crests, which rise from the condyles, to give origin to muscles for the movement of the fore foot and toes.* The ulna (l) is extremely broad and powerful at its upper extremity, affording large space for the origin of muscles, concerned in the movements of the foot. The radius (m) revolves freely on the ulna, as in the Sloths and Ant-eaters, both of which make much use of the fore leg, though for different purposes ; it has a cavity at its upper end, which turns upon a spherical portion of the lower part of the humerus, and a large apophysis (n), projecting from its longitudinal crest, indicates great power in the muscles that gave rotatory motion.

The entire fore-foot must have been about a yard in length, and more than twelve inches wide ; forming a most efficient instrument for moving the earth, from that depth within which succulent roots are usually most abundant. This great length of the fore-foot, when resting upon the ground, though unfavourable to progressive motion, must have enabled one fore-leg, when acting in conjunction with the two hind legs and tail, to support the entire weight of the body ; leaving the other fore-leg at liberty to be employed exclusively in the operation of digging food.†

* There is a similar expansion of the lower part of the Humerus in the Ant-eater, which employs its fore feet in digging up the solid hills of the Termite Ants.

† At Pl. 5, beneath Fig. 1, are represented the fore-foot of an Armadillo (Dasypus Peba), and the fore-foot of the Chlamyphorus, each adapted, like that of the Megatherium, to form an

The toes of the fore-foot are terminated by large and powerful claws of great length; the bones, supporting these claws, are composed partly of an axis, or pointed core, (o,) which filled the internal cavity of the horny claw; and partly of a bony sheath, that formed a strong case to receive and support its base. These claws were set obliquely to the ground, like the digging claws of the Mole, a position which made them instruments of greater power for the purpose of excavation.

Posterior Extremities.

The *pelvis* of the Megatherium (Pl. 5, Fig. 2. p.) is of vast solidity and expanse; and the enormous bones of the ilium (r) are set nearly at right angles to the spine of the back, and at their outer margin, or crest, are more than five feet asunder, very much exceeding the diameter across the haunches of the largest Elephant: the crest of the ilium, (s,) is much flattened, as if by the pressure of the armour. This enormous size of the pelvis would be disproportionate and inconvenient to an animal of ordinary stature and

instrument of peculiar power for the purpose of digging; and each presenting an extraordinary enlargement and elongation of the extreme bones of the toes, for the support of long and massive claws. At Pl. 5, Figs. 18, 19, the anterior parts of these animals are represented, and show how large a proportion the claws bear to the other parts of the body.

functions; but was probably attended with much advantage to the Megatherium, in relation to its habit of standing great part of its time on three legs, whilst the fourth was occupied in digging.

The pelvis being thus, unusually wide and heavy, presents a further deviation from other animals, as to the place and direction of the acetabulum, or socket which articulates with the head of the thigh bone (u). This cavity, in other animals, is usually set more or less obliquely outwards, and by this obliquity facilitates the movement of the hind leg; but in the Megatherium it is set perpendicularly downwards, over the head of the femur, and is also nearer than usual to the spine; deriving from this position increase of strength for supporting vertical pressure, but attended with a diminished capability of rapid motion.*

From the enormous width of the pelvis, it

---

* There is also a further peculiarity for the increase of strength in the manner in which that part, which, in most other animals, is an open space, called the *ischiatic notch* (Pl. 5, Fig. 2 c.), is nearly closed with solid bone by the union of the spines of the ischia with the elongated transverse processes of the sacral vertebræ, (a).

Further evidence of the enormous size and power in the muscles of the thigh and leg is afforded by the magnitude of the cavity in the sacrum, (Pl. 5. d,) for the passage of the spinal marrow: this cavity being about four inches in diameter, the spinal marrow must have been a foot in circumference. The extraordinary magnitude also of the nerves which proceeded from it to supply the leg, is indicated by the prodigious size of the sacral foramina.

follows also that the abdominal cavity was extremely large, and the viscera voluminous, and adapted to the digestion of vegetable food.

The form and proportions of the thigh bone, (v) are not less extraordinary than those of the pelvis, being nearly three times the thickness of the femur of the largest Elephant. Its breadth is nearly half its entire length, and its head is united to the body of the bone by a neck of unusual shortness and strength, twenty-two inches in circumference. Its length is two feet four inches, and its circumference at the smallest part two feet two inches; and at the largest part, three feet two inches. Its body is also flattened; and by means of this flatness, expanded outwards to a degree of which Nature presents no other example. These peculiarities in the femur appear to be subservient to a double purpose: first, to give extraordinary strength by the shortness and solidity of all its proportions; and secondly, to afford compensation, by its flatness outwards; for the debility which would otherwise have followed from the inward position of the sockets, (t,) by which the femur, (u,) articulates with the pelvis.

The two bones of the leg (x, y,) are also extremely short, and on a scale of solidity and strength, commensurate with that of the femur that rests upon them. This strength is much increased by their being united at both extremities; an union which is said by Cuvier to

occur in no other animals except the Armadillo
and Chlamyphorus; both of which are continually occupied in digging for their food.

The articulation of the leg with the hind foot
is admirably contrived for supporting the enormous pressure of downward weight; the astra-
galus (z), or great bone of the instep, being nine
inches broad, and nine inches high, is in due
proportion to the lower extremity of the tibia, or
leg bone, with which it articulates; and rests
upon a heel bone, of the extraordinary length of
seventeen inches, with a circumference of twenty-
eight inches. This enormous bone, pressing on
the ground, gives a firm bearing and solid support to the continuous accumulation of weight,
which we have been tracing down from the
pelvis through the thigh and leg: in fact the
heel bone occupies nearly one-half of the entire
length of the hind foot; the bones of the toes
are all short, excepting the extreme joint, which
forms an enormous claw-bone; larger than
the largest of those in the fore-foot, measuring
thirteen inches in circumference, and having
within its sheath a core, ten inches long, for the
support of the horny claw with which it was
invested. The chief use of this large claw was
probably to keep the hind foot fixed steadily
upon the ground.*

* It is probable that the large thick claw, Pl. 5 5', was placed
on the second toe of the hind foot. Its size approaches nearly
to that of the first toe of this foot, and both of these differ mate-

Feet and legs thus heavily constructed, must have been very inefficient organs of rapid locomotion, and may consequently seem imperfect, if considered in relation to the ordinary functions of other quadrupeds; but, viewed as instruments adapted for supporting an almost stationary creature, of unusual weight, they claim our admiration equally with every other piece of animal mechanism, when its end and uses are understood. The perfection of any instrument can only be appreciated by looking to the work it is intended to perform. The hammer and anvil of an anchorsmith, though massive, are neither clumsy nor imperfect; but bear the same proportionate relation to the work in which they are employed, as the light and fine tools of the watchmaker bear to the more delicate wheels of his chronometer.

Bony Armour.

Another remarkable character of the Megatherium, in which it approaches most nearly to the Armadillo, and Chlamyphorus, consists, in its hide having probably been covered with a bony coat of armour; varying from three-fourths of an inch, to an inch and half in thickness, and

rially in form and proportions, from the three more elongated and flatter claw bones of the fore foot, the oblique form of which is peculiarly adapted for digging.

resembling the armour which covers these living inhabitants, of the same warm and sandy regions of South America. Fragments of this armour are represented at Pl. 5, Figs. 12, 13.*

A covering of such enormous weight, would have been consistent with the general structure of the Megatherium ; its columnar hind legs and colossal tail, were calculated to give it due support ; and the strength of the loins and ribs, being very much greater than in the Elephant, seems to have been necessary for carrying so ponderous a cuirass as that which we suppose to have covered the body.†

* The resemblance between some parts of this fossil armour, and of the armour of an Armadillo, (Dasypus Peba) is extended even to the detail of the patterns of the tuberculated compartments into which they are divided, see Pl. 5, Figs. 12, 14. The increase of size in the entire shield is in both cases provided for, by causing the centre of every plate to form a centre of growth, around which the margin receives continual additions, as the increasing bulk of the body requires an increase in the dimensions of the bony case, by which it is invested. Figs. 15, 16, 17, represent portions of the armour of the head, body, and tail piece of the Chlamyphorus. Figs. 18, 19, represent the manner in which the armour is disposed over the head and anterior part of the body of the Chlamyphorus, and Dasypus Peba. The body of the Megatherium, when covered with its corresponding coat of armour, must in some degree have resembled a tilted waggon.

† In the Transactions of the Academy of Berlin, 1830, Professor Weis has published an account of some Bones of the Megatherium, discovered near Monte Video, accompanied by several fragments of bony armour. Much of this armour he

It remains to consider, of what use this cuirass could have been to the gigantic animal on which it probably was placed. As the locomotive organs of the Megatherium indicate very slow power of progression, the weight of a cuirass would have afforded little impediment to such tardy movements; its use was probably defensive, not only against the tusks and claws of beasts of prey, but also, against the myriads of insects, that usually swarm in such climates as those wherein its bones are found; and to which an animal that obtained its food by digging beneath a broiling sun, would be in a peculiar

refers without doubt to the Megatherium; other portions of it, and also many bones from the same district, he assigns to other animals. A similar admixture of bones and armour, derived from more than one species of animal, bearing a bony cuirass, is found in the collection made at several and distant points of the country above Buenos Ayres, by Mr. Parish. Although no armour was found with the fragments of the large skeleton, in the bed of the Salado, the rough broad flattened surface of a part of the crest of the ileum of this skeleton, (see Pl. 5, Fig. 2. r, s,) and the broad condition of the summit of the spinous processes of many vertebræ, and also of the superior convex portion of certain ribs on which the armour would rest, afford evidence of pressure, similar to that we find on the analogous parts of the skeleton of the Armadillo, from which we might have inferred that the Megatherium also was covered with heavy armour, even had no such armour been discovered near bones of this animal in other parts of the same level district of Paraguay. In all these flattened bones the effects of pressure are confined to those parts of the skeleton, on which the armour would rest, and are such as occur in a remarkable degree in the Armadillo.

G. M

degree exposed. We may also conjecture it to
have had a further use in the protection afforded
by it to the back, and upper parts of the body;
not only against the sun and rain, but against
the accumulations of sand and dust, that might
otherwise have produced irritation and disease.*

Conclusion.

We have now examined in detail the skeleton
of an extinct quadruped of enormous magnitude;
every bone of which presents peculiarities, that
at first sight appear imperfectly contrived, but
which become intelligible when viewed in their

---

* To animals that dig only occasionally, like Badgers, Foxes,
and Rabbits, to form a habitation beneath the ground, but seek
their food upon the surface, a defence of this kind would not
only have been unnecessary but inconvenient.

The Armadillo and Chlamyphorus are the only known animals
that have a coat of armour composed of thick plates of bone,
like that of the Megatherium. As this peculiar covering is confined to these quadrupeds, we can hardly imagine its use to be
solely for protection against other beasts and insects; but as the
Armadillo obtains its food by digging in the same dry and sandy
plains, which were once inhabited by the Megatherium, and
the Chlamyphorus lives almost entirely in burrows beneath the
surface of the same sandy regions; they both probably receive
from their cuirass the same protection to the upper parts of their
bodies from sand and dust, which we suppose to have been
afforded by its cuirass to the Megatherium. The Pangolins are
covered with a different kind of armour, composed of horny
moveable scales, in which there is no bony matter.

relations to one another, and to the functions of the animal in which they occur.

The size of the Megatherium exceeds that of the existing Edentata, to which it is most nearly allied, in a greater degree than any other fossil animal exceeds its nearest living congeners. With the head and shoulders of a Sloth, it combined in its legs and feet, an admixture of the characters of the Ant-eater, the Armadillo, and the Chlamyphorus; it probably also still further resembled the Armadillo and Chlamyphorus, in being cased with a bony coat of armour. Its haunches were more than five feet wide, and its body twelve feet long and eight feet high; its feet were a yard in length, and terminated by most gigantic claws; its tail was probably clad in armour, and much larger than the tail of any other beast, among extinct or living terrestrial Mammalia. Thus heavily constructed, and ponderously accoutred, it could neither run, nor leap, nor climb, nor burrow under the ground, and in all its movements must have been necessarily slow; but what need of rapid locomotion to an animal, whose occupation of digging roots for food was almost stationary? and what need of speed for flight from foes, to a creature whose giant carcase was encased in an impenetrable cuirass, and who by a single pat of his paw, or lash of his tail, could in an instant have demolished the Couguar or the

Crocodile? Secure within the panoply of his bony armour, where was the enemy that would dare encounter this Leviathan of the Pampas? or, in what more powerful creature can we find the cause that has effected the extirpation of his race?

- His entire frame was an apparatus of colossal mechanism, adapted exactly to the work it had to do; strong and ponderous, in proportion as this work was heavy, and calculated to be the vehicle of life and enjoyment to a gigantic race of quadrupeds; which, though they have ceased to be counted among the living inhabitants of our planet, have, in their fossil bones, left behind them imperishable monuments of the consummate skill with which they were constructed. Each limb, and fragment of a limb, forming co-ordinate parts of a well adjusted and perfect whole; and through all their deviations from the form and proportion of the limbs of other quadrupeds, affording fresh proofs of the infinitely varied, and inexhaustible contrivances of Creative Wisdom.

SECTION III.

FOSSIL SAURIANS.

In those distant ages that elapsed during the formation of strata of the secondary series, so large a field was occupied by reptiles, referrible to the order of Saurians, that it becomes an important part of our enquiry to examine the history and organization of these curious relics of ancient creations, which are known to us only in a fossil state. A task like this may appear quite hopeless to persons unaccustomed to the investigation of subjects of such remote antiquity; yet Geology, as now pursued, with the aid of comparative anatomy, supplies abundant evidence of the structure and functions of these extinct families of reptiles; and not only enables us to infer from the restoration of their skeletons, what may have been the external form of their bodies; but instructs us also as to their economy and habits, the nature of their food, and even of their organs of digestion. It further shows their relations to the then existing condition of the world, and to the other forms of organic life with which they were associated.

The remains of these reptiles bear a much greater resemblance to one another, than to those of any animals we discover in deposits preceding or succeeding the secondary series.*

The species of fossil Saurians are so numerous, that we can only select a few of the most remarkable among them, for the purpose of exemplifying the prevailing conditions of animal life, at the periods when the dominant class of animated beings were reptiles; attaining, in many cases, a magnitude unknown among the living orders of that class, and which seems to have been peculiar to those *middle ages* of geological chronology, that were intermediate between the transition and tertiary formations.

During these *ages of reptiles,* neither the carnivorous nor lacustrine Mammalia of the tertiary periods had begun to appear; but the most formidable occupants, both of land and water, were Crocodiles, and Lizards; of various forms, and often of gigantic stature, fitted to endure the turbulence, and continual convulsions of the unquiet surface of our infant world.

When we see that so large and important a range has been assigned to reptiles among the

---

* The oldest strata in which any reptiles have yet been found are those connected with the magnesian-limestone formation. (Pl. 1, Sec. 16). The existence of reptiles allied to the Monitor in the cupriferous slate and zechstein of Germany, has long been known. In 1834, two species of reptiles, allied to the Iguana and Monitor, were discovered in the dolomitic conglomerate, on Durdham Down, near Bristol.

former population of our planet, we cannot but
regard with feelings of new and unusual interest,
the comparatively diminutive existing orders
of that most ancient family of quadrupeds, with
the very name of which we usually associate a
sentiment of disgust. We shall view them with
less contempt, when we learn from the records of
geological history, that there was a time when
reptiles not only constituted the chief tenants,
and most powerful possessors of the earth, but
extended their dominion also over the waters of
the seas; and that the annals of their history
may be traced back through thousands of years,
antecedent to that latest point in the progressive
stages of animal creation, when the first parents
of the human race were called into existence.

Persons to whom this subject may now be
presented for the first time, will receive, with
much surprise, perhaps almost with incredulity,
such statements as are here advanced. It
must be admitted, that they at first seem much
more like the dreams of fiction and romance,
than the sober results of calm and deliberate
investigation; but to those who will examine the
evidence of facts upon which our conclusions
rest, there can remain no more reasonable doubt
of the former existence of these strange and
curious creatures, in the times and places we
assign to them; than is felt by the antiquary,
who, finding the catacombs of Egypt stored

with the mummies of Men, and Apes, and Crocodiles, concludes them to be the remains of
mammalia and reptiles, that have formed part
of an ancient population on the banks of the
Nile.

SECTION IV.

ICHTHYOSAURUS.

NEARLY at the head of the surprising discoveries, which have been made relating to the
family of Saurians, we may rank the remains
of many extraordinary species, which inhabited
the sea; and which present almost incredible
combinations of form, and structure; adapting
them for modes of life, that do not occur among
living reptiles. These remains are most abundant throughout the lias and oolite formations
of the secondary series.*   In these deposits we

* The chief repository in which these animals have been found
is the lias, at Lyme Regis ; but they abound also along the whole
extent of this formation throughout England, e. g. from the
coast of Dorset, through Somerset and Leicestershire, to the
coast of Yorkshire: they are found also in the lias of Germany
and France. The range of the genus Ichthyosaurus seems to have
begun with the Muschelkalk, and to have extended through the
whole of the oolitic period into the cretaceous formation. The
most recent stratum in which any remains of this genus have yet
been found is the chalk marl at Dover, where they have been
discovered by Mr. Mantell: I have found them in the gault,
near Benson, Oxon.

find not only animals allied to Crocodiles, and nearly approaching to the Gavial of the Ganges; but also still more numerous gigantic Lizards, that inhabited the then existing seas and estuaries.

Some of the most remarkable of these reptiles have been arranged under the genus Ichthyosaurus, (or Fish Lizard), in consequence of the partial resemblance of their vertebræ to those of fishes. (See Plate 1, Fig. 51, and Plates 7, 8, 9.) If we examine these creatures with a view to their capabilities of locomotion, and the means of offence and defence, which their extraordinary structure afforded to them; we shall find combinations of form and mechanical contrivances, which are now dispersed through various classes and orders of existing animals, but are no longer united in the same genus. Thus, in the same individual, the snout of a Porpoise is combined with the teeth of a Crocodile, the head of a Lizard with the vertebræ of a Fish, and the sternum of an Ornithorhynchus with the paddles of a Whale. The general outline of an Ichthyosaurus must have most nearly resembled the modern Porpoise, and Grampus. It had four broad feet, or paddles, (Pl. 7), and terminated behind in a long and powerful tail. Some of the largest of these reptiles must have exceeded thirty feet in length.

There are seven or eight known species of the genus Ichthyosaurus, all agreeing with one

another in the general principles of their construction, and the possession of those peculiar
organs, in which I shall endeavour to point out
the presence of mechanism and contrivance,
adapted to their habits and state of life. As
it will be foreign to our purpose to enter on
details respecting species, I shall content myself
with referring to the figures of the four most
common forms (Plates 7, 8, 9.)*

Head.

The head, which in all animals forms the most
important and characteristic part, (see Pl. 10,

* Pl. 7, is a large and nearly perfect specimen of the Ichthyosaurus Platyodon, from the lias at Lyme Regis, being one of
the splendid series of Saurians, purchased in 1834 of Mr.
Hawkins by the British Museum. Portions of the paddles, and
many lost fragments, are restored from the corresponding parts
which are preserved; a few vertebræ, and the extremity of
the tail are also restored conjecturally. Beautiful and accurate
lithographed figures of this specimen, and of the greater part of
this collection, are published in Mr. Hawkins's Memoirs of
Ichthyosauri and Plesiosauri, London, 1834. Pl. 8. Fig. 1, is a
small specimen of the Ichthyosaurus Communis, from the lias
at Lyme-Regis, belonging to the Geol. Soc. of London. Pl. 8,
Fig. 2, a small Ichthyosaurus Intermedius, from the lias at Lyme
Regis belonging to Sir Astley Cooper. Pl. 9, Fig. 1, an Ichthyosaurus Tenuirostris, from the lias of Street, near Glastonbury, in
the collection of Rev. D. Williams. Fig. 2 is the continuation
of the tail, and Fig. 3, the reverse of the head. The teeth in
this species are small, and in due proportion to the slender
character of the snout.

Figs. 1, 2), at once shows that the Ichthyosauri were Reptiles, partaking partly of the characters of the modern Crocodiles, but more allied to Lizards. They approach nearest to Crocodiles in the form and arrangement of their teeth. The position of the nostril is not, as in Crocodiles, near the point of the snout; it is set, as in Lizards, near the anterior angle of the orbit of the eye. The most extraordinary feature of the head is the enormous magnitude of the eye, very much exceeding that of any living animal.* The expansion of the jaws must have been prodigious; their length in the larger species, (Ichthyosaurus Platyodon), sometimes exceeding six feet; the voracity of the animal was doubtless in proportion to its powers of destruction. The neck was short, as in fishes.

Teeth.

The teeth of the Ichthyosaurus (Pl. 11, B, C,) are conical, and much like those of the Crocodiles, but considerably more numerous, amounting in some cases to a hundred and eighty; they vary in each species; they are not enclosed in deep and separate sockets, as the teeth of Croco-

---

* In the collection of Mr. Johnson at Bristol is a skull of Ichthyosaurus Platyodon, in which the longer diameter of the orbital cavity measures fourteen inches.

diles, but are ranged in one long continuous
furrow, (Pl. 11, B, c), of the maxillary bone, in
which the rudiments of a separation into distinct
alveoli may be traced in slight ridges extending between the teeth, along the sides and
bottom of the furrow. The contrivance by
which the new tooth replaces the old one, is
very nearly the same in the Ichthyosauri as in
the Crocodiles (Pl. 11, A, B, c); in both, the
young tooth begins its growth at the base of
the old tooth, where, by pressure on one side,
it causes first a partial absorption of the base,
and finally a total removal of the body of the
older tooth, which it is destined to replace.*

As the predaceous habits of the Ichthyosauri
exposed them, like modern Crocodiles, to frequent loss of their teeth, an abundant provision
has in each case been made for their continual
renewal.

---

* In Pl. 11, Fig. A, shews the manner in which the older
tooth in the Crocodile becomes absorbed, by pressure of a younger
tooth rising within the cavity of its hollow base. Fig. c, represents a transverse section of the left side of the lower jaw of an
Ichthyosaurus, shewing two teeth in their natural place, within
the furrow of the jaw; the younger tooth, by lateral pressure,
has caused absorption of the inside portion of the base of the
older tooth. Fig. B, represents a transverse section of the entire
snout of an Ichthyosaurus, in which the lower jaw exhibits on
both sides, a small tooth (a), which has caused partial absorption
of the base of the larger tooth, (c). In the upper jaw, the
bases of two large teeth (d, d,) are seen in their respective
furrows.

Eyes.

The enormous magnitude of the eye of the Ichthyosaurus (Pl. 10, Fig 1, 2), is among the most remarkable peculiarities in the structure of this animal. From the quantity of light admitted in consequence of its prodigious size, it must have possessed very great powers of vision; we have also evidence that it had both microscopic and telescopic properties. We find on the front of the orbital cavity in which this eye was lodged, a circular series of petrified thin bony plates, ranged around a central aperture, where once was placed the pupil; the form and thickness of each of these plates very much resembles that of the scales of an artichoke (Pl. 10, Fig. 3). This compound circle of bony plates, does not occur in fishes; but is found in the eyes of many birds,* as well as of Turtles,

---

* The bony sclerotic of the Ichthyosaurus approaches to the form of the bony circle in the eye of the Golden Eagle (Pl. 10, Fig. 5); one of its uses in each case being to vary the sphere of distinct vision, in order to descry their prey at long or short distances. These bony plates also assist to maintain the prominent position of the front of the eye, which is so remarkable in birds. In Owls, whose nocturnal habits render distant vision impossible, Mr. Yarrel observes, that the bony circle (Pl. 10, Fig. 4), is concave, and elongated forwards, so that the front of the eye is placed at the end of a long tube, and thus projects beyond the loose and downy feathers of the head; he adds; " The *extent* of vision enjoyed by the Falcons is probably denied to the Owls, but their more spherical lens and corresponding cornea give them an *intensity* better suited to the opacity of the medium in which they are required to exercise this

Tortoises, and Lizards; and in a less degree in Crocodiles. (Pl. 10, Figs. 4. 5. 6.)

In living animals these bony plates are fixed in the exterior or sclerotic coat of the eye, and vary its scope of action, by altering the convexity of the cornea : by their retraction they press forward the front of the eye and convert it into a microscope; in resuming their position, when the eye is at rest, they convert it into a telescope. The soft parts of the eyes of the Ichthyosauri have of course entirely perished; but the preservation of this curiously constructed hoop of bony plates, shews that the enormous eye, of which they formed the front, was an optical instrument of varied and prodigious power, enabling the Ichthyosaurus to descry its prey at great or little distances, in the obscurity of night, and in the depths of the sea ; it also tends to associate the animal, in which it existed, with the family of Lizards, and exclude it from that of fishes.*

power. They may be compared to a person near-sighted, who sees objects with superior magnitude and brilliancy when within the prescribed limits of his natural powers of vision, from the increased angle these objects subtend." Yarrel on the Anatomy of Birds of Prey, Zool. Journal, v. 3, p. 188.

* There are analogous contrivances for the purpose of resisting pressure, and maintaining the form of the eye in fishes, by the partial or total ossification of the exterior capsule; but in fishes, this ossification is usually simple, though carried to a different extent in different species; and the bone is never divided transversely into many plates, as in Lizards and Birds; these capsules of the eye are often preserved in the heads of fossil fishes : they abound in the London clay; and occasionally occur in chalk.

A further advantage resulting from this curious apparatus of bony plates, was to give strength to the surface of so large an eye-ball, enabling it the better to resist the pressure of deep water, to which it must often have been exposed ; it would also have protected this important organ from injury by the waves of the sea, to which an eye, sometimes larger than a man's head, must frequently have been subject, when the nose was brought to the surface, for the necessary purpose of breathing air : the position of the nostrils, close to the anterior angle of the eye, rendered it impossible for the Ichthyosaurus to breathe without raising its eye to the surface of the water.

Jaws.

The Jaws of the Ichthyosauri, like those of Crocodiles and Lizards, which are all more or less elongated into projecting beaks, are composed of many thin plates, so arranged as to combine strength with elasticity and lightness, in a greater degree than could have been effected by single bones, like those in the jaws of Mammalia. It is obvious that an under jaw so slender, and so much elongated as that of a Crocodile or Ichthyosaurus, and employed in seizing and retaining the large and powerful animals which formed their prey, would have been comparatively weak and liable to fracture

if composed of a single bone. Each side of the
lower jaw was therefore made up of six separate
pieces, set together in a manner that will be
best understood by reference to the Figures in
Pl. 11.*

This contrivance in the lower jaw, to combine
the greatest elasticity and strength with the
smallest weight of materials, is similar to that
adopted in binding together several parallel
plates of elastic wood, or steel, to make a cross-
bow; and also in setting together thin plates of
steel in the springs of carriages. As in the
carriage spring, or compound bow, so also
in the compound jaw of the Ichthyosaurus,
the plates are most numerous and strong, at
the parts where the greatest strength is required
to be exerted; and are thinner, and fewer,
towards the extremities, where the service to
be performed is less severe. Those who have

---

* These figures are selected from various plates by Mr. Conybeare and Mr. De la Beche. Fig. 1 is a restoration of the entire head
of an Ichthyosaurus, in which each component bone is designated
by the letters appropriated by Cuvier to the equivalent bones in
the head of the Crocodile. In the lower jaw, u, marks the dental
bone; v, the angular bone; x, superangular or coronoid; y, articular bone; z, complementary; ξ, opercular. Fig. 2, is part
of an under jaw of an Ichthyosaurus, shewing the manner in
which the flat bones, v, x, u, are applied to each other, towards
the posterior part of the jaw. Figs. 3, 4, 5, 6, 7, shew the
manner in which these bones overlap, and lock into each other,
at the transverse sections, indicated by the lines immediately
above them in Fig. 2. Fig. 8, shews the composition of the
bones in the lower jaw, as seen from beneath.

witnessed the shock given to the head of a
Crocodile, by the act of snapping together its
thin long jaws, must have seen how liable to
fracture the lower jaw would be, were it composed of one bone only on each side: a similar
inconvenience would have attended the same
simplicity of structure in the jaw of the Ichthyosaurus. In each case, therefore, the splicing
and bracing together of six thin flat bones of
unequal length, and of varying thickness, on both
sides of the lower jaw, affords compensation for
the weakness and risk of fracture, that would
otherwise have attended the elongation of the
snout.

Mr. Conybeare points out a further beautiful
contrivance in the lower jaw of the Ichthyosaurus,
analogous to the cross bracings lately introduced
in naval architecture, (see Pl. 11, Fig. 2.)*

Vertebræ.

The vertebral column in the Ichthyosaurus was
composed of more than one hundred joints; and

---

* The coronoid bone, (x) is interposed between the dental, (a),
and opercular (&), its fibres have a slanting direction, whilst
those of the two latter bones are disposed horizontally; thus,
the strength of the part is greatly increased by a regular dia-
gonal bracing, without the least addition of weight or bulk;
a similar structure may be noticed in the overlapping bones of
the heads of fish, and in a less degree, in those of Turtles.—
Geol. Trans. Lond. Vol. V. p. 565, and Vol. I. N. S. p. 112.

although united to a head nearly resembling
that of a Lizard, assumed, in the leading principles of its construction, the character of the
vertebræ of fishes. As this animal was constructed for rapid motion through the sea, the
mechanism of hollow vertebræ, which gives facility of movement in water to fishes, was
better calculated for its functions than the
solid vertebræ of Lizards and Crocodiles.*
(See Plate 12, A. and B.) This hollow conical form would be inapplicable to the vertebræ of land quadrupeds, whose back, being
nearly at right angles to the legs, requires a
succession of broad and nearly flat surfaces,
which press with considerable weight against

* The sections of the vertebræ of a fish (A c. c.) present two
hollow cones, united at their apex in the centre of each vertebra, in the form of an hour-glass; but the base of each cone,
(b. b.) instead of terminating in a broad flat surface, like the base
of the hour-glass, is bounded by a thin edge, like the edge of a
wine glass, and by this alone touches the corresponding edge of
the adjacent vertebra. Between these hollow vertebræ, a soft
and flexible intervertebral substance, in the form of a double
solid cone (e. e.) is so placed that each hollow cone of bone plays
on the cone of elastic substance contained within it, with a
motion in every direction ; thus forming a kind of universal joint,
and giving to the entire column great strength, and power of
rapid flexion in the water. But as the inflections in the perpendicular direction are less necessary than in the lateral, they are
limited by the overlapping, or contiguity of the spines.

This mode of articulation gives mechanical advantage to animals like fishes, whose chief organ of progressive motion is the
tail; and the weight of whose bodies being always suspended in
water, creates little or no pressure on the edges, by which alone
the vertebræ touch each other.

each other. It is quite certain, therefore, that such large and bulky creatures as the Ichthyosauri, having their vertebræ constructed after the manner of fishes, had they been furnished with legs instead of paddles, could not have moved on land without injury to their backs.*

Ribs.

The ribs were slender, and most of them bifurcated at the top: they were also continuous along the whole vertebral column, from the head to the pelvis, (see Plates 7, 8, 9); and in this respect agree with the structure of modern Lizards. A considerable number of them were united in front across the chest: their mode of articulation may be seen in Pl. 14.

* Sir E. Home has further remarked a peculiarity of the spinal canal, which exists in no other animals; the annular part (Pl. 12, D a. and E a.) being neither consolidated with the body of the vertebra, as in quadrupeds ; nor connected by a suture, as in Crocodiles ; but remaining always distinct, and articulating by a peculiar joint, resembling a compressed oval ball and socket joint, (D g. and E g.). And Mr. Conybeare adds, that this mode of articulation co-operates with the cup-shaped form of the intervertebral joints, in giving flexibility to the vertebral column, and assisting its vibratory motions; for, had these parts been consolidated, as in quadrupeds, their articulating processes must have locked the whole column together, so as to render such a motion of its parts impossible ; but by means of this joint every part yields to that motion. The tubercle by which the transverse apophysis of the head of the rib articulates with the vertebra, is seen at d.

The ribs of the right side were united to those of the left, by *intermediate* bones, analogous to the cartilaginous *intermediate* and *sternal* portions of the ribs in Crocodiles; and to the bones which, in the Plesiosaurus, form what Mr. Conybeare has called the *sterno-costal* arcs. (See Pl. 17.) This structure was probably subservient to the purpose of introducing to their bodies an unusual quantity of air; the animal by this means being enabled to remain long beneath the water, without rising to the surface for the purpose of breathing.*

* The sterno-costal ribs probably formed part of a condensing apparatus, which gave these animals the power of compressing the air within its lungs, before they descended beneath the water. In the Lond. and Edin. Phil. Mag. Oct. 1833, Mr. Faraday has noticed a method of preparing the organs of respiration in man, so as considerably to extend the time of holding the breath in an impure atmosphere; or under water, as practised by pearl-fishers; and illustrated by experiments of Sir Graves C. Houghton. If a person inspires deeply, and ceasing with his lungs full of air, holds his breath as long as he is able, the time during which he can remain without breathing will be double, or more than double, that which he could do if he held his breath without such deep inspiration. When Mr. Brunel, jun. and Mr. Gravatt descended in a diving-bell to examine the hole where the Thames had broken into the tunnel at Rotherhithe, at the depth of about thirty feet of water, Mr. Brunel, having inspired deeply the compressed air within the diving-bell, descended into the water below the bell; and found that he could remain twice as long under water, going into it from the diving-bell at that depth, as he could under ordinary circumstances.

Mr. Gravatt has also informed me that he is able to dive, and remain three minutes under water, after inflating his lungs with

Sternum.

To a marine animal that breathed air, it was essential to possess an apparatus whereby its ascent and descent in the water may have been easily accomplished ; accordingly we find such an apparatus, constructed with prodigious strength, in the anterior paddles of the Ichthyosaurus ; and in the no less extraordinary combination of bones that formed the sternal arch, or that part of the chest, on which these paddles rested. Pl. 12, Fig. 1.

It is a curious fact, that the bones composing the sternal arch are combined nearly in the same manner as in the Ornithorhynchus * of New Holland ; which seeks its food at the bottom of lakes and rivers, and is obliged, like the

the largest possible quantity of common air, by a succession of strong and rapid inspirations, and immediately compressing the lungs thus filled with air, by muscular exertion, and contraction of the chest, before he plunges into the water. By this compression of the lungs, the specific gravity of the body is also increased, and the descent is consequently much facilitated.

All these advantages were probably united in the mode of respiration of the Ichthyosaurus, and also in the Plesiosaurus.

* In this anomalous animal the Ornithorhynchus or Platypus, we have a quadruped clothed with fur, having a bill like a duck, with four webbed feet, suckling its young, and most probably ovoviviparous : the male is furnished with spurs.—See Mr. R. Owen's Papers on the Ornithorhynchus Paradoxus, in the Phil. Trans. London, 1832, Part II. and 1834, Part II. See also Mr. Owen's Paper on the same subject in Trans. Zool. Soc. Lond. Part III. 1835, in which he points out many approximations in the generative and other systems of this animal to the organization of reptiles.

Ichthyosaurus, to be continually rising to the surface to breathe air.*

Here then we have a race of animals that became extinct at the termination of the secondary series of geological formations, presenting, in their structure, a series of contrivances, the same in principle, with those employed at the present day to effect a similar purpose in one of the most curiously constructed aquatic quadrupeds of New Holland.†

Paddles.

In the form of its extremities, the Ichthyosaurus deviates from the Lizards, and approaches the Whales. A large animal, moving rapidly through the sea, and breathing air, must have

---

* In both these animals there is superadded to the ordinary type of bones in quadrupeds, an enlargement of the coracoid bone (c), and a peculiar form of sternum, resembling the furcula of birds. In Pl. 12, Fig. 1, a. represents the peculiar sternum or furcula; b. b. the clavicles; c. c. the coracoid bones; d. d. the scapulæ; e. e. the humeri; f. g. the radius and ulna. At Fig 2, the same letters are attached to the corresponding bones of the Ornithorhynchus.

The united power of all these bones imparts to the chest and paddles peculiar strength for an unusual purpose; not so much to effect progressive motion (which, in the Ichthyosaurus, was produced with much greater facility and power by the tail), as to ascend and descend vertically in quest of air and food.

† The Echidna, or spiny Ant-eater, of New Holland, is the only known land quadruped that has a similar furcula and clavicles. As this animal feeds on Ants, and takes refuge in deep burrows, this structure may be subsidiary to its great power of digging. A cartilaginous rudiment of a furcula occurs also in the Dasypus; and seems subservient to the same purpose.

required great modification of the fore-leg and foot of the Lizard, to fit it for such cetaceous habits. The extremities were to be converted into fins instead of feet, and as such we shall find them to combine even a still greater union of elasticity with strength, than is presented by the fin or paddle of the Whale. Plate 12, Fig. 1, shows the short and strong bones of the arm (e), and those of the fore arm (f, g); and beyond these the series of polygonal bones that made up the phalanges of the fingers. These polygonal bones vary in number in different species, in some exceeding one hundred; they differ also in form from the phalanges both of Lizards and Whales; and derive, from their increase of number, and change of dimensions, an increase of elasticity and power. The arm and hand thus converted into an elastic oar or paddle, when covered with skin, must have much resembled externally the undivided paddle of a Porpoise or Whale. The position also of the paddles on the anterior part of the body was nearly the same; to these were super-added posterior extremities, or hind fins, which are wanting in the cetacea, and which possibly make compensation for the absence of their flat horizontal tail: these hind paddles in the Ichthyosaurus are nearly by one half smaller than the anterior paddles.*

* In the Ornithorhynchus, also, the membranous expansion, or web of the hind feet, is very much less than that on the fore foot.

Mr. Conybeare remarks, with his usual acumen, that "the reasons of this variation from the proportions of the posterior extremities of quadrupeds in general, are the same which lead to a similar diminution of the analogous parts in Seals, and their total disappearance in the cetacea, namely, the necessity of placing the centre of the organs of motion, when acting laterally, before the centre of gravity. For the same reason, the wings of birds are placed in the fore part of their body, and the centre of the moving forces given to ships by their sails, and to steam-boats by their paddles, is similarly placed. The great organ of motion in fishes, the tail, is indeed posteriorly placed, but this by its mode of action generates a *vis a tergo*, which impels the animal straight forwards, and does not therefore operate under the same conditions with organs laterally applied." G. T. V. 5, p. 579.

I shall conclude this detailed review of the peculiarities of one of the most curious, as well as the most ancient, among the many genera of extinct reptiles presented to us by Geology, with a few remarks on the final causes of those deviations from the normal structure of its proper type, the Lizard; under which the Ichthyosaurus combines in itself the additional characters of the fish, the Whale, and Ornithorhynchus. As the form of vertebræ by which it is associated with the class of fishes, seems to have been

introduced for the purpose of giving rapid motion in the water to a Lizard inhabiting the element of fishes; so the further adoption of a
structure in the legs, resembling the paddles of
a Whale, was superadded in order to convert
these extremities into powerful fins. The still
further addition of a furcula and clavicles, like
those of the Ornithorhynchus, offers a third and
not less striking example of selection of contrivances, to enable animals of one class to live in
the element of another class.

If the laws of co-existence are less rigidly
maintained in the Ichthyosaurus, than in other
extinct creatures which we discover amid the
wreck of former creations, still these deviations
are so far from being fortuitous, or evidencing
imperfection; that they present examples of
perfect appointment and judicious choice, pervading and regulating even the most apparently anomalous aberrations.

Having the vertebræ of a fish, as instruments of rapid progression; and the paddles of
a Whale, and sternum of an Ornithorhynchus,
as instruments of elevation and depression; the
reptile Ichthyosaurus united in itself a combination of mechanical contrivances, which are
now distributed among three distinct classes of
the animal kingdom. If, for the purpose of
producing vertical movements in the water, the
sternum of the living Ornithorhynchus assumes

forms and combinations that occur but in one
other genus of Mammalia, they are the same
that co-existed in the sternum of the Ichthyosaurus of the ancient world; and thus, at
points of time, separated from each other by the
intervention of incalculable ages, we find an
identity of objects effected by instruments so
similar, as to leave no doubt of the unity of the
design in which they all originated.

It was a necessary and peculiar function in the
economy of the fish-like Lizard of the ancient
seas, to ascend continually to the surface of the
water in order to breathe air, and to descend
again in search of food; it is a no less peculiar
function in the Duck-billed Ornithorhynchus of
our own days, to perform a series of similar movements in the lakes and rivers of New Holland.

The introduction to these animals, of such
aberrations from the type of their respective
orders, to accommodate deviations from the usual
habits of these orders, exhibits an union of com-
pensative contrivances, so similar in their relations, so identical in their objects, and so perfect
in the adaptation of each subordinate part, to
the harmony and perfection of the whole; that
we cannot but recognise throughout them all,
the workings of one and the same eternal principle of Wisdom and Intelligence, presiding from
first to last over the total fabric of Creation.

SECTION V.

INTESTINAL STRUCTURE OF ICHTHYOSAURUS AND OF FOSSIL FISHES.

FROM the teeth and organs of locomotion, we come next to consider those of digestion in the Ichthyosaurus. If there be any point in the structure of extinct fossil animals, as to which it should have seemed hopeless to discover any kind of evidence, it is the form and arrangement of the intestinal organs; since these soft parts, though of prime importance in the animal economy, yet being suspended freely within the cavity of the body, and unconnected with the skeleton, would leave no traces whatever upon the fossil bones.

It is impossible to have seen the large apparatus of teeth, and strength of jaws, which we have been examining in the Ichthyosauri, without concluding that animals furnished with such powerful instruments of destruction, must have used them freely in restraining the excessive population of the ancient seas. This inference has been fully confirmed by the recent discovery within their skeletons, of the half digested remains of fishes and reptiles, which they had devoured, (see Pl. 13, 14,), and by the further discovery of Coprolites, (see Pl. 15,) i. e. of fœcal remains in a state of petrifaction, dispersed through the same strata in which these skeletons are buried. The state of preservation of these very curious petrified bodies is often so perfect, as to indicate not only the food of the animals from which they were derived, but also the dimensions, form, and structure of their stomach, and intestinal canal.*

On the shore at Lyme Regis, these Coprolites are so abundant, that they lie in some parts of the lias like potatoes scattered in the ground ; still

---

* The following description of these Coprolites, is given in my memoir on this subject, published in the Transactions of the Geological Society of London, 1829, (vol. iii. N. s. part i. p. 224. with three plates.)

" In variety of size and external form, the Coprolites resemble oblong pebbles or kidney-potatoes. They, for the most part, vary from two to four inches in length, and from one to two inches in diameter. Some few are much larger, and bear a due proportion to the gigantic calibre of the largest Ichthyosauri ; others are small, and bear a similar ratio to the more infantine individuals of the same species, and to small fishes : some are flat and amorphous, as if the substance had been voided in a semifluid state ; others are flattened by pressure of the shale. Their usual colour is ash-grey, sometimes interspersed with black, and sometimes wholly black. Their substance is of a compact earthy texture, resembling indurated clay, and having a conchoidal and glossy fracture. The structure of the Coprolites at Lyme Regis is in most cases tortuous, but the number of coils is very unequal ; the most common number is three : the greatest I have seen is six : these variations may depend on the various species of animals from which they are derived ; I find analogous variations in the tortuous intestines of modern Skates, Sharks, and Dog-fish. Some

more common are they in the lias of the Estuary of the Severn, where they are similarly disposed in strata of many miles in extent, and mixed so abundantly with teeth and rolled fragments of the bones of reptiles and fishes, as to show that this region, having been the bottom of an ancient sea, was for a long period the receptacle of the bones and fœcal remains of its inhabitants. The occurrence of Coprolites is not however peculiar to the places just mentioned, they are found in greater or less abundance throughout the lias of England; they occur also in strata, of all

Coprolites, especially the small ones, show no traces at all of contortion.

" The sections of these fœcal balls, (*see Pl.* 15, *Figs.* 4, *and* 6,) show their interior to be arranged in a folded plate, wrapped spirally round from the centre outwards, like the whorls of a turbinated shell; their exterior also retains the corrugations and minute impressions, which, in their plastic state, they may have received from the intestines of the living animals. (*See Pl.* 15, *Figs.* 3, *and* 10 *to* 14.) Dispersed irregularly and abundantly throughout these petrified fœces, are the scales, and occasionally the teeth and bones of fishes, that seem to have passed undigested through the bodies of the Saurians; just as the enamel of teeth and sometimes fragments of bone, are found undigested both in the recent and fossil album græcum of hyænas. These scales are the hard bright scales of the *Dapedium politum*, and other fishes which abound in the lias, and which thus appear to have formed no small portion of the food of the Saurians. The bones are chiefly vertebræ of fishes and of small Ichthyosauri; the latter are less frequent than the bones of fishes, but still are sufficiently numerous, to show that these monsters of the ancient deep, like many of their successors in our modern oceans, may have devoured the small and weaker individuals of their own species."

ages that contain the remains of carnivorous reptiles, and have been recognized in many and distant regions both of Europe and America.*

The certainty of the origin of these Coprolites is established, by their frequent presence in the abdominal region of fossil skeletons of Ichthyosauri found in the lias of Lyme Regis. One of the most remarkable of these is represented in Pl. 13; the coprolitic matter loaded with fish scales, within the ribs of these and similar specimens, is identical in appearance and chemical composition with the insulated coprolites that occur in the same strata with the skeletons.†

---

* Professor Jæger has recently discovered many Coprolites in the alum slate of Gaildorf in Wirtemberg; a formation which he considers to be in the lower region of that part of the new red sandstone formation which in Germany is called Keuper; and which contains the remains of two species of Saurians.

In the United States Dr. Dekay has also discovered Coprolites in the Green-sand formation of Monmouth, in New Jersey, see Pl. 15, Fig. 13.

† This specimen has been presented by Viscount Cole to the Geological Collection of the University of Oxford. It affords decisive proof that the substances in question cannot be referred to adventitious matter, placed accidentally in contact with the fossil body, inasmuch as the large coprolitic mass is enclosed between the back bone and the right and left series of the ribs, of which the greater number remain nearly in their natural position. The quantity of this coprolite is prodigious, when compared with the size of the animal in which it occurs; and if we were not acquainted with the powers of the digestive organs of reptiles and fishes, and their capacity of gorging the larger animals that form their prey; the great space within these fossil skeletons of Ichthyosauri, which is occasionally filled with coprolitic matter, would appear inexplicable.

The preservation of such fœcal matter, and its conversion to the state of stone, result from the imperishable nature of the phosphate of lime, of which both bones, and the products of digested bones are equally composed.

The skeleton of another Ichthyosaurus in the Oxford Museum, from the lias at Lyme Regis, (Pl. 14) shews a large mass of fish scales, chiefly referrible to the Pholidophorus limbatus,* intermixt with coprolite throughout the entire region of the ribs; this mass is overlaid by many ribs, and although, in some degree perhaps, extended by pressure, it shows that the length

* According to Professor Agassiz, the scales of Pholidophorus limbatus, a species very frequent among the fossils of the lias, are more abundant than those of any other fish in the Coprolites found in that formation at Lyme Regis; and shew that this species was the principal food of these reptiles. In Coprolites from the coal formation, near Edinburgh, he has also recognised the scales of Palæoniscus, and of other fishes that are often found entire in strata that accompany the coal of that neighbourhood. Scales of the Zeus Lewisiensis, a fish discovered by Mr. Mantell, in the chalk, occur in Coprolites derived from voracious fishes during the deposition of this formation.

A Coprolite from the lias, (Pl. 15, Fig. 3), remarkable for its spiral convolutions, and vascular impressions, affords a striking example of the minute accuracy with which investigations are now conducted by naturalists, and of the kind of evidence which comparative anatomy contributes in aid of geological enquiry. On one side of this Coprolite, there is a small scale, (Fig. 3, a,) which I could only refer to some unknown fish, of the numerous species that occur in the lias. The instant I shewed it to M. Agassiz, he not only pronounced its species to be the Pholidophorus limbatus; but at once declared the precise place which this scale had occupied upon the body of the fish. A minute

of the stomach was nearly co-extensive with the trunk.

Among living voracious reptiles we have examples of stomachs equally capacious; we know that whole human bodies have been found within the stomachs of large Crocodiles ; we know also, from the form of their teeth, that the Ichthyosauri, like the Crocodiles, must have gorged their prey entire ; and when we find, imbedded in Coprolites derived from the larger Ichthyosauri, bones of smaller Ichthyosauri, of such dimensions, (see Pl. 15, Fig. 18. And Geol. Trans. 2, S. vol. iii, Pl. 29, Figs. 2, 3, 4, 5,) that the individuals from which they were derived, must have measured several feet in length ; we infer that the stomach of these animals formed a pouch, or sac, of prodigious size, extending through nearly the entire cavity of the body, and of capacity duly proportioned to the jaws and teeth with which it co-operated.

tube upon its inner surface, (Pl. 15, Fig. 3′,) scarcely visible without a microscope, shewed it to have been one of those which form the *lateral line* of perforated scales, that pass from the head towards the tail, one on each side of every fish; and convey a tube for the transmission of lubricating mucus from glands in the head, to the extremity of the body. The place of the scale in this line, had been on the left side, not far from the head. Fig. 3″ is the upper surface of a similar scale, shewing at e the termination of the mucous duct.

Spiral Disposition of Small Intestines.

As the more solid parts of animals alone, are usually susceptible of petrifaction, we cannot demonstrate by direct evidence the form and size of the small intestines of the Ichthyosauri, but the contents of these viscera are preserved in such perfection in a fossil state, as to afford circumstantial evidence that the bowels in which they were moulded, were formed in a manner resembling the spiral intestines of some of the swiftest and most voracious of our modern fishes.

We shall best understand the structure of these intestines by examining the corresponding organs of Sharks and Dog-fish, animals not less peculiarly rapacious among the inhabitants of our modern seas, than the Ichthyosauri were in those early periods to which our considerations are carried back. We find in the intestines of these fishes, (see Pl. 15, Figs. 1, and 2,) and also in those of Rays, an arrangement resembling that of the interior of an Archimedes screw, admirably adapted to increase the extent of internal surface for the absorption of nutriment from the food, during its passage through a tube containing within it a continuous spiral fold, coiled in such a manner, as to afford the greatest

G. O

possible extent of surface in the smallest space. A similar contrivance is shown by the Coprolites to have existed in the Ichthyosaurus. See Pl. 15, Figs. 3, 4, 6.*

Impressions of the Mucous Membrane on Coprolites.

Besides the spiral structure and consequent shortness of the small intestine, we have additional evidence to shew even the form of the minute vessels and folds of the mucous mem-

---

* These cone-shaped bodies are made up of a flat and continuous plate of digested bone, coiled round itself whilst it was yet in a plastic state. The form is nearly that which would be assumed by a piece of riband, forced continually forward into a cylindrical tube, through a long aperture in its side. In this case, the riband moving onwards, would form a succession of involuted cones, coiling one round the other, and after a certain number of turns within the cylinder, (the apex moving continually downwards,) these cones would emerge from the end of the tube in a form resembling that of the Coprolites, Pl. 15, Figs. 3, 5, 7, 10, 11, 12, 13, 14. In the same manner, a lamina of coprolitic matter would be coiled up spirally into a series of successive cones, in the act of passing from a small spiral vessel into the adjacent large intestine. Coprolites thus formed fell into soft mud, whilst it was accumulating at the bottom of the sea, and together with this mud, (which has subsequently been indurated into shale and stone,) they have undergone so complete a process of petrifaction, that in hardness, and beauty of the polish of which they are susceptible, they rival the qualities of ornamental marble.

Fig. 6, shews a longitudinal section through the axis of

brane, by which it was lined. This evidence consists in a series of vascular impressions and corrugations on the surface of the Coprolite, which it could only have received during its passage through the windings of this flat tube.* Specimens thus marked are engraved at Pl. 15, Figs. 3, 5, 7, 10, 12, 13, 14.

If we attempt to discover a final cause for these curious provisions in the bowels of the extinct reptile inhabitants of the seas of a former world, we shall find it to be the same that explains the existence of a similar structure

a coprolite, from the inferior chalk, in which this involute conical form is well defined. Fig. 4, is the transverse section of another Coprolite from the lias, shewing the manner in which the plate coils round itself, till it terminates externally in a broken edge, (at b). In all the figures the letter b, marks the transverse section of this plate, where it is broken off near the termination of its outer coil; the sections at b, shew also the size and form of the flattened passage through the interior of the screw.

A lamina of tenacious plastic substance pressed continually forwards from the interior of such a screw, into the cavity of the large intestine, would coil up spirally within it, until it attained the largest size admitted by its diameter; from this coil successive portions would be broken off abruptly, (at b,) and descending into the cloaca would be thence discharged into the sea.

* These impressions cannot have been derived from the membrane of the inferior large intestine, because they are continued along those surfaces of the inner coils of the Coprolite, which became permanently covered by its outer coils, in the act of passing from the spiral tube into this large intestine.

in the modern voracious tribes of Sharks and Dog-fish.*

As the peculiar voracity of all these animals required the stomach to be both large and long, there would remain but little space for the smaller viscera; these are therefore reduced, as we have seen, nearly to the state of a flattened tube, coiled like a corkscrew around itself; their bulk is thus materially diminished, whilst the amount of absorbing surface remains almost the same, as if they had been circular. Had a large expansion of intestines been superadded to the enormous stomach and lungs of the Ichthyosaurus, the consequent enlargement of the body would have diminished the power of progressive motion, to the great detriment of an

---

* Paley, in his chapter on mechanical compensations in the structure of animals, mentions a contrivance similar to that which we attribute to the Ichthyosaurus, as existing in a species of Shark, (the Alopecias, Squalus Vulpes, or Sea Fox). " In this animal, he says, the intestine is straight from one end to the other: but, in this straight, and consequently short intestine, is a winding, cork-screw, spiral passage, through which the food, not without several circumvolutions, and in fact by a long route, is conducted to its exit. Here the shortness of the gut is *compensated* by the obliquity of the perforation."

Dr. Fitton has called my attention to a passage in Lord King's Life of Locke, 4°. p. 166, 167, from which it appears that the importance of a spiral disposition within the intestinal canal, which he observed in many preparations in the collection of anatomy at Leyden, was duly appreciated by that profound philosopher.

animal which depended on its speed for the capture of its prey.

The above facts which we have elicited from the coprolitic remains of the Ichthyosauri, afford a new and curious contribution to our knowledge both of the anatomy and habits of the extinct inhabitants of our planet. We have found evidence which enables us to point out the existence of beneficial arrangements and compensations, even in those perishable, yet important parts which formed their organs of digestion. We have ascertained the nature of their food, and the form and structure of their intestinal canal; and have traced the digestive organs through three distinct stages of descent, from a large and long stomach, through the spiral coils of a compressed ileum, to their termination in a cloaca; from which the Coprolites descended into the mud of the nascent lias. In this lias they have been interred during countless ages, until summoned from its deep recesses by the labours of the Geologist, to give evidence of events that passed at the bottom of the ancient seas, in ages long preceding the existence of man.

Intestinal Structure of Fossil Fishes.

Discoveries have recently been made of Coprolites derived from fossil fishes.   Mr. Mantell
has found them within the body of the Macropoma Mantellii, from the chalk of Lewes, placed
in contact with the long stomach of this voracious fish : the coats of its stomach are also well
preserved.*   Miss Anning also has discovered
them within the bodies of several species of
fossil fish, from the lias at Lyme Regis.
Dr. Hibbert has shown that the strata of
fresh-water limestone, in the lower region of
the coal formation, at Burdie House, near Edinburgh, are abundantly interspersed with Coprolites, derived from fishes of that early era; and
Sir Philip Egerton has found similar fœcal
remains, mixed with scales of the Megalichthys, and fresh-water shells, in the coal formation of Newcastle-under-Lyne.   In 1832,
Mr. W. C. Trevelyan recognized Coprolites in

* See Mantell's Geol. of Sussex, Pl. 38.   I learn from Mr.
Mantell, that the form of the Coprolites within the Macropoma
most nearly resemble those engraved, Pl. 15, Figs. 8, 9, of the
present work: he also conjectures that the more tortuous kinds,
(Pl. 15, Figs. 5, 7), long known by the name of Juli, and supposed to be fossil fir cones, may have been derived from fishes
of the Shark family, (Ptychodus) whose large palatal teeth (Pl.
27. f) abound in the same localities of the chalk formation with
them, at Steyning and Hamsey.

the centre of nodules of clay ironstone, that abound in a low cliff composed of shale, belonging to the coal formation at Newhaven, near Leith. I visited the spot, with this gentleman and Lord Greenock, in September, 1834, and found these nodules strewed so thickly upon the shore, that a few minutes sufficed to collect more specimens than I could carry; many of these contained a fossil fish, or fragment of a plant, but the greater number had for their nucleus, a Coprolite, exhibiting an internal spiral structure; they were probably derived from voracious fishes, whose bones are found in the same stratum. These nodules take a beautiful polish, and have been applied by the lapidaries of Edinburgh to make tables, letter presses, and ladies' ornaments, under the name of Beetle stones, from their supposed insect origin. Lord Greenock has discovered, between the laminæ of a block of coal, from the neighbourhood of Edinburgh, a mass of petrified intestines distended with Coprolite, and surrounded with the scales of a fish, which Professor Agassiz refers to the Megalichthys.

This distinguished naturalist has recently ascertained that the fossil worm-like bodies, so abundant in the lithographic slate of Solenhofen, and described by Count Münster in the Petrefacten of Goldfuss, under the name of Lumbricaria, are either the petrified intestines of fishes, or the contents of their intestines, still retaining the form of the tortuous tube in which they were lodged. To these remarkable fossils he has given the name of *Cololites*. (Pl. 15′ is copied from one of a series that are engraved in Goldfuss. Petrefacten, Pl. 66.) He has also found similar tortuous petrifactions within the abdominal cavity of fossil fishes, belonging to several species of the genus Thrissops and Leptolepis, occupying the ordinary position of the intestines between the ribs.*   (See Agassiz Poissons Fossiles, liv. 2, Appendix, p. 15.)

* As these Cololites are most frequently found insulated in the lithographic limestone, M. Agassiz has ingeniously explained this fact by observing the process of decomposition of dead fishes in the lakes of Switzerland. The dead fish floats on the surface, with its belly upwards, until the abdomen is so distended with putrid gas, that it bursts: through the aperture thus formed the bowels come forth into the water, still adhering together in their natural state of convolution. This intestinal mass is soon torn from the body by the movement of the waves; the fish then sinks, and the bowels continue a long time floating on the water: if cast on shore, they remain many days upon the sand before they are completely decomposed. The small bowels only are thus detached from the body, the stomach and other viscera remain within it.

We owe this illustration of the nature of these fossil bodies, whose origin has hitherto been inexplicable, to the author of a most important work on fossil fishes, now under publication at Neuchatel. His qualifications for so great and difficult a task are abundantly guaranteed by the fact, that Cuvier, on seeing the progress he had made, at once placed at the disposal of Professor Agassiz the materials he had himself collected towards a similar work.

It is probable that to many persons inexperienced in anatomy, any kind of information on a subject so remote, and apparently so inaccessible, as the intestinal structure of an extinct reptile or a fossil fish, may at first appear devoid of the smallest possible importance; but it assumes a character of high value, in the investigation of the proofs of creative wisdom and design, that are unfolded by the researches of Geology; and supplies a new link to that important chain, which connects the lost races that formerly inhabited our planet, with species that are actually living and moving around ourselves.* The systematic recurrence, in animals of such distant eras, of the same contrivances, similarly disposed to effect similar purposes, with analogous adaptations to peculiar conditions of existence, shews that they all originated in the same Intelligence.

When we see the body of an Ichthyosaurus, still containing the food it had eaten just before its death, and its ribs still surrounding the remains of fishes, that were swallowed ten thou-

* Le temps qui répand de la dignité sur tout ce qui échappe à son pouvoir destructeur, fait voir ici un exemple singulier de son influence: ces substances si viles dans leur origine, étant rendues à la lumière après tant de siècles, deviennent d'une grande importance puis qu'elles servent à remplir un nouveau chapitre dans l'histoire naturelle du globe.—Bulletin Soc. Imp. de Moscow, No. VI. 1833, p. 23.

sand, or more than ten times ten thousand years ago, all these vast intervals seem annihilated, time altogether disappears, and we are almost brought into as immediate contact with events of immeasurably distant periods, as with the affairs of yesterday.

SECTION VI.

PLESIOSAURUS.*

WE come next to consider a genus of extinct animals, nearly allied in structure to the Ichthyosaurus, and co-extensive with it through the middle ages of our terrestrial history. The discovery of this genus forms one of the most important additions that Geology has made to comparative anatomy. It is of the Plesiosaurus, that Cuvier asserts the structure to have been the most heteroclite, and its characters altogether the most monstrous, that have been yet found amid the ruins of a former world.† To the head of a Lizard, it united the teeth of a Crocodile; a neck of enormous length, resembling the body of a Serpent: a trunk and tail having the proportions of an ordinary quadruped, the ribs of a

---

* See Pl. 16, 17, 18, 19.

† Cet habitant de l'ancien monde peut-être la plus hétéroclite et celui de tous qui paroît le plus mériter le nom de monstre.— Oss. Foss. V. Pt. 2, p. 476.

Camelion, and the paddles of a Whale. Such are the strange combinations of form and structure in the Plesiosaurus —a genus, the remains of which, after interment for thousands of years amidst the wreck of millions of extinct inhabitants of the ancient earth, are at length recalled to light by the researches of the Geologist, and submitted to our examination, in nearly as perfect a state as the bones of species that are now existing upon the earth.

The Plesiosauri appear to have lived in shallow seas and estuaries, and to have breathed air like the Ichthyosauri, and our modern Cetacea. We are already acquainted with five or six species, some of which attained a prodigious size and length; but our present observations will be chiefly limited to that which is the best known, and perhaps the most remarkable of them all, viz. the P. Dolichodeirus.*

* The first specimens of this animal were discovered in the lias of Lyme Regis, about the year 1823, and formed the foundation of that admirable paper (Geol. Trans. Lond. vol. 5, Pt. 2.) in which Mr. Conybeare and M. De la Beche established and named this genus. Other examples have since been recognised in the same formations in different parts of England, Ireland, France, and Germany, and in formations of various ages, from the muschel kalk upwards to the chalk. The first specimen discovered in a state approaching to perfection, was that in the collection of the Duke of Buckingham, (figured in the Geol. Trans. Lond. N. S. Vol. 1, Pt. 2, Pl. 48). Another specimen, nearly entire, in the collection of the British Museum, eleven feet in length, is figured in our second volume, (Pl. 16); and at Pl. 17, a still more per-

Head.*

The head of the P. Dolichodeirus exhibits a combination of the characters of the Ichthyosaurus, the Crocodile, and the Lizard, but most nearly approaches to the latter. It agrees with the Ichthyosaurus in the smallness of its nostrils, and also in their position near the anterior angle of the eye; it resembles the Crocodile, in having the teeth lodged in distinct alveoli; but differs from both, in the form and shortness of its head, many characters of which approach closely to the Iguana.†

---

fect fossil skeleton, also in the British Museum, discovered by Mr. Hawkins, in the lias at Street, near Glastonbury. At Pl. 16 is also copied Mr. Conybeare's restoration of this animal, from dislocated fragments, before any entire skeletons were found. The near approach of this restoration to the character of the perfect skeletons, affords a striking example of the sure grounds on which comparative anatomy enables us to reconstruct the bodies of fossil animals, from a careful combination of insulated parts. The soundness of the reasoning of Cuvier, on the fossil quadrupeds of Montmartre, was established by the subsequent discovery of skeletons, such as he had conjecturally restored from insulated bones. Mr. Conybeare's restoration of the Plesiosaurus Dolichodeirus, (Pl. 16,) was not less fully confirmed by the specimens above mentioned.

     * See Pl. 16, 17, 18.

     † Mr. Conybeare, in the Geol. Trans. second series, vol. 1, part 1, Pl. 19, has published figures of the superior and lateral view of a nearly perfect head of this animal. Our figure, Pl. 18, Fig. 2, represents the head of the specimen in the British Mu-

Neck.

The most anomalous of all the characters of P. Dolichodeirus is the extraordinary extension of the neck, to a length almost equalling that of the body and tail together, and surpassing in the number of its vertebræ (about thirty-three) that of the most long-necked bird, the Swan : it thus deviates in the greatest degree from the almost

seum, of which the entire figure, on a smaller scale, is given in Pl. 16. The head is in a supine position; the upper jaw is distorted, and shows several of the separate alveoli that contained the teeth, and also the posterior portion of the palate. The under jaw is but little disturbed.

A figure of another lower jaw is given at Pl. 18, Fig. 1, taken from a specimen also in the British Museum, found by Mr. Hawkins, at Street.

Pl. 19, Fig. 3, represents the extremity of the dental bone of another lower jaw, in the same collection, retaining several teeth in the anterior sockets, and also exhibiting a series of new teeth, rising within an interior range of small cavities. This arrangement for the formation of new teeth, in cells within the bony mass that contains the older teeth, from which they shoot irregularly forwards through the substance of the bone, forms an important point of resemblance whereby the Plesiosaurus assumes, in the renovation of its teeth, the character of Lizards, combined with the position of the perfect teeth in distinct alveoli, after the manner of Crocodiles.

The number of teeth in the lower jaw was fifty-four, which, if met by a corresponding series in the upper jaw, must have made the total number to exceed one hundred. The anterior part of the extremity of the jaw enlarges itself like the bowl of a spoon, to allow space for the reception of the six first teeth on each side, which are the largest of all.

universal law, which limits the cervical vertebræ of quadrupeds to a very small number. Even in the Camelopard, the Camel, and Lama, their number is uniformly seven. In the short neck of the Cetacea the type of this number is maintained. In Birds it varies from nine to twenty-three; and in living Reptiles from three to eight.* We shall presently find in the habits of the Plesiosaurus a probable cause for this extraordinary deviation from the normal character of the Lizards.

* To compensate for the weakness that would have attended this great elongation of the neck, the Plesiosaurus had an addition of a series of hatchet-shaped processes, on each side of the lower part of the cervical vertebræ. (Pl. 17, and Pl. 19, 1, 2.) Rudiments and modifications of these processes exist in birds, and in long-necked quadrupeds. In the Crocodiles they assume a form, most nearly approaching that which they bear in the Plesiosaurus.

The bodies of the vertebræ also more nearly resemble those of certain fossil Crocodiles, than of Ichthyosauri or Lizards; they agree further with the Crocodile, in having the annular part attached to the body by sutures; so that we have in the neck of the P. Dolichodeirus a principle of construction resembling that of the vertebræ of Crocodiles; combined with an elongation very much exceeding that of the longest neck in birds, and such as occurs in no other known animal of the extinct or living creations. The length of the neck in P. Dolichodeirus is nearly five times that of the head; that of the trunk four times the length of the head, and of the tail three times; the head itself being one-thirteenth part of the whole body.—See Geol. Trans. Lond. Vol. 5, p. 559, and Vol. I. N. S. p. 103, et seq.

Back and Tail.

The vertebræ of the back were not disposed in
hollow cones, like those of fishes, but presented
to each other nearly flat surfaces, giving to the
column a stability, like that which exists in the
back of terrestrial quadrupeds. The articulating processes, also, were locked into one another
in such manner as to give strength, rather than
that peculiar kind of flexibility, which admitted
of the same quick progressive motion in the
Ichthyosauri that we find in fishes: but as rapid
motion was incompatible with the structure of
the other parts of the Plesiosaurus, the combination of strength, rather than of speed with
flexibility, was more important.

The tail, being comparatively short, could not
have been used like the tail of fishes, as an
instrument of rapid impulsion in a forward direction; but was probably employed more as a
rudder, to steer the animal when swimming on
the surface, or to elevate or depress it in ascending and descending through the water. The
same consequence as to slowness of motion
would follow from the elongation of the neck,
to so great a distance in front of the anterior
paddles. The total number of vertebræ in the
entire column was about ninety. From all these
circumstances we may infer that this animal,
although of considerable size, had to seek its food,

as well as its safety, chiefly by means of artifice
and concealment.

Ribs.*

The ribs are composed of two parts, one vertebral and one ventral; the ventral portions of
one side, (Pl. 18, 3, b,) uniting with those on
the opposite side by an intermediate transverse
bone, (a, c,) so that each pair of ribs encircled
the body with a complete belt, made up of five
parts.† Cuvier observes that the similarity of
this structure to that of the ribs of Cameleons
and two species of Iguana, (Lacerta Marmorata,
Lin. and Anolius, Cuvier,) seems to shew that
the lungs of the Plesiosaurus Dolichodeirus,
(as in these three subgenera of living Saurians,)
were very large; and *possibly* that the colour
of its skin also was changeable, by the varied
intensity of its inspirations.‡   Oss. Foss. Vol. V.
Pt. 2. p. 280.

* See Pl. 16, 17, 18.

† The ventral portion of each rib, (Pl. 17, and Pl. 18, 3,
b,) appears to have been composed of three slender bones fitted
to one another by oblique grooves, allowing of great expansive
movement during the inflation of the lungs: the manner in
which these triple bones were folded over one another, is best
seen in a single series between a, and b, the upper ends of the
ventral portions of the ribs (b) have been separated by pressure,
from the lower ends of the vertebral portions. (d.)

‡ We have no means to verify this ingenious conjecture, that
the Plesiosaurus may have been a kind of *sub-marine* Cameleon,

This hypothesis of Cuvier is but conjectural, respecting the power of the Plesiosaurus to change the colour of its skin; and to the un-experienced in comparative anatomy, it may seem equally conjectural, to deduce any other conclusions respecting such perishable organs as the lungs, from the discovery of peculiar con-trivances, and unusual apparatus in the ribs; yet we argue on similar grounds, when from the form and capabilities of these fossil ribs, we infer that they were connected, as in the cameleon, with vast and unusual powers of expansion and contraction in the lungs; and when, on finding the ribs and wood-work of a worn-out bellows, near the ruins of a blacksmith's forge, we con-clude that these more enduring parts of the

possessing the power of altering the colour of its skin; it must however be admitted that such a power would have been of much advantage to this animal, in defending it by concealment from its most formidable enemy the Ichthyosaurus, with which, its diminutive head and long slender neck, must have rendered it a very unequal combatant, and from whose attacks its slow locomotive powers must have made escape by flight impossible: the enlarged condition of the lungs, would also have been of great advantage in diminishing the frequency of its ascents to the surface, to inspire air; an operation that must have been attended with constant danger, in a sea thickly swarming with Ichthyosauri. Dr. Stark has recently observed that certain fishes, especially minnows, have a tendency to assume the colour of the vessel in which they are kept. (Proceedings Zool. Soc. Lond. July, 1833.) As in animals of this class there are no lungs, this change of colour must arise from other cause than that to which it has been attributed in the Cameleon.

G.   P

frame of this instrument, have been connected with a proportionable expansion of leather.

The compound character of the ribs, probably also gave to the Plesiosaurus the same power of compressing air within its lungs, and in that state taking it to the bottom, which we have considered as resulting from the structure of the steno-costal apparatus of the Ichthyosauri.

Extremities.*

As the Plesiosaurus breathed air, and was therefore obliged to rise often to the surface for inspiration, this necessity was met by an apparatus in the chest and pelvis, and in the bones of the arms and legs, enabling it to ascend and descend in the water after the manner of the Ichthyosauri and Cetacea; accordingly the legs were converted into paddles, longer and more powerful than those of the Ichthyosaurus, thus compensating for the comparatively small assistance which it could have derived from its tail.†

Comparing these extremities with those of other vertebrated animals, we trace a regular

* See Pl. 16, 17, 19.

† The number of joints representing the phalanges of the fingers and toes exceeds that in the Lizards and Birds, and also in all Mammalia, excepting the Whales, some of which present a similar increase of number to accommodate them to the corresponding office of a paddle. The mode of connection between the joints was (like that in the Whales,) by *synchondrosis*. The phalanges of the Plesiosaurus present a link, between the

series of links and gradations, from the corresponding parts of the highest mammalia, to their least perfect form in the fins of fishes. In the fore paddle of the Plesiosaurus, we have all the essential parts of the fore leg of a quadruped, and even of a human arm; first the scapula, next the humerus, then the radius and ulna, succeeded by the bones of the carpus and metacarpus, and these followed by five fingers, each composed of a continuous series of phalanges. (see Pl. 16, 17, 19.) The hind paddle also offers precisely the same analogies to the leg and foot of the Mammalia; the pelvis and femur are succeeded by a tibia and fibula, which articulate with the bones of the tarsus and metatarsus, followed by the numerous phalanges of five long toes.

From the consideration of all its characters, Mr. Conybeare has drawn the following inferences with respect to the habits of the Plesiosaurus Dolichodeirus, " That it was aquatic is evident, from the form of its paddles; that it was marine is almost equally so, from the remains with which it is universally associated;

still more numerous and angular joints of the paddle of the Ichthyosaurus, and the phalanges of land quadrupeds, which are more or less cylindrical; in these sea Lizards they were flattened, for the purpose of giving breadth to the extremities as organs of swimming. As its paddles give no indication of having carried even such imperfect claws, as those of the Turtles and Seals, the Plesiosaurus apparently could have made little or no progress in any other element than water.

that it may have occasionally visited the shore,
the resemblance of its extremities to those of
the Turtle may lead us to conjecture; its motion
however must have been very awkward on land;
its long neck must have impeded its progress
through the water; presenting a striking contrast to the organization which so admirably
fits the Ichthyosaurus to cut through the waves.
May it not therefore be concluded (since, in
addition to these circumstances, its respiration
must have required frequent access of air,) that
it swam upon, or near the surface; arching
back its long neck like the swan, and occasionally darting it down at the fish which happened
to float within its reach. It may perhaps have
lurked in shoal water along the coast, concealed
among the sea-weed, and raising its nostrils to a
level with the surface from a considerable depth,
may have found a secure retreat from the as-
saults of dangerous enemies; while the length
and flexibility of its neck may have compensated for the want of strength in its jaws, and
its incapacity for swift motion through the water,
by the suddenness and agility of the attack
which they enabled it to make on every animal
fitted for its prey, which came within its reach,"
—Geol. Trans. N. s. vol. i. part ii. p. 388.

We began our account of the Plesiosaurus
with quoting the high authority of Cuvier, for
considering it as one of the most anomalous and
monstrous productions of the ancient systems of

creation; we have seen in proceeding through
our examination of its details, that these apparent anomalies consist only in the diversified
arrangement, and varied proportion, of parts
fundamentally the same as those that occur in
the most perfectly formed creatures of the present world.

Pursuing the analogies of construction, that
connect the existing inhabitants of the earth
with those extinct genera and species which
preceded the creation of our race, we find an
unbroken chain of affinities pervading the entire
series of organized beings, and connecting all
past and present forms of animal existence by
close and harmonious ties. Even our own
bodies, and some of their most important organs,
are brought into close and direct comparison
with those of reptiles, which, at first sight, appear the most monstrous productions of creation; and in the very hand and fingers with
which we write their history, we recognise the
type of the paddles of the Ichthyosaurus and
Plesiosaurus.

Extending a similar comparison through the
four great classes of vertebral animals, we find
in each species a varied adaptation of analogous parts, to the different circumstances and
conditions in which it was intended to be
placed. Ascending from the lower orders, we
trace a gradual advancement in structure and
office, till we arrive at those whose functions

are the most exalted: thus, the fin of the fish becomes the paddle of the reptile Plesiosaurus and Ichthyosaurus; the same organ is converted into the wing of the Pterodactyle, the bird and bat; it becomes the fore-foot, or paw, in quadrupeds that move upon the land, and attains its highest consummation in the arm and hand of rational man.

I will conclude these observations in the words and with the feelings of Mr. Conybeare, which must be in unison with those of all who have had the pleasure to follow him through his masterly investigations of this curious subject, from which great part of our information respecting the genus Plesiosaurus has been derived:

" To the observer actually engaged in tracing the various links that bind together the chain of organised beings, and struck at every instant by the development of the most beautiful analogies, almost every detail of comparative anatomy, however minute, acquires an interest, and even a charm; since he is continually presented with fresh proof of the great general law, which Scarpa himself, one of its most able investigators, has so elegantly expressed : ' Usque adeo natura, una eadem semper atque multiplex, disparibus etiam formis effectus pares, admirabili quadam varietatum simplicitate conciliat.' "

SECTION VII.

MOSASAURUS, OR GREAT ANIMAL OF MAESTRICHT.

THE Mosasaurus has been long known by the name of the great animal of Maestricht, occuring near that city, in the calcareous freestone which forms the most recent deposit of the cretaceous formation, and contains Ammonites, Belemnites, Hamites, and many other shells belonging to the chalk, mixt with numerous remains of marine animals that are peculiar to itself. A nearly perfect head of this animal was discovered in 1780, and is now in the Museum at Paris. This celebrated head during many years baffled all the skill of Naturalists; some considered it to be that of a Whale, others of a Crocodile; but its true place in the animal kingdom was first suggested by Adrian Camper, and at length confirmed by Cuvier. By their investigations it is proved to have been a gigantic marine reptile, most nearly allied to the Monitor.*
The geological epoch at which the Mosasaurus

---

* The Monitors form a genus of Lizards, frequenting marshes and the banks of rivers in hot climates; they have received this name from the prevailing, but absurd, notion that they give warning by a whistling noise, of the approach of Crocodiles and Caymans. One species, the Lacerta nilotica, which devours the eggs of Crocodiles, has been sculptured on the monuments of ancient Egypt.

first appeared, seems to have been the last of the long series, during which the oolitic and cretaceous groupes were in process of formation. In these periods the inhabitants of our planet seem to have been principally marine, and some of the largest creatures were Saurians of gigantic stature, many of them living in the sea, and controlling the excessive increase of the then existing tribes of fishes.

From the lias upwards, to the commencement of the chalk formation, the Ichthyosauri and Plesiosauri were the tyrants of the ocean ; and just at the point of time when their existence terminated, during the deposition of the chalk, the new genus Mosasaurus appears to have been introduced, to supply for a while their place and office,* being itself destined in its turn to give place to the Cetacea of the tertiary periods. As no Saurians of the present world are inhabitants of the sea, and the most powerful living representatives of this order, viz. the Crocodiles, though living chiefly in water, have recourse to stratagem rather than speed, for the capture of their prey, it may not be unprofitable to examine the mechanical contrivances, by which a reptile, most nearly allied to the Monitor, was so constructed, as to possess the power of moving in the sea, with sufficient velocity to

* Remains of the Mosasaurus have been discovered by Mr. Mantell in the upper chalk near Lewes, and by Dr. Morton in the green sand of Virginia.

overtake and capture such large and powerful
fishes, as from the enormous size of its teeth
and jaws, we may conclude it was intended to
devour.

The head and teeth, (Pl. 20.) point out the
near relations of this animal to the Monitors;
and the proportions maintained throughout all the
other parts of the skeleton warrant the conclusion, that this monstrous Monitor of the ancient
deep was five and twenty feet in length, although
the longest of its modern congeners does not
exceed five feet. The head here represented
measures four feet in length, that of the largest
Monitor does not exceed five inches. The most
skilful Anatomist would be at a loss to devise a
series of modifications, by which a Monitor
could be enlarged to the length and bulk of a
Grampus,* and at the same time be fitted to
move with strength and rapidity through the
waters of the sea; yet in the fossil before us,
we shall find the genuine characters of a Monitor maintained throughout the whole skeleton,
with such deviations only as tended to fit the
animal for its marine existence.

The Mosasaurus had scarcely any character
in common with the Crocodile, but resembled
the Iguanas, in having an apparatus of teeth
fixed on the pterygoid bone, (Pl. 20, k.) and
placed in the roof of its mouth, as in many

* The Grampus is from 20 to 25 feet long, and very fero-
cious, feeding on seals and porpoises as well as on fishes.

serpents and fishes, where they act as barbs to prevent the escape of their prey.*

The other parts of the skeleton follow the character indicated by the head. The vertebræ are all concave in front, and convex behind; being fitted to each other by a ball and socket joint, admitting easy and universal flexion. From the centre of the back to the extremity of the tail, they are destitute of articular apophyses, which are essential to support the back of animals that move on land : in this respect, they agree with the vertebræ of Dolphins, and were calculated to facilitate the power of swimming ; the vertebræ of the neck allowed to that part also more flexibility than in the Crocodiles.

The tail was flattened on each side, but high and deep in the vertical direction, like the tail of a Crocodile; forming a straight oar of immense strength to propel the body by horizontal

* The teeth have no true roots and are not hollow, as in the Crocodiles, but when full grown, are entirely solid, and united to the sockets by a broad and firm base of bone, formed from the ossification of the pulpy matter which had secreted the tooth, and still further attached to the jaw by the ossification of the capsule that had furnished the enamel. This indurated capsule, passed like a circular buttress around its base, tending to make the tooth an instrument of prodigious strength. The young tooth first appeared in a separate cell in the bone of the jaw, (Pl. 20, h.) and moved irregularly across its substance, until it pressed against the base of the old tooth ; causing it gradually to become detached, together with its base by a kind of *necrosis*, and to fall off like the horns of a Deer. The teeth, in the roof of the mouth, are also constructed on the same principle with those in the jaw, and renewed in like manner.

movements, analogous to those of skulling. Although the number of caudal vertebræ was nearly the same as in the Monitor, the proportionate length of the tail was much diminished by the comparative shortness of the body of each vertebra; the effect of this variation being to give strength to a shorter tail as an organ for swimming; and a rapidity of movement, which would have been unattainable by the long and slender tail of the Monitor, which assists that animal in climbing. There is a further provision to give strength to the tail, by the *chevron* bones being soldered firmly to the body of each vertebra, as in fishes.

The total number of vertebræ was one hundred and thirty-three, nearly the same as in the Monitors, and more than double the number of those in the Crocodiles. The ribs had a single head, and were round, as in the family of Lizards. Of the extremities, sufficient fragments have been found to prove that the Mosasaurus, instead of legs, had four large paddles, resembling those of the Plesiosaurus and the Whale: one great use of these was probably to assist in raising the animal to the surface, in order to breathe, as it apparently had not the horizontal tail, by means of which the Cetacea ascend for this purpose. All these characters unite to show that the Mosasaurus was adapted to live entirely in the water, and that although it was of such vast proportions compared with

the living genera of these families, it formed
a link intermediate between the Monitors and
the Iguanas. However strange it may appear
to find its dimensions so much exceeding those
of any existing Lizards, or to find marine
genera in the order of Saurians, in which there
exists at this time no species capable of living
in the sea; it is scarcely less strange than the
analogous deviations in the Megalosaurus and
Iguanodon, which afford examples of still greater
expansion of the type of the Monitor and Iguana,
into colossal forms adapted to move upon the
land. Throughout all these variations of proportion, we trace the persistence of the same laws,
which regulate the formation of living genera,
and from the combinations of perfect mechanism
that have, in all times, resulted from their operation, we infer the perfection of the wisdom
by which all this mechanism was designed, and
the immensity of the power by which it has ever
been upheld.

Cuvier asserts of the Mosasaurus that before
he had seen a single vertebra, or a bone of any
of its extremities, he was enabled to announce
the character of the entire skeleton, from the examination of the jaws and teeth alone, and even
from a single tooth. The power of doing this
results from those magnificent laws of co-existence, which form the basis of the science of comparative anatomy, and which give the highest
interest to its discoveries.

SECTION VIII.

PTERODACTYLE.*

AMONG the most remarkable disclosures made
by the researches of Geology, we may rank the
flying reptiles, which have been ranged by
Cuvier under the genus Pterodactyle; a genus
presenting more singular combinations of form,
than we find in any other creatures yet discovered amid the ruins of the ancient earth.†

The structure of these animals is so exceedingly anomalous, that the first discovered Pterodactyle (Pl. 21) was considered by one naturalist to be a bird, by another as a species of
bat, and by a third as a flying reptile.

This extraordinary discordance of opinion
respecting a creature whose skeleton was almost
entire, arose from the presence of characters
apparently belonging to each of the three classes
to which it was referred. The form of its head,
and length of neck, resembling that of birds, its
wings approaching to the proportion and form of

* See Pl. 1, Figs. 42, 43, and Plates 21, 22.

† Pterodactyles have hitherto been found chiefly in the quarries of lithographic limestone of the jura formation at Aichstadt
and Solenhofen; a stone abounding in marine remains, and also
containing Libellulæ, and other insects. They have also been
discovered in the lias at Lyme Regis, and in the oolitic slate of
Stonesfield.

those of bats, and the body and tail approximating to those of ordinary Mammalia. These
characters, connected with a small skull, as is
usual among reptiles, and a beak furnished with
not less than sixty pointed teeth, presented a
combination of apparent anomalies which it was
reserved for the genius of Cuvier to reconcile.
In his hands, this apparently monstrous production of the ancient world, has been converted into
one of the most beautiful examples yet afforded
by comparative anatomy, of the harmony that
pervades all nature, in the adaptation of the
same parts of the animal frame, to infinitely
varied conditions of existence.

In the case of the Pterodactyle we have an
extinct genus of the Order Saurians, in the class
of Reptiles, (a class that now moves only on
land or in the water), adapted by a peculiarity
of structure to fly in the air. It will be interesting to see how the anterior extremity, which in
the fore leg of the modern Lizard and Crocodiles
is an organ of locomotion on land, became converted into a membraniferous wing; and how
far the other parts of the body are modified so as
to fit the entire animal machine for the functions of flight. The details of this enquiry will
afford such striking examples of numerical agreement in the component bones of every limb, with
those in the corresponding limbs of living
Lizards, and are at the same time so illustrative
of contrivances for the adjustment of the same

organ to effect different ends, that I shall select
for examination a few points, from the long and
beautiful analysis which Cuvier has given of the
structure of this animal.

The Pterodactyles are ranked by Cuvier
among the most extraordinary of all the extinct
animals that have come under his consideration;
and such as, if we saw them restored to life,
would appear most strange, and most unlike to
any thing that exists in the present world.—
" Ce sont incontestablement de tous les êtres
dont ce livre nous révele l'ancienne existence, les
plus extraordinaires, et ceux qui, si on les voyait
vivans, paroîtroient les plus étrangers à toute la
nature actuelle." (Cuv. Oss. Foss. Vol. V. Pt. 11,
p. 379.)

We are already acquainted with eight species of this genus, varying from the size of a
Snipe to that of a Cormorant.*

In external form, these animals somewhat
resemble our modern Bats and Vampires:
most of them had the nose elongated, like the
snout of a Crocodile, and armed with conical

* In Pl. 21, I have given an engraving of the Pterodactylus
longirostris, which was first published by Collini, and formed the
basis on which this genus was established.

At Pl. 22, O. is engraved the smallest known species, P. Brevirostris, from Solenhofen, described by Professor Soemmering.

A figure and description of a third species, P. macronyx, from
the lias at Lyme Regis, have been published by myself, (Geol.
Trans. Lond. second series, Vol. 3, Pt. 1). This species was
about the size of a Raven, and its wings, when expanded, must

teeth. Their eyes were of enormous size, apparently enabling them to fly by night. From their wings projected fingers, terminated by long hooks, like the curved claw on the thumb of the Bat. These must have formed a powerful paw, wherewith the animal was enabled to creep or climb, or suspend itself from trees.

It is probable also that the Pterodactyles had the power of swimming, which is so common in reptiles, and which is now possessed by the Pteropus Pselaphon, or Vampire Bat of the island of Bonin. (See Zool. Journ. No. 16, p. 458.) " Thus, like Milton's fiend, all qualified for all services and all elements, the creature was a fit companion for the kindred reptiles that swarmed in the seas, or crawled on the shores of a turbulent planet.

> " The Fiend,
> O'er bog, or steep, through strait, rough, dense, or rare,
> With head, hands, wings, or feet, pursues his way,
> And swims, or sinks, or wades, or creeps, or flies."
>
> Paradise Lost, Book II. line 947.

With flocks of such-like creatures flying in the

---

have been about four feet from tip to tip. A fourth species, P. crassirostris, has been described by Professor Goldfuss. In Pl. 22, N. I have given a reduced copy of his plate of the specimen ; and in Pl. 22, A. a copy of his restoration of the entire animal. Count Munster has described another species, P. medius. Cuvier describes some bones of a species, P. grandis, four times as large as P. longirostris, which latter was about the size of a Woodcock. Professor Goldfuss has described a seventh species from Solenhofen, P. Munsteri ; and has proposed the name P. Bucklandi, for the eighth undescribed species found at Stonesfield.

air, and shoals of no less monstrous Ichthyosauri and Plesiosauri swarming in the ocean,
and gigantic Crocodiles, and Tortoises crawling
on the shores of the primæval lakes and rivers,
air, sea, and land must have been strangely
tenanted in these early periods of our infant
world." *

As the most obvious feature of these fossil
reptiles is the presence of organs of flight, it
is natural to look for the peculiarities of the
Bird or Bat, in the structure of their component
bones. All attempts, however, to identify them
with Birds are stopped at once by the fact of
their having teeth in the beak, resembling those
of reptiles: the form of a single bone, the os
quadratum, enabled Cuvier to pronounce at
once that the creature was a Lizard: but a
Lizard possessing wings exists not in the present creation, and is to be found only among the
Dragons of romance and heraldry;† while a
moment's comparison of the head and teeth

* Geol. Trans. Lond. N. S. Vol. III. part. 1.

† One diminutive living species of Lizard, (the Draco volans,
see Pl. 22, L.) differs from all other Saurians, in having an appearance of imperfect wings, produced by a membranous expansion of the skin over the false ribs which project almost horizontally from the back ; the membrane expanded by these false ribs,
acts like a parachute to support the animal in leaping from tree
to tree, but has no power to beat the air, or become an instrument of true flight, like the arm or wing of Birds and Bats ; the
arm or fore leg of the Draco volans differs not from that of
common Lizards.

G.   Q

with those of Bats (Pl. 21, and Pl. 22, M.) shows that the fossil animals in question cannot be referred to that family of flying Mammalia.

The vertebræ of the neck are much elongated, and are six or seven only in number, whereas they vary from nine to twenty-three in birds.* In birds the vertebræ of the back also vary from seven to eleven, whilst in the Pterodactyles there are nearly twenty; the ribs of the Pterodactyles are thin and thread-shaped, like those of Lizards, those of birds are flat and broad, with a still broader recurrent apophysis, peculiar to them. In the foot of birds, the metatarsal bones are consolidated into one: in the Pterodactyles all the metatarsal bones are distinct; the bones of the pelvis also differ widely from those of a bird, and resemble those of a Lizard;

* In one species of Pterodactyle, viz. the P. macronyx, Geol. Trans. n. s. V. iii. pl. 27, page 220, from the lias at Lyme Regis, there is an unusual provision for giving support and movement to a large head at the extremity of a long neck, by the occurrence of bony tendons running parallel to the cervical vertebræ, like the tendons that pass along the back of the Pigmy Musk (Moschus pygmæus,) and of many birds. This provision does not occur in any modern Lizards, whose necks are short, and require no such aid to support the head. In the compensation which these tendons afforded for the weakness arising from the elongation of the neck, we have an example of the same mechanism in an extinct order of the most antient reptiles, which is still applied to strengthen other parts of the vertebral column, in a few existing species of mammalia and birds.

all these points of agreement, with the type of Lizards, and of difference from the character of birds, leave no doubt as to the place in which the Pterodactyles must be ranged, among the Lizards, notwithstanding the approximation which the possession of wings seems to give them to Birds or Bats.

The number and proportions of the bones in the fingers and toes in the Pterodactyle, require to be examined in some detail, as they afford coincidences with the bones in the corresponding parts of Lizards, from which important conclusions may be derived.

As an insulated fact, it may seem to be of little moment, whether a living Lizard or a fossil Pterodactyle, might have four or five joints in its fourth finger, or its fourth toe; but those who have patience to examine the minutiæ of this structure, will find in it an exemplification of the general principle, that things apparently minute and trifling in themselves, may acquire importance, when viewed in connexion with others, which, taken singly, appear equally insignificant. Minutiæ of this kind, viewed in their conjoint relations to the parts and proportions of other animals, may illustrate points of high importance in physiology, and thereby become connected with the still higher considerations of natural theology. If we examine the fore-foot of the existing Lizards, (Pl. 22, B.) we

find the number of joints regularly increased by
the addition of one, as we proceed from the first
finger, or thumb, which has two joints, to the
third, in which there are four; this is precisely
the numerical arrangement which takes place in
the three first fingers of the hand of the Pterodactyle; (Pl. 22, c. d. e. n. o. Figs. 30—38.)
thus far the three first fingers of the fossil
reptile agree in structure with those of the fore-
foot of living Lizards; but as the hand of the
Pterodactyle was to be converted into an organ
of flight, the joints of the fourth, or fifth finger
were lengthened, to become expansors of a
membranous wing.*

* Thus in the P. Longirostris (Pl. 21, 39—42.) and P.
Brevirostris, (Pl. 22, Fig. O, 39—42,) the fourth finger is
stated by Cuvier to have four elongated joints, and the fifth or
ungual joint to be omitted, as its presence is unnecessary.
In the P. Crassirostris, according to Goldfuss (Pl. 22, Figs. a,
n,) this claw is present upon the fourth finger, (43) which thus
has five bones, and the fifth finger is elongated to carry the wing.
Throughout all these arrangements in the fore foot, the normal
numbers of the type of Lizards are maintained.

If, as appears from the specimen engraved by Goldfuss, of
P. Crassirostris, (Pl. 22, n, 44, 45,) the fifth finger was elongated to expand the wing, we should infer from the normal
number of joints in the fifth finger of Lizards being only three,
that this wing finger had but three joints. In the fossil itself
the two first joints only are preserved, so that his conjectural
addition of a fourth joint to the fifth finger, in the restored
figure, (Pl. 22, a, 47,) seems inconsistent with the analogies,
that pervade the structure of this, and of every other species of
Pterodactyle, as described by Cuvier.

As the bones in the wing of the Pterodactyle
thus agree in number and proportion with those
in the fore foot of the Lizard, so do they differ
entirely from the arrangement of the bones
which form the expansors of the wing of the
Bat.*

The total number of toes in the Pterodactyles
is usually four; the exterior, or little toe, being
deficient; if we compare the number and proportion of the joints in these four toes with
those of Lizards, (Pl. 22, F, G, H, I,) we find the
agreement as to number, to be not less perfect
than it is in the fingers; we have, in each case,
two joints in the first, or great toe, three in the
second, four in the third, and five in the fourth.
As to proportion also, the penultimate joint is
always the longest, and the antepenultimate, or
last but two, the shortest; these relative proportions are also precisely the same, as in the feet
of Lizards.†   The apparent use of this disposi-

---

* The Bat, see Pl. 22, M, 30, 31, the first finger or thumb
alone, is free, and applied to the purpose of suspension and
creeping; the expansors of the wing are formed by the meta-
carpal bones, (26—29,) much elongated and terminated by the
minute phalanges of the other four fingers, 32—45, thus
presenting an adaptation of the hand of the mammalia to the
purposes of flight, analogous to that which in the fossil world,
the Pterodactyle affords with respect to the hand of Lizards.

† According to Goldfuss the P. Crassirostris had one more
toe than Cuvier assigns to the other species of Pterodactyles; in this respect it is so far from violating the analogies

tion of the shortest joints in the middle of the toes of Lizards, is to give greater power of flexion for bending round, and laying fast hold on twigs and branches of trees of various dimensions, or on inequalities of the surface of the ground or rocks, in the act of climbing, or running.*

All these coincidences of number and proportion, can only have originated in a premeditated adaptation of each part to its peculiar office; they teach us to arrange an extinct animal under an existing family of reptiles; and when we find so many other peculiarities of this tribe in almost every bone of the skeleton of the Pterodactyle, with such modifications, and such only as were necessary to fit it for the purposes of flight, we perceive unity of design pervading every part, and adapting to motion in the air, organs which in other genera

we are considering, that it adds another approximation to the character of the living Lizards; we have seen that it also differs from the other Pterodactyles, in having the fifth, instead of the fourth finger elongated, to become the expansor of the wing.

It is however probable that the fifth toe had only three joints, for the same reasons that are assigned respecting the number of joints in the fifth finger. In the P. Longirostris, Cuvier considers the small bone, (Pl. 21, 5, 6,) to be a rudimentary form of the fifth toe.

* A similar numerical disposition prevails also in the toes of birds, attended by similar advantages.

are calculated for progression on the ground, or in the water.

If we compare the foot of the Pterodactyle with that of the Bat, (see Pl. 22, ĸ,) we shall find that the Bat, like most other mammalia, has three joints in every toe, excepting the first, which has only two; still these two, in the Bat, are equal in length to the three bones of the other toes, so that the five claws of its foot range in one strait line, forming altogether the compound hook, by which the animal suspends itself in caves, with its head downwards, during its long periods of hybernation; the weight of its body being, by this contrivance, equally divided between each of the ten toes. The unequal length of the toes of the Pterodactyle must have rendered it almost impossible for its claws to range uniformly in line, like those of the Bat, and as no single claw could have supported for a long time the weight of the whole body, we may infer that the Pterodactyles did not suspend themselves after the manner of the Bats. The size and form of the foot, and also of the leg and thigh, show that they had the power of standing firmly on the ground, where, with their wings folded, they possibly moved after the manner of birds; they could also perch on trees, and climb on rocks and cliffs, with their hind and fore feet conjointly, like Bats and Lizards.

With regard to their food, it has been conjectured by Cuvier, that they fed on insects, and from the magnitude of their eyes that they may also have been noctivagous. The presence of large fossil Libellulæ, or Dragon-flies, and many other insects, in the same lithographic quarries with the Pterodactyles at Solenhofen, and of the wings of coleopterous insects, mixed with bones of Pterodactyles, in the oolitic slate of Stonesfield, near Oxford, proves that large insects existed at the same time with them, and may have contributed to their supply of food. We know that many of the smaller Lizards of existing species are insectivorous; some are also carnivorous, and others omnivorous, but the head and teeth of two species of Pterodactyle, are so much larger and stronger than is necessary for the capture of insects, that the larger species of them may possibly have fed on fishes, darting upon them from the air after the manner of Sea Swallows and Solan Geese. The enormous size and strength of the head and teeth of the P. Crassirostris, would not only have enabled it to catch fish, but also to kill and devour the few small marsupial mammalia which then existed upon the land.

The entire range of ancient anatomy, affords few more striking examples of the uniformity of the laws, which connect the extinct animals of the fossil creation with existing organized beings, than those we have been examining in the case of the Pterodactyle. We find the details of parts which, from their minuteness should seem insignificant, acquiring great importance in such an investigation as we are now conducting; they shew not less distinctly, than the colossal limbs of the most gigantic quadrupeds, a numerical coincidence, and a concurrence of proportions, which it seems impossible to refer to the effect of accident ; and which point out unity of purpose, and deliberate design, in some intelligent First Cause, from which they were all derived. We have seen that whilst all the laws of existing organization in the order of Lizards, are rigidly maintained in the Pterodactyles; still, as Lizards modified to move like birds and Bats in the air, they received, in each part of their frame, a perfect adaptation to their state. We have dwelt more at length on the minutiæ of their mechanism, because they convey us back into ages so exceedingly remote, and show that even in those distant eras, the same care of a common Creator, which we witness in the mechanism of our own bodies, and those of the myriads of inferior creatures that move around us, was extended to the structure of creatures, that at first sight seem made up only of monstrosities.

SECTION IX.

MEGALOSAURUS.*

THE Megalosaurus, as its name implies, was a Lizard, of great size, of which, although no skeleton has yet been found entire, so many perfect bones and teeth have been discovered in the same quarries, that we are nearly as well acquainted with the form and dimensions of its limbs, as if they had been found together in a single block of stone.*

From the size and proportions of these bones, as compared with existing Lizards, Cuvier concludes the Megalosaurus to have been an enormous reptile, measuring from forty to fifty feet in length, and partaking of the structure of the Crocodile and the Monitor.

---

* This genus was established by the Author, in a Memoir, published in the Geol. Trans. of London, (Vol. I., N. S. Pt. 2, 1824), and was founded upon specimens discovered in the oolitic slate of Stonesfield, near Oxford, the place in which these bones have as yet chiefly occurred. Mr. Mantell has discovered remains of the same animal in the Wealden fresh-water formation of Tilgate Forest; and from this circumstance we infer that it existed during the deposition of the entire series of oolitic strata. The author, in 1826, saw fragments of a jaw, containing teeth, and of some other bones of Megalosaurus, in the museum at Besançon, from the oolite of that neighbourhood.

As the femur and tibia measure nearly three feet each, the entire hind leg must have attained a length of nearly two yards: a metatarsal bone, thirteen inches long, indicates a corresponding length in the foot.* The bones of the thigh and leg are not solid at the centre, as in Crocodiles, and other aquatic quadrupeds, but have large medullary cavities, like the bones of terrestrial animals. We learn from this circumstance, added to the character of the foot, that the Megalosaurus lived chiefly upon the land.

In the internal condition of these fossil bones, we see the same adaptation of the skeleton to its proper element, which now distinguishes the bones of terrestrial, from those of aquatic Saurians.† In the Ichthyosauri and Plesiosauri, whose paddles were calculated exclusively to move in water, even the largest bones of the arms and legs were solid throughout. Their weight would in no way have embarrassed their action in the fluid medium they inhabited ; but in the huge Megalosaurus, and still more gigantic Iguanodon, which are shown by the character of their feet to have been fitted to move on land, the larger bones of the legs were diminished

* See Geol. Trans. 2nd series, Vol. 3, p. 427, Pl. 41.

† I learn from Mr. Owen that the long bones of land Tortoises have a close cancellous internal structure, but not a medullary cavity.

in weight, by being internally hollow, and having their cavities filled with the light material of marrow, while their cylindrical form tended also to combine this lightness with strength.*

The form of the teeth shews the Megalo-

---

* The medullary cavities in the fossil bones of Megalosaurus, from Stonesfield, are usually filled with calcareous spar.   In the Oxford Museum there is a specimen from the Wealden fresh-water formation at Langton, near Tunbridge Wells, which is perhaps unique amongst organic remains : it presents the curious fact of a perfect cast of the interior of a large bone, appa-rently the femur of a Megalosaurus, exhibiting the exact form and ramifications of the marrow, whilst the bone itself has entirely perished.   The substance of this cast is fine sand, cemented by oxide of iron, and its form distinctly represents all the minute reticulations, with which the marrow filled the intercolumniations of the cancelli, near the extremity of the bone.   It exhibits also casts of the perforations along the internal parietes, whereby the vessels entered obliquely from the exterior of the bone, to communicate with the marrow. A mould of the exterior of the same bone has been also formed by the sandstone in which it was imbedded : hence, although the bone itself has perished, we have precise representations both of its external form and internal cavities, and a model of the mar-row that filled this femur, nearly as perfect as could be made by pouring wax into an empty marrow-bone, and corroding away the bone with acid.   The sand which formed this cast must have entered the medullary cavity by a fracture across the other ex-tremity of the bone, which was wanting in the specimen.

From this natural preparation of ancient anatomy we learn that the disposition of marrow, and its connection with the reti-culated extremities of the interior of the femur, were the same in these gigantic Lizards of a former world, as in medullary cavi-ties of existing species.

saurus to have been in a high degree carnivorous: it probably fed on smaller reptiles, such as Crocodiles and Tortoises, whose remains abound in the same strata with its bones. It may also have taken to the water in pursuit of Plesiosauri and fishes.*

The most important part of the Megalosaurus yet found, consists of a fragment of the lower jaw, containing many teeth, (Pl. 23, Figs. 1'—2'). The form of this jaw shows that the head was terminated by a straight and narrow snout, compressed laterally like that of the Delphinus Gangeticus.

As in all animals, the jaws and teeth form the most characteristic parts, I shall limit my present observations to a few striking circumstances in the dentition of the Megalosaurus. From these we learn that the animal was a reptile, closely allied to some of our modern Lizards; and viewing the teeth as instruments for providing food to a carnivorous creature of enormous magnitude, they appear to have been admirably adapted to the destructive office for which they were designed. Their form and

---

* Mr. Broderip informs me that a living Iguana (I. Tuberculata), in the gardens of the Zoological Society of London, in the summer of 1834, was observed frequently to enter the water, and swim across a small pond, using its long tail as the instrument of progression, and keeping its fore feet motionless.

mechanism will best be explained by reference to the figures in Pl. 23.*

In the structure of these teeth, (Pl. 23, Figs. 1, 2, 3), we find a combination of mechanical contrivances analogous to those which are adopted in the construction of the knife, the sabre, and the saw. When first protruded above the gum, (Pl. 23, Figs. 1'. 2'.) the apex of each tooth presented a double cutting edge of serrated enamel. In this stage, its position and line of action were nearly vertical, and its form like that of the two-edged point of a sabre, cutting equally on each side. As the tooth advanced in growth, it became curved

* The outer margin of the jaw (Pl. 23, Fig. 1'. 2'.) rises nearly an inch above its inner margin, forming a continuous lateral parapet to support the teeth on the exterior side, where the greatest support was necessary; whilst the inner margin (Pl. 23, Fig. 1') throws up a series of triangular plates of bone, forming a zig-zag buttress along the interior of the alveoli. From the centre of each triangular plate, a bony partition crosses to the outer parapet, thus completing the successive alveoli. The new teeth are seen in the angle between each triangular plate, rising in reserve to supply the loss of the older teeth, as often as progressive growth, or accidental fracture, may render such renewal necessary; and thus affording an exuberant provision for a rapid succession and restoration of these most essential implements. They were formed in distinct cavities, by the side of the old teeth, towards the interior surface of the jaw, and probably expelled them by the usual process of pressure and absorption; insinuating themselves into the cavities thus left vacant. This contrivance for the renewal of teeth is strictly analogous to that which takes place in the dentition of many species of existing Lizards.

backwards, in the form of a pruning knife,
(Pl. 23, Figs. 1. 2. 3.), and the edge of serrated
enamel was continued downwards to the base of
the inner and cutting side of the tooth, (Fig. 1,
B. D.), whilst, on the outer side, a similar
edge descended, but to a short distance from the
point (Fig. 1, B. to C.), and the convex portion
of the tooth (A.) became blunt and thick, as the
back of a knife is made thick, for the purpose
of producing strength. The strength of the
tooth was further increased by the expansion
of its sides, (as represented in the transverse
section, Fig. 4, A. D). Had the serrature
continued along the whole of the blunt and
convex portion of the tooth, it would, in this
position, have possessed no useful cutting power;
it ceased precisely at the point (C.), beyond
which it could no longer be effective. In a
tooth thus formed for cutting along its concave
edge, each movement of the jaw combined the
power of the knife and saw; whilst the apex,
in making the first incision, acted like the two-
edged point of a sabre. The backward curvature of the full-grown teeth, enabled them to
retain, like barbs, the prey which they had
penetrated. In these adaptations, we see contrivances, which human ingenuity has also
adopted, in the preparation of various instruments of art.

In a former chapter (Ch. XIII.) I endea-

voured to show that the establishment of carnivorous races throughout the animal kingdom
tends materially to diminish the aggregate
amount of animal suffering. The provision of
teeth and jaws, adapted to effect the work of
death most speedily, is highly subsidiary to the
accomplishment of this desirable end. We act
ourselves on this conviction, under the impulse
of pure humanity, when we provide the most
efficient instruments to produce the instantaneous, and most easy death, of the innumerable
animals that are daily slaughtered for the supply of human food.

SECTION X.

IGUANODON.*

As the reptiles hitherto considered appear from
their teeth to have been carnivorous, so we find
extinct species of the same great family, that
assumed the character and office of herbivora.
For our knowledge of this genus, we are indebted to the scientific researches of Mr. Mantell. This indefatigable historian of the Wealden fresh-water formation, has not only found

---

* See Pl. 1, Fig. 45, and Pl. 24; and Mantell's Geology of
Sussex, and of the South-east of England.

the remains of the Plesiosaurus, Megalosaurus, Hylæosaurus,* and several species of Crocodiles and Tortoises in these deposits, of a period intermediate between the oolitic and cretaceous series, but has also discovered in Tilgate Forest the remains of the Iguanodon, a reptile much more gigantic than the Megalosaurus, and which, from the character of its teeth, appears to have been herbivorous.† The teeth of the Iguanodon are so precisely similar, in the principles of their construction, to the teeth of the modern Iguana, as to leave no

* The Hylæosaurus, or Lizard of the Weald, was discovered in Tilgate Forest, in Sussex, in 1832. This extraordinary Lizard was probably about twenty-five feet long. Its most peculiar character consists in the remains of a series of long, flat, and pointed bones, which seems to have formed an enormous dermal fringe, like the horny spines on the back of the modern Iguana. These bones vary in length from five to seventeen inches, and in width from three to seven inches and a half at the base. Together with them were found the remains of large dermal bones, or thick scales, which were probably lodged in the skin.

† The Iguanodon has hitherto been found only, with one exception, in the Wealden fresh-water formation of the south of England, (Pl. 1, sec. 22.), intermediate between the marine oolitic deposits of the Portland stone and those of the green-sand formation in the cretaceous series. The discovery, in 1834, (Phil. Mag. July 1834, p. 77), of a large proportion of the skeleton of one of these animals, in strata of the latter formation, in the quarries of Kentish Rag, near Maidstone, shews that the duration of this animal did not cease with the completion of the Wealden series. The individual from which this skeleton was derived had probably been drifted to sea, as those which afforded the bones found in the fresh-water depo-

doubt of the near connection of this most gigantic extinct reptile with the Iguanas of our own time. When we consider that the largest living Iguana rarely exceeds five feet in length, whilst the congenerous fossil animal must have been nearly twelve times as long, we cannot but be impressed by the discovery of a resemblance, amounting almost to identity, between such characteristic organs as the teeth, in one of the most enormous among the extinct reptiles of the fossil world, and those of a genus whose largest species is comparatively so diminutive. According to Cuvier, the common Iguana inhabits all the warm regions of America: it lives chiefly upon trees, eating fruits, and seeds, and leaves. The female occasionally visits the water, for the purpose of laying in the sand its eggs, which are about the size of those of a pigeon.*

sits subjacent to this marine formation, had been drifted into an estuary. This unique skeleton is now in the museum of Mr. Mantell, and confirms nearly all his conjectures respecting the many insulated bones which he had referred to the Iguanodon.

* In the Appendix to a paper in the Geol. Trans. Lond. (N. S. Vol. III. Pt. 3) on the fossil bones of the Iguanodon, found in the Isle of Wight and Isle of Purbeck, I have mentioned the following facts, illustrative of the herbivorous habits of the living Iguana.

In the spring of 1829, " Mr. W. J. Broderip saw a living Iguana, about two feet long, in a hothouse at Mr. Miller's nursery gardens, near Bristol. It had refused to eat insects, and

As the modern Iguana is found only in the warmest regions of the present earth, we may reasonably infer that a similar, if not a still warmer climate, prevailed at the time when so huge a Lizard as the Iguanodon inhabited what are now the temperate regions of the southern coasts of England. We know from the fragment of a femur, in the collection of Mr. Mantell, that the thigh-bone of this reptile much exceeded in bulk that of the largest Elephant: this fragment presents a circumference of twenty-two inches in its smallest part, and the entire length must have been between four and five feet. Comparing the proportions of this monstrous bone with those of the fossil teeth with which it is associated, it appears that they bear to one another nearly the same ratio that the femur of the Iguana bears to the similarly constructed and peculiar teeth of that animal.*

other kinds of animal food, until happening to be near some kidney-bean plants that were in the house for forcing, it began to eat of their leaves, and was from that time forth supplied from these plants." In 1828, Captain Belcher found, in the island of Isabella, swarms of Iguanas, that appeared omnivorous; they fed voraciously on the eggs of birds, and the intestines of fowls and insects.

* From a careful comparison of the bones of the Iguanodon with those of the Iguana, made by taking an average from the proportions of different bones from eight separate parts of the respective skeletons, Mr. Mantell has arrived at these dimensions as being the proportionate measures of the following parts of this extraordinary reptile :

It has been stated, in the preceding section, that the large medullary cavities in the femur of the Iguanodon, and the form of the bones of the feet, show that this animal, like the Megalosaurus, was constructed to move on land.

A further analogy between the extinct fossil and the recent Iguana is offered by the presence in both of a horn of bone upon the nose, (Pl. 24, Fig. 14). The concurrence of peculiarities so remarkable as the union of this nasal horn with a mode of dentition of which there is no example, except in the Iguanas, affords one of the many proofs of the universality of the laws of co-existence, which prevailed no less constantly throughout the extinct genera and species of the fossil world, than they do among the living members of the animal kingdom.

	Feet.
Length from snout to the extremity of the tail....	70
Length of tail	$52\frac{1}{2}$
Circumference of body	$14\frac{1}{2}$

Mr. Mantell calculates the femur of the Iguanodon to be twenty times the size of that of a modern Iguana; but as animals do not increase in length in the same ratio as in bulk, it does not follow that the Iguanodon attained the enormous length of one hundred feet, although it approached perhaps nearly to seventy feet.

As the Iguanodon, from its enormous bulk, must have been unable to mount on trees, it could not have applied its tail to the same purpose as the Iguana, to assist in climbing; and the longitudinal diameter of its caudal vertebræ is much less in proportion than in the Iguana, and shews the entire tail to have been comparatively shorter.

Teeth.

As the teeth are the most characteristic and important parts of the animal, I shall endeavour to extract from them evidence of design, both in their construction and mode of renewal, and also in their adaptation to the office of consuming vegetables, in a manner peculiar to themselves. They are not lodged in distinct sockets, like the teeth of Crocodiles, but fixed, as in Lizards, along the internal face of the dental bone, to which they adhere by one side of the bony substance of their root. (Pl. 24, Fig. 13.)

The teeth of most herbivorous quadrupeds, (exclusively of the defensive tusks), are divided into two classes of distinct office, viz. incisors and molars; the former destined to collect and sever vegetable substances from the ground, or from the parent plant; the latter to grind and masticate them on their way towards the stomach. The living Iguanas, which are in great part herbivorous, afford a striking exception to this economy: as their teeth are little fitted for grinding, they transmit their food very slightly comminuted into the stomach.

Our giant Iguanodon, also, had teeth resembling those of the Iguana, and of so herbivorous a character, that at first sight they were supposed by Cuvier to be the teeth of a Rhinoceros.

The examination of these teeth will lead us to the discovery of remarkable contrivances, adapting them to the function of cropping tough vegetable food, such as the Clathraria, and similar plants, which are found buried with the Iguanodon, might have afforded. We know the form and power of iron pincers to gripe and tear nails from their lodgment in wood: a still more powerful kind of pincers, or nippers, is constructed for the purpose of cutting wire, which yields to them nearly as readily as thread to a pair of scissors. Our figures (Pl. 24, Figs. 6, 7, 8, 12) show the place of the cutting edges, and form of curvature, and points of enlargement and contraction, in the teeth of the Iguanodon, to be nearly the same as in the corresponding parts of these powerful metallic tools; and the mechanical advantages of such teeth, as instruments for tearing and cutting, must have been similar.*

The teeth exhibit also two kinds of provisions to maintain sharp edges along the cutting surface, from their first protrusion, until they were worn down to the very stump. The first

---

* Fig. 2. represents the front of a young tooth; and Figs. 5, 6, 7, 8, the front of four other teeth, thrown slightly into profile. In all of these we recognise a near approach to the form of the nipping pincers, with a sharp cutting edge at the upper margin of the enamel. The enamel is here expressed by wavy lines, which represent its actual structure: it is placed only in front, like the enamel in front of the incisors of Rodentia.

of these is a sharp and serrated edge, extending on each side downwards, from the point to the broadest portion of the body of the tooth. (See Figs. 1, 2, 6, 8, 12, &c.)

The second provision is one of compensation for the gradual destruction of this serrated edge, by substituting a plate of thin enamel, to maintain a cutting power in the anterior portion of the tooth, until its entire substance was consumed in service.*

Whilst the crown of the tooth was thus gradually diminishing above, a simultaneous absorption of the root went on below, caused by the pressure of a new tooth rising to replace the old one, until by this continual consumption at both extremities, the middle portion of the older tooth was reduced to a hollow stump, (Figs. 10, 11), which fell from the jaw to make room for a

---

* This perpetual edge resulted from the enamel being placed only on the front of the tooth, like that on the incisors of Rodentia. As the softer material of the tooth itself must have worn away more readily than this enamel, and most readily at the part remotest from it, an oblique section of the crown was thus perpetually maintained, with a sharp cutting edge in front, like that of the nippers. (See Figs. 7. 8. 12.)

The younger tooth, (Fig. 1), when first protruded, was lancet-shaped, with a serrated edge, extending on each side downwards, from the point to its broadest portion, as in the living Iguana. (Pl. 24. f. 13, and Fig. 4.) This serrature ceased at the broadest diameter of the tooth, i. e. precisely at the line, below which, had they been continued, they would have had no effect in cutting. (Pl. 24. f. 2. 6. 8. 9. 12.) As these saws were gradually worn away, the cutting power was transferred to the enamel in front,

more efficient successor.*   In this last stage the
form of the tooth had entirely changed, and the
crown had become flat, like the crown of worn-
out human incisors, and capable of performing
imperfect mastication after the cutting powers
had diminished.   There is, I believe, no other
example of teeth which possess the same mechanical advantages as instruments of cutting
and tearing portions of vegetable matter from
tough and rigid plants.   In this curious piece
of animal mechanism, we find a varied adjustment of all parts and proportions of the tooth,
to the exercise of peculiar functions ; attended
by compensations adapted to shifting conditions
of the instrument, during different stages of its

and here we find a provision of another kind to give efficacy and
strength.   The front was traversed longitudinally by alternate
ridges and furrows, (Pl. 24, Figs. 2, 5, 6, 7, 8), the ridges
serving as ribs or buttresses to strengthen and prevent the
enamel from scaling off, and forming, together with the furrows,
an edge slightly wavy, and disposed in a series of minute googes,
or fluted chisels ; hence the tooth became an instrument of greater
power to cut tough vegetables under the action of the jaw, than
if the enamel had been in a continuous straight line.   By these
contrivances, also it continued effective during every stage
through which it passed, from the serrated lancet-point of the
new tooth, (Fig. 1), to its final consumption.   (Fig. 10, 11.)

    * In Pl. 24, Fig. 13, the jaw of a recent Iguana exhibits the
commencement of this process, and a number of young teeth are
seen forcing their way upwards, and causing absorption at the
base of the older teeth.   Figs. 10, 11, exhibit the effect of similar absorption upon the residuary stump of the fossil tooth of an
Iguanodon.

consumption. And we must estimate the works of nature by a different standard from that which we apply to the productions of human art, if we can view such examples of mechanical contrivance, united with so much economy of expenditure, and with such anticipated adaptations to varying conditions in their application, without feeling a profound conviction that all this adjustment has resulted from design and high intelligence.

SECTION XI.

AMPHIBIOUS SAURIANS ALLIED TO CROCODILES.

THE fossil reptiles of the Crocodilean family do not deviate sufficiently from living genera, to require any description of peculiar and discontinued contrivances, like those we have seen in the Ichthyosaurus, Plesiosaurus, and Pterodactyle; but their occurrence in a fossil state is of high importance, as it shows that whilst many forms of vertebrated animals have one after another been created, and become extinct, during the successive geological changes of the surface of our globe; there are others which have survived all these changes and revolutions, and still retain the leading features under which they first appeared upon our planet.

If we look to the state of the earth, and the character of its population, at the time when Crocodilean forms were first added to the number of its inhabitants, we find that the highest class of living beings were reptiles, and that the only other vertebrated animals which then existed were fishes; the carnivorous reptiles at this early period must therefore have fed chiefly upon them, and if in the existing family of Crocodiles there be any, that are in a peculiar degree piscivorous, their form is that we should expect to find in those most ancient fossil genera, whose chief supply of food must have been derived from fishes.

In the living sub-genera of the Crocodilean family, we see the elongated and slender beak of the Gavial of the Ganges, constructed to feed on fishes; whilst the shorter and stronger snout of the broad-nosed Crocodiles and Alligators gives them the power of seizing and devouring quadrupeds, that come to the banks of rivers in hot countries to drink. As there were scarcely any mammalia* during the secondary periods, whilst the waters were abundantly stored with

---

* The small Opossums in the oolite formation at Stonesfield, near Oxford, are the only land mammalia whose bones have been yet discovered in any strata more ancient than the tertiary.

fishes, we might, *à priori*, expect that if any Crocodilean forms had then existed they would most nearly have resembled the modern Gavial. And we have hitherto found only those genera which have elongated beaks, in formations anterior to, and including the chalk ; whilst true Crocodiles, with a short and broad snout, like that of the Cayman and the Alligator, appear for the first time in strata of the tertiary periods, in which the remains of mammalia abound.*

During these grand periods of lacustrine mammalia, in which but few of the present genera of terrestrial carnivora had been called into existence, the important office of controlling the excessive increase of the aquatic herbivora appears to have been consigned to the Crocodiles, whose habits fitted them, in a peculiar degree, for such a service. Thus, the past history of the Crocodilean tribe presents another example of the well regulated workings of a

* One of these, found by Mr. Spencer in the London clay of the Isle of Sheppy, is engraved, Pl. 25', Fig 1. Crocodiles of this kind have been found in the chalk of Meudon, in the plastic clay of Auteuil, in the London clay, in the gypsum of Mont Martre, and in the lignites of Provence.

The modern broad-nosed Crocodileans, though they have the power to capture mammalia, are not limited to this kind of prey ; they feed largely also on fishes, and occasionally on birds. This omnivorous character of the existing Crocodilean family, seems adapted to the present general diffusion of more varied kinds of food, than existed when the only form of the beak in this family was fitted, like that of the Gavial, to feed chiefly on Fishes.

consistent plan in the economy of animated nature, under which each individual, whilst following its own instinct, and pursuing its own good, is instrumental in promoting the general welfare of the whole family of its co-temporaries.

Cuvier observes, that the presence of Croco-dilean reptiles, which are usually inhabitants of fresh water, in various beds, loaded with the remains of other reptiles and shells that are decidedly marine, and the further fact of their being, in many cases, accompanied by fresh-water Tortoises, shows that there must have existed dry land, watered by rivers, in the early periods when these strata were deposited, and long before the formation of the lacustrine ter-tiary strata of the neighbourhood of Paris.* The living species of the Crocodile family are twelve in number, namely, one Gavial, eight true Crocodiles, and three Alligators. There are also many fossil species: no less than six of these have been made out by Cuvier, and several

* M. Geoffroy St. Hilaire has arranged the fossil Saurians with long and narrow beaks, like that of the Gavial, under the two new genera, Teleosaurus and Steneosaurus. In the Teleo-saurus, (Pl. 25′, Fig. 2.) the nostrils form almost a vertical section of the anterior extremity of the beak ; in the Steneo-saurus, (Pl. 25′, Fig. 3.) this anterior termination of the nasal canal had nearly the same arrangement as in the Gavial, opening upwards, and being almost semi-circular on each side.—Recher-ches sur les grands Sauriens, 1831.

others, from the secondary and tertiary formations in England remain to be described.*

It would be foreign to our present purpose, to enter into a minute comparison of the osteology of living and fossil genera and species of this family. We may simply observe, with respect to their similar manner of dentition, that they all present the same examples of provision for extraordinary expenditure of teeth, by an unusually abundant store of these most essential organs.† As Crocodiles increase to no less than four hundred times their original bulk,

---

* One of the finest specimens of fossil Teleosauri yet discovered, (see Pl. 25, Fig. 1), was found in the year 1824, in the alum shale of the lias formation at Saltwick, near Whitby, and is engraved in Young and Bird's Geological Survey of the Yorkshire Coast, 2d Ed. 1828 : its entire length is about eighteen feet, the breadth of the head twelve inches, the snout was long and slender, as in the Gavial, the teeth, one hundred and forty in number, are all small and slender, and placed in nearly a straight line. The heads of two other individuals of the same species, found near Whitby, are represented in the same plate, Figs. 2. 3.

Some of the ungual phalanges, which are preserved on the hind feet of this animal, Fig. 1, show that these extremities were terminated by long and sharp claws, adapted for motion upon land, from which we may infer that the animal was not exclusively marine; from the nature of the shells with which they are associated, in the lias and oolite formations, it is probable that both the Steneosaurus and Teleosaurus frequented shallow seas. Mr. Lyell states that the larger Alligator of the Ganges sometimes descends beyond the brackish water of the delta into the sea.

† This mode of dentition has been already exemplified in speaking of the dentition of the Ichthyosaurus, P. 172, and Pl. 11. A.

between the period at which they leave the egg
and their full maturity, they are provided with
a more frequent succession of teeth than the
mammalia, in order to maintain a duly proportioned supply during every period of their life.
As the predaceous habits of these animals cause
their teeth, placed in so long a jaw, to be peculiarly liable to destruction, the same provision serves also to renew the losses which must
often be occasioned by accidental fracture.

The existence of these remedial forces, thus
uniformly adapted to supply anticipated wants,
and to repair foreseen injuries, affords an example of those supplementary contrivances,
which give double strength to the argument
from design, in proof of the agency of Intelligence, in the construction and renovation of
the animal machinery wherein such contrivances are introduced.

The discovery of Crocodilean forms so nearly
allied to the living Gavial, in the same early
strata that contains the first traces of the Ichthyosaurus and the Plesiosaurus, is a fact which
seems wholly at variance with every theory that
would derive the race of Crocodiles from Ichthyosauri and Plesiosauri, by any process of gradual
transmutation or developement. The first appearance of all these three families of reptiles
seems to have been nearly simultaneous; and
they all continued to exist together until the ter-

mination of the secondary formations; when the Ichthyosauri and Plesiosauri became extinct, and forms of Crocodiles, approaching to the Cayman and the Alligator, were for the first time introduced.

SECTION XII.

FOSSIL TORTOISES, OR TESTUDINATA.

AMONG the existing animal population of the warmer regions of the earth, there is an extensive order of reptiles, comprehended by Cuvier under the name of Chelonians, or Tortoises. These are subdivided into four distinct families; one inhabiting salt water, two others fresh water lakes and rivers, and a fourth living entirely upon the land. One of the most striking characters of this Order consists in the provision that is made for the defence of creatures, whose movements are usually slow and torpid, by inclosing the body within a double shield or cuirass, formed by the expansion of the vertebræ, ribs and sternum, into a broad bony case.

The small European Tortoise, Testudo Græca, and the eatable Turtle, Chelonia Mydas, are familiar examples of this peculiar arrangement both in terrestrial and aquatic reptiles; in each

case the shield affords compensation for the
want of rapidity of motion to animals that have
no ready means of escape by flight or concealment from their enemies.   We learn from Geology that this Order began to exist nearly at the
same time with the Order of Saurians, and has
continued coextensively with them through the
secondary and tertiary formations, unto the present time: their fossil remains present also the
same threefold divisions that exist among modern Testudinata, into groups respectively adapted to live in salt and fresh water, and upon the
land.

Animals of this Order have yet been found
only in strata more recent than the carboniferous
series.*   The earliest example recorded by Cuvier, (Oss. Foss. Vol. 5, Pt. 2, p. 525), is that of a
very large species of Sea Turtle, the shell of
which was eight feet long, occurring in the Muschelkalk at Luneville.   Another marine species
has been found at Glaris, in slate referrible to
the lower cretaceous formation.   A third occurs
in the upper cretaceous freestone at Maestricht.
All these are associated with the remains of other
animals that are marine; and though they differ
both from living Turtles and from one another,
they still exhibit such general accordance in

---

* The fragment from the Caithness slate, engraved in the Geol.
Trans. Lond. V. iii. Pl. 16, Fig. 6, as portions of a trionyx, is
pronounced by M. Agassiz to be part of a fish.

the principles of their construction, with the conditions by which existing Turtles are fitted for their marine abode, that Cuvier was at once enabled to pronounce these fossil species to have been indubitably inhabitants of the sea.*

The genera Trionyx and Emys, present their fossil species in the Wealden freshwater formations of the Secondary series; and still more abundantly in the Tertiary lacustrine deposits; all these appear to have lived and died, under circumstances analogous to those which attend their cognate species in the lakes and rivers of the present tropics. They have also been found

* Plate 25', Fig. 4, represents a Turtle from the slate of Glaris: it is shewn to have been marine by the unequal elongation of the toes in the anterior paddle; because, in freshwater Tortoises, all the toes are nearly equal, and of moderate length; and in land Tortoises, they are also nearly equal, and short; but in all marine species they are very long, and the central toe of the anterior paddle, is by much the longest of all. The accordance with this latter condition in the specimen before us, is at once apparent; and both in this respect and in general structure, it approaches very nearly to living genera. This figure is copied from Vol. 5, Pt. 2, Tab. 14, *f.* 4, of the Oss. Foss. of Cuvier. M. Agassiz has favoured me with the following details respecting important parts which are imperfectly represented in the drawing from which Cuvier's engraving was taken. " The ribs show evidently that it is nearly connected with the genera Chelonia and Sphargis, but referrible to no known species; the fingers of the left fore paddle are five in number; the two exterior are the shortest, and have each three articulations; and the three internal fingers, of which the middle one is the longest, have each four articulations, as in the existing genera, Chelonia and Sphargis.

G.   S

in marine deposits, where their admixture with the remains of Crocodilean animals shows that they were probably drifted, together with them, into the sea, from land, at no great distance.*

In the close approximation of the generic characters of these fossil Testudinata, of various and ancient geological epochs, to those of the present day, we have a striking example of the unity of design which has pervaded the construction of animals, from the most distant periods in which these forms of organized beings were also called into existence. As the paddle of the Turtle has at all times been adapted to move in the waves of the sea, so have the feet of the Trionyx and Emys ever been constructed for a more quiescent life in freshwater, whilst those of the Tortoise have been no less uniformly fitted to creep and burrow upon land.

The remains of land Tortoises have been more rarely observed in a fossil state. Cuvier mentions but two examples, and these in very recent formations at Aix, and in the Isle of France.

Scotland has recently afforded evidence of the existence of more than one species of these ter-

---

* Thus two large extinct species of Emys occur, together with marine shells, in the jura limestone at Soleure. The Emys also and Crocodiles, are found in the marine deposits of the London clay at Sheppy and Harwich; and the former is associated with marine exuviæ at Brussels. Very perfect impressions of small horny scales of Testudinata, occur in the Oolite slate of Stonesfield, near Oxford.

restrial reptiles, during the period of the New red, or Variegated sandstone formation. (See Pl. 1, Sec. 17). The nature of this evidence is almost unique in the history of organic remains.*

It is not uncommon to find on the surface of sandstone, tracks which mark the passage of small Crustacea and other marine animals, whilst

---

* See Dr. Duncan's account of tracks and footmarks of animals, impressed on sandstone in the quarry of Corn Cockle Muir, Dumfries-shire, Trans. Royal Society of Edinburgh, 1828.

Dr. Duncan states that the strata which bear these impressions lie on each other like volumes on the shelf of a library, when all inclining to one side : that the quarry has been worked to the depth of forty-five feet from the top of the rock ; throughout the whole of this depth similar impressions have been found, not on a single stratum only, but on many successive strata ; i. e. after removing a large slab which contained foot-prints, they found perhaps the very next stratum at the distance of a few feet, or it might be less than an inch, exhibiting a similar phenomenon. Hence it follows that the process by which the impressions were made on the sand, and subsequently buried, was repeated at successive intervals.

I learn, by a letter from Dr. Duncan, dated October, 1834, that similar impressions, attended by nearly the same circumstances, have recently been discovered about ten miles south of Corn Cockle Muir, in the Red sandstone quarries of Craigs, two miles east of the town of Dumfries. The inclination of the strata of this place is about 45° S.W. like that of almost all the sandstone strata of the neighbourhood. One of these tracks extended from twenty to thirty feet in length : in this place also, as at Corn Cockle Muir, no bones of any kind have yet been discovered.

Sir William Jardine has informed Dr. Duncan that tracks of animals have been found also in other quarries near Corn Cockle Muir.

this stone was in a state of loose sand at the bottom of the sea. Laminated sandstones are also often disposed in minute undulations, resembling those formed by the ripple of agitated water upon sand.*

The same causes, which have so commonly preserved these undulations, would equally preserve any impressions that might happen to have been made on beds of sand, by the feet of animals; the only essential condition of such preservation being, that they should have become covered with a further deposit of earthy matter, before they were obliterated by any succeeding agitations of the water.

The nature of the impressions in Dumfries-

* In 1831, Mr. G. P. Scrope, after visiting the quarries of Dumfries, found rippled markings, and abundant foot tracks of small animals on the Forest marble beds north of Bath. These were probably tracks of Crustacea.—See Phil. Mag. May, 1831, p. 376.

We find on the surface of slabs both of the calcareous grit, and Stonesfield slate, near Oxford, and on sandstones of the Wealden formation, in Sussex and Dorsetshire, perfectly preserved and petrified castings of marine worms, at the upper extremity of holes bored by them in the sand, while it was yet soft at the bottom of the water; and within the sandstones, traces of tubular holes in which the worms resided. The preservation of these tubes and castings shews the very quiet condition of the bottom, and the gentle action of the water, which brought the materials that covered them over, without disturbing them.

Cases of this kind add to the probability of the preservation of footsteps of Tortoises on the Red sandstone, and also afford proof of the alternation of intervals of repose with periods of violence, during the destructive processes by which derivative strata were formed.

shire may be seen by reference to Pl. 26. They traverse the rock in a direction either up or down, and not across the surfaces of the strata, which are now inclined at an angle of 38°. On one slab there are twenty-four continuous impressions of feet, forming a regular track, with six distinct repetitions of the mark of each foot, the fore-foot being differently shaped from the hind-foot; the marks of claws are also very distinct.*

Although these footsteps are thus abundant in the extensive quarries of Corn Cockle Muir, no trace whatever has been found of any portion of the bones of the animals whose feet they represent. This circumstance may perhaps be explained by the nature of the siliceous sandstone having been unfavourable to the preservation of organic remains. The conditions which would admit of the entire obliteration of bones,

---

* On comparing some of these impressions with the tracks which I caused to be made on soft sand, and clay, and upon unbaked pie-crust, by a living Emys and Testudo Græca, I found the correspondence with the latter sufficiently close, allowing for difference of species, to render it highly probable that the fossil footsteps were also impressed by the feet of land Tortoises.

In the bed of the Sapey and Whelpley brooks near Tenbury, circular markings occur in the Old Red Sandstone, which are referred by the natives to the tracks of Horses, and the impressions of Patten-rings, and a legendary tale has been applied to explain their history. They are caused by concretions of Marlstone and Iron, disposed in spherical cases around a solid core of sandstone, and intersected by these water courses.

would in no way interfere with the preservation of impressions made by feet, and speedily filled up by a succeeding deposit of sand, which would assume, with the fidelity of an artificial plaster mould, the precise form of the surface to which it was applied.

Notwithstanding this absence of bones from the rocks which are thus abundantly impressed with footsteps, the latter alone suffice to assure us both of the existence and character of the animals by which they were made. Their form is much too short for the feet of Crocodiles, or any other known Saurians; and it is to the Testudinata, or Tortoises, that we look, with most probability of finding the species to which their origin is due.*

The Historian or the Antiquary may have traversed the fields of ancient or of modern

---

* This evidence of footsteps, on which we are here arguing, is one which all mankind appeal to in every condition of society. The thief is identified by the impression which his shoe has left near the scene of his depredations. Captain Parry found the tracks of human feet upon the banks of the stream in Possession Bay, which appeared so fresh, that he at first imagined them to have been recently made by some natives : on examination they were distinctly ascertained to be the marks of the shoes of some of his own crew, eleven months before. The frozen condition of the soil had prevented their obliteration. The American savage not only identifies the Elk and Bison by the impression of their hoofs, but ascertains also the time that has elapsed since each animal had passed. From the Camel's track upon the sand, the Arab can determine whether it was heavily or lightly laden, or whether it was lame.

battles ; and may have pursued the line of march of triumphant Conquerors, whose armies trampled down the most mighty kingdoms of the world. The winds and storms have utterly obliterated the ephemeral impressions of their course. Not a track remains of a single foot, or a single hoof, of all the countless millions of men and beasts whose progress spread desolation over the earth. But the Reptiles, that crawled upon the half-finished surface of our infant planet, have left memorials of their passage, enduring and indelible. No history has recorded their creation or destruction; their very bones are found no more among the fossil relics of a former world. Centuries, and thousands of years, may have rolled away, between the time in which these footsteps were impressed by Tortoises upon the sands of their native Scotland, and the hour when they are again laid bare, and exposed to our curious and admiring eyes. Yet we behold them, stamped upon the rock, distinct as the track of the passing animal upon the recent snow ; as if to show that thousands of years are but as nothing amidst Eternity—and, as it were, in mockery of the fleeting perishable course of the mightiest Potentates among mankind.*

---

* A similar discovery of fossil footsteps has recently been made in Saxony, at the village of Hessberg, near Hildburg-hausen, in several quarries of grey quartzose sandstone, alter-

SECTION XIII.

FOSSIL FISHES.

THE history of Fossil Fishes is the branch of Paleontology which has hitherto received least attention, in consequence of the imperfect

nating with beds of red sandstone, nearly of the same age with that of Dumfries. (See Pl. 26'. 26''. 26'''.)

The following account of them is collected from notices by Dr. Hohnbaum and Professor Kaup. " The impressions of feet are partly hollow, and partly in relief; all the depressions are upon the *upper* surfaces of slabs of sandstone, whilst the reliefs are only upon the *lower* surfaces, covering those which bear the depressions. These reliefs are natural casts, formed in the subjacent footsteps as in moulds. On one slab (see Pl. 26'), six feet long by five feet wide, there occur many footsteps of more than one animal, and of various sizes. The larger impressions, which seem to be of the hind foot, are eight inches long, and five wide. (See Pl. 26''.) One was twelve inches long. Near to each large footstep, and at the regular distance of an inch and a half before it, is a smaller print of a fore foot, four inches long and three inches wide. These footsteps follow one another in pairs, at intervals of fourteen inches from pair to pair, each pair being in the same line. Both large and small steps have the great toes alternately on the right and left side; each has the print of five toes, and the first, or great toe is bent inwards like a thumb. The fore and hind foot are nearly similar in form, though they differ so greatly in size.

On the same slabs are other tracks, of smaller and diffe-

state of our knowledge of existing Fishes. The inaccessible recesses of the waters they inhabit, renders the study of their nature and habits much more difficult than that of terrestrial animals. The arrangement of this large and important class of Vertebrata was the last great work undertaken by Cuvier, not long before his lamented death, and nearly eight thousand species of living Fishes had come under his observation. The full development of their history

rently shaped feet, armed with nails. Many of these (Pl. 26') resemble the impressions on the sandstone of Dumfries, and are apparently the steps of Tortoises.

Professor Kaup has proposed the provisional name of Chirotherium for the great unknown animal that formed the larger footsteps, from the distant resemblance, both of the fore and hind feet, to the impression of a human hand; and he conjectures that they may have been derived from some quadruped allied to the Marsupialia. The presence of two small fossil mammalia related to the Opossum, in the Oolite formation of Stonesfield, and the approximation of this order to the class of Reptiles, which has already been alluded to, (page 73, *note*), are circumstances which give probability to such a conjecture. In the Kangaroo, the first toe of the fore foot is set obliquely to the others, like a thumb, and the disproportion between the fore and hind feet is also very great.

A further account of these footseps has been published by Dr. Sickler, in a letter to Blumenbach, 1834. Our figure (Pl. 26'), is copied from a plate that accompanies this letter; on comparing it with a large slab, covered with similar footmarks, from the same quarries, lately placed in the British Museum, (1835) I find that the representations, both of the large and small footsteps, correspond most accurately. The hind foot (Pl. 26''), is drawn from one on this slab. Pl. 26''' is drawn from a plaster

and numbers, and of the functions they discharge in the economy of nature, he has left to his able successors.

The fact of the formation of so large a portion of the surface of the earth beneath the water, would lead us to expect traces of the former existence of Fishes, wherever we have the remains of aquatic Mollusca, Articulata, and Radiata. Although a few remarkable places have long been celebrated as the repositories of fossil Fishes, even of these there are some, whose geological relations have scarcely yet been ascertained, while the nature of their Fishes remains in still greater obscurity.*

The task of arranging all this disorder has

cast in the British Museum, taken from another slab found in the same quarries, and impressed with footsteps of some small aquatic Reptile.

Some fragments of bones were found in the same quarries with these footsteps, but were destroyed.

A thin deposit of Green Marl, which lay upon the inferior bed of sand, at the time when the footsteps were impressed, causes the slabs above and below it to part readily, and exhibit the casts that were formed by the upper sand, in the prints that the animals had made on the lower stratum, through the marl, while soft, and sufficiently tenacious to retain the form of the footsteps.

* The most celebrated deposits of fossil Fishes in Europe are the coal formation of Saarbrück, in Lorraine; the bituminous slate of Mansfeld, in Thuringia; the calcareous lithographic slate of Solenhofen; the compact blue slate of Glaris; the limestone of Monte Bolca, near Verona; the marlstone of Oeningen, in Switzerland; and of Aix, in Provence.

Every attempt that has yet been made at a systematic arrange-

at length been undertaken by an individual, to whose hands Cuvier at once consigned the materials he had himself collected for this important work. The able researches of Professor Agassiz have already extended the number of fossil Fishes to two hundred genera, and more than eight hundred and fifty species.* The results of his enquiry throw a new and most important light on the state of the earth, during each of the great periods into which its past history has been divided. The study of fossil Ichthyology is therefore of peculiar importance to the geologist, as it enables him to follow an entire Class of animals, of so high a Division as the vertebrate, through the whole series of geological formations; and to institute comparisons between their various conditions during successive Periods of the earth's formation, such as Cuvier could carry only to a much more limited extent in the classes of Reptiles, Birds, and Mammifers, for want of adequate materials.

ment of these Fishes has been more or less defective, from an endeavour to arrange them under existing genera and families. The imperfection of his own, and of all preceding classifications of Fishes, is admitted by Cuvier; and one great proof of this imperfection is that they have led to no general results, either in Natural History, Physiology, or Geology.

* No existing genus is found among the fossil Fishes of any stratum older than the Chalk formation. In the inferior chalk there is one living genus, Fistularia; in the true chalk, five; and in the Tertiary strata of M. Bolca, thirty-nine living genera, and thirty-eight which are extinct.—Agassiz.

The system upon which M. Agassiz has established his classification of recent Fishes is in a peculiar degree applicable to fossil Fishes, being founded on the character of the external coverings, or Scales. This character is so sure and constant, that the preservation, even of a single scale, will often announce the genus and even the species of the animal from which it was derived ; just as certain feathers announce to a skilful ornithologist the genus or species of a Bird. It follows still further, that as the nature of their outward covering indicates the relations of all animals to the external world, we derive from their scales certain indications of the relations of Fishes;* the scales forming a kind of external skeleton, analogous to the crustaceous or

---

* The foundation of this character is laid upon the dermal covering, the skin being that organ which, more than any other part of the body, shews the relations of every animal to the element in which it moves.

The form and conditions of the feathers and down show the relation of Birds to the air in which they fly, or the water in which they swim or dive. The varied forms of fur and hair and bristles on the skins of Beasts are adapted to their respective place, and climate, and occupations upon the land. The scales of Fishes show a similar adaptation to their varied place and occupations beneath the waters.

Mr. Burchell informs me that he has observed, both in Africa and South America, that in the order of Serpents a peculiar character of the scales appears to indicate a natural subdivision ; and that in that tribe, to which the Viper, and nearly all the venomous Snakes belong, an acute ridge, or *carina*, along each dorsal scale may be considered as a distinctive mark.

horny coverings of Insects, to the feathers of
Birds, and the fur of Quadrupeds, which shows
more directly than the internal bones, their adaptation to the medium in which they lived.

A further advantage arises from the fact that
the enamelled condition of the scales of most
Fishes, which existed during the earlier geological epochs, rendered them much less destructi-
ble than their internal skeleton; and cases frequently occur where the entire scales and figure
of the Fish are perfectly preserved, whilst the
bones within these scales have altogether disappeared; the enamel of the scales being less
soluble than the more calcareous material of the
bone.*

* The following are the new Orders into which M. Agassiz
divides the Class of Fishes.

First Order, *PLACOIDIANS.* (Pl. 27, Figs. 1, 2, Etym.
πλαξ, *a broad plate.*) Fishes of this Order are characterized
by having their skin covered irregularly with plates of enamel,
often of considerable dimensions, and sometimes reduced to
small points, like the shagreen on the skins of many Sharks,
and the prickly, tooth-like tubercles on the skin of Rays. It
comprehends all the cartilaginous fishes of Cuvier, excepting the
Sturgeon.

The enamelled prickly tubercles on the skin of Sharks and
Dog-Fishes are well known, from the use made of them in rasp-
ing and polishing wood, and for shagreen.

Second Order, *GANOIDIANS.* (Pl. 27, 3, 4. Etym.
γανος, *splendour,* from the bright surface of their enamel.) The
families of this Order are characterized by angular scales, composed of horny or bony plates, covered with a thick plate of
enamel. The bony Pike (Lepidosteus Osseus, Pl. 27ª, Fig. 1);

It must be obvious that another and most important branch of natural history is enlisted in aid of Geology, as soon as the study of the character of fossil Fishes has been established on any footing, which admits of such general application as the system now proposed. We introduce an additional element into geological calculations; we bring an engine of great power, hitherto unapplied, to bear on the field of our enquiry, and seem almost to add a new sense to our powers of geological perception. The general result is, that fossil Fishes approximate

and Sturgeons are of this Order. It contains more than sixty genera, of which fifty are extinct.

Third Order, *CTENOIDIANS*. (Pl. 27, Figs. 5, 6, Etym. κτεις, a *comb*.) The Ctenoïdians have their scales jagged or pectinated, like the teeth of a comb, on their posterior margin. They are formed of laminæ of horn or bone, but have no enamel. The Perch affords a familiar example of scales constructed on this principle.

Fourth Order, *CYCLOIDIANS*. (Pl. 27, Figs. 7, 8. Etym. κυκλος, a *circle*.) Families of this Order have their scales smooth, and simple at their margin, and often ornamented with various figures on the upper surface : these scales are composed of laminæ of horn or bone, but have no enamel. The Herring and Salmon are examples of Cycloidians.

Each of these Orders contains both cartilaginous and bony Fishes : the representatives of each prevailed in different proportions during different epochs; only the two first existed before the commencement of the Cretaceous formations; the third and fourth Orders, which contain three-fourths of the eight thousand known species of living Fishes, appear for the first time in the Cretaceous strata, when all the preceding fossil genera of the two first Orders had become extinct.

nearest to existing genera and species, in the most recent Tertiary deposits; and differ from them most widely in strata whose antiquity is the highest; and that strata of intermediate age are marked by intermediate changes of ichthyological condition.

It appears still further, that all the great changes in the character of fossil Fishes take place simultaneously with the most important alterations in the other classes of fossil animals, and in fossil vegetables; and also in the mineral condition of the strata.*

It is satisfactory to find that these conclusions are in perfect accordance with those to which geologists had arrived from other data. The details that lead to them, will be described by M. Agassiz, in a work of many volumes, and will form a continuation of the Ossemens Fossiles of Cuvier. From the parts of this work already published, and from communications by the author, I select a few examples, illustrating

* The genera of Fishes which prevail in strata of the Carboniferous order are found no more after the deposition of the Zechstein, or Magnesian limestone. Those of the Oolitic series were introduced after the Zechstein, and ceased suddenly at the commencement of the Cretaceous formations. The genera of the Cretaceous formations are the first that approximate to existing genera. Those of the lower Tertiary deposits of London, Paris, and Monte Bolca, are still more nearly allied to existing forms; and the fossil Fishes of Oeningen and Aix approximate again yet closer to living genera, although every one of their species appears to be extinct.

the character of some of the most remarkable families of fossil Fishes.

It appears that the character of fossil Fishes does not change *insensibly* from one formation to another, as in the case of many Zoophytes and Testacea; nor do the same genera, or even the same families, pervade successive series of great formations; but their changes take place *abruptly*, at certain definite points in the vertical succession of the strata, like the sudden changes that occur in fossil Reptiles and Mammalia.* Not a single species of fossil Fishes has yet been found that is common to any two great geological formations; or living in our present seas.†

One important geological result has already attended the researches of M. Agassiz, viz. that the age and place of several formations hitherto unexplained by any other character, have been made clear by a knowledge of the fossil Fishes which they contain.‡

---

* M. Agassiz observes that fossil Fishes in the same formation present greater variations of species at distant localities, than we find in the species of shells and Zoophytes, in corresponding parts of the same formation; and that this circumstance is readily explained by the greater locomotive powers of this higher class of animals.

† The nodules of clay stone on the coast of Greenland, containing fishes of a species now living in the adjacent seas, (Mallotus Villosus) are probably modern concretions.

‡ Thus the slate of Engi, in the canton of Glaris, in Swit-

Sauroid Fishes in the Order Ganoid.

The voracious family of Sauroid, or Lizard-like Fishes, first claims our attention, and is highly important in the physiological consideration of the history of Fishes, as it combines in the structure both of the bones, and some of the soft parts, characters which are common to the class of reptiles. M. Agassiz has already ascer-

zerland, has long been one of the most celebrated, and least understood localities of fossil Fishes in Europe, and the mineral character of this slate had till lately caused it to be referred to the early period of the Transition series. M. Agassiz has found that among its numerous fishes, there is not one belonging to a single genus, that occurs in any formation older than the Cretaceous series; but that many of them agree with fossil species found in Bohemia, in the lower Cretaceous formation, or Pläner kalk; hence he infers that the Glaris slate is an altered condition of an argillaceous deposit, subordinate to the great Cretaceous formations of other parts of Europe, probably of the Gault.

Another example of the value of Ichthyology, in illustration of Geology, occurs in the fact, that as the fossil Fishes of the Wealden estuary formation are referrible to genera that characterize the strata of the Oolitic series, the Wealden deposits are hereby connected with the Oolitic period that preceded their commencement, and are separated from the Cretaceous formations that followed their termination. A change in the condition of the higher orders of the inhabitants of the waters seems to have accompanied the changes that occurred in the genera and species of inferior animals at the commencement of the Cretaceous formations.

A third example occurs, in the fact that M. Agassiz has, by resemblances in the character of their fossil Fishes, identified the hitherto unknown periods of the freshwater deposits of Oeningen, and of Aix in Provence, with that of the Molasse of Switzerland.

tained seventeen genera of Sauroid Fishes.
Their only living representatives are the genus
Lepidosteus,* or bony Pike (Pl. 27ᵃ Fig. 1.),
and the genus Polypterus (Agass. Poiss. Foss.
Vol. 2. Tab. C.), the former containing five
species, and the latter two.   Both these genera
are found only in fresh waters, the Lepidosteus
in the rivers of North America, and the Polypterus in the Nile, and the waters of Senegal.†

The teeth of the Sauroid Fishes are striated
longitudinally towards the base, and have a
hollow cone within. (See Pl. 27ᵃ, 2, 3, 4 ; and
Pl. 27. 9, 10, 11, 12, 13, 14.) The bones of the
palate also are furnished with a large apparatus of teeth. ‡

---

* Lepidosteus *Agassiz*—Lepisosteus *Lacépède*.

† The bones of the skull, in Sauroid Fishes, are united by
closer sutures than those of common Fishes. The vertebræ articulate with the spinous processes by sutures, like the vertebræ of
Saurians; the ribs also articulate with the extremities of the
spinous processes. The caudal vertebræ have distinct chevron
bones, and the general condition of the skeleton is stronger and
more solid than in other Fishes: the air-bladder also is bifid and
cellular, approaching to the character of lungs, and in the throat
there is a glottis, as in Sirens and Salamanders, and many Saurians.—See Report of Proceedings of Zool. Soc. London, October, 1834.

‡ The object of the extensive apparatus of teeth, over the
whole interior of the mouth of many of the most voracious
Fishes, appears not to be for mastication, but to enable them to
hold fast, and swallow the slippery bodies of other Fishes that
form their prey. No one who has handled a living Trout or Eel
can fail to appreciate duly the importance of the apparatus in
question.

Pl. 27, Figs. 11, 12, 13, 14, represent teeth of the largest Sauroid Fishes yet discovered, equalling in size the teeth of the largest Crocodiles: they occur in the lower region of the Coal formation near Edinburgh, and are referred by M. Agassiz to a new genus, Megalicthys. Pl. 27, Fig. 9, and Pl. 27ª, Fig. 4, are fragments of jaws, containing many smaller teeth of the same kind. The external form of all these teeth is nearly conical, and within them is a conical cavity, like that within the teeth of many Saurians; their base is fluted, like the base of the teeth of the Ichthyosaurus. Their prodigious size shows the magnitude which Fishes of this family attained at a period so early as that of the Coal formation :* their structure coincides

* We owe the discovery of these very curious teeth, and much valuable information on the Geology of the neighbourhood of Edinburgh, to the zeal and discernment of Dr. Hibbert, in the spring of 1834. The limestone in which these Fishes occur lies near the bottom of the Coal formation, and is loaded with Coprolites, derived apparently from predaceous Fishes. It is abundantly charged also with ferns, and other plants of the coal formation; and with the crustaceous remains of Cypris, a genus known only as an inhabitant of fresh water. These circumstances, and the absence of Corals and Encrinites, and of all species of marine shells, render it probable that this deposit was formed in a freshwater lake, or estuary. It has been recognized in various and distant places, at the bottom of the carboniferous strata near Edinburgh.

In the Transactions of the Royal Society of Edinburgh, Vol. XIII. Dr. Hibbert has published a most interesting description of the recent discoveries made in the limestone of Burdie House,

entirely with that of the teeth of the living Lepidosteus osseus. (Pl. 27ª, Figs. 1, 2, 3.)

Smaller Sauroid Fishes only have been noticed in the Magnesian limestone, forming about
one fifth of the total number yet observed in
this formation. Very large bones of this voracious family occur in the lias of Whitby and
Lyme Regis, and its genera abound throughout
the Oolite formation.* In the Cretaceous formations they become extremely rare.† They

illustrated with engravings, from which the larger teeth in our
plate are copied. (Pl. 27, Fig. 11, 12, 13, 14). The smaller
figures, Pl. 27, Fig. 9, and Pl. 27ª, Fig. 4, are drawn from
specimens belonging to Dr. Hibbert and the Royal Society of
Edinburgh.

In this memoir, Dr. Hibbert has also published figures of some
curious large scales, found at Burdie House, with the teeth of
Megalicthys, and referred by M. Agassiz to that Fish. Similar
scales have been noticed in various parts of the Edinburgh Coal
field, and also in the Coal formation of Newcastle-on-Tyne.
Unique specimens of the heads of two similar Fishes, and part
of a body covered with scales, from the Coal field near Leeds,
are preserved in the museum of that town.

Sir Philip Grey Egerton has recently discovered scales of the
Megalicthys, with teeth and bones of some other Fishes, and also
Coprolites, in the Coal formation of Silverdale, near Newcastle-
under-Line. These occur in a stratum of shale, containing shells
of three species of Unio, with balls of argillaceous iron ore and
plants.

* The Aspidorhynchus, from the Jurassic limestone of Solenhofen, (Pl. 27ª, Fig. 5), represents the general character of the
sauroid Fishes.

† The Macropoma is the only genus of Sauroid Fishes yet
found in the Chalk of England.

have not yet been discovered in any of the
Tertiary strata; and in the waters of the present world are reduced to the two genera, Lepidosteus and Polypterus.

Thus we see that this family of Sauroids
holds a very important place in the history of
fossil Fishes. In the waters of the Transition
period, the Sauroids and Sharks constituted the
chief voracious forms, destined to fulfil the
important office of checking excessive increase
of the inferior families. In the Secondary
strata, this office was largely shared by Ichthyosauri and other marine Saurians, until the commencement of the Chalk. The cessation of these
Reptiles and of the semi-reptile Sauroid Fishes
in the Tertiary formations made room for the
introduction of other predaceous families, approaching more nearly to those of the present
creation.*

* Much light has been thrown on the history of Fishes in the
Old red sandstone at the base of the Carboniferous series, by the
discoveries of Professor Sedgwick and Mr. Murchison, in the
bituminous schist of Caithness, (Geol. Trans. Lond. N. s. Vol. 3,
part 1.); and those of Dr. Traile, in the same schist in Orkney.
Dr. Fleming also has made important observations on Fishes in
the old red sandstone of Fifeshire. Further discoveries have
been made by Mr. Murchison of Fishes in the old red sandstone
of Salop and Herefordshire. The general conditions of all these
Fishes accord with those in the carboniferous series, but their
specific details present most interesting peculiarities. Many of
them will be figured by Mr. Murchison in his splendid Illustrations
of the Geology of the Border Counties of England and Wales.

Fishes in Strata of the Carboniferous Order.

I select the genus Amblypterus (Pl. 27[b].), as an example of Fishes whose duration was limited to the early periods of geological Formations; and which are marked by characters that cease after the deposition of the Magnesian limestone.

This genus occurs only in strata of the Carboniferous order, and presents four species at Saarbrück, in Lorraine;* it is found also in Brazil. The character of the teeth in Amblypterus, and most of the genera of this early epoch, shews the habit of these Fishes to have been to feed on decayed sea-weed, and soft animal substances at the bottom of the water: they are all small and numerous, and set close together like a brush. The form of the body, being not calculated for rapid progression, accords with this habit.

* The Fishes at Saarbrück are usually found in balls of clay ironstone, which form nodules in strata of bituminous coal shale. Lord Greenock has recently discovered many interesting examples of this, and other genera of Fishes in the coal formation at Newhaven, and Wardie, near Leith. The shore at Newhaven is strewed with nodules of ironstone, washed out by the action of the tide, from shale beds of the coal formation. Many of these ironstones have for their nucleus a fossil Amblypterus, or some other Fish; and an infinitely greater number contain Coprolites, apparently derived from a voracious species of Pygopterus, that preyed upon the smaller Fishes.

The vertebral column continues into the upper lobe of the tail, which is much longer than the lower lobe, and is thus adapted to sustain the body in an inclined position, with the head and mouth nearest to the bottom.

Among existing cartilaginous Fishes, the vertebral column is prolonged into the caudal fin of Sturgeons and Sharks: the former of these perform the office of scavengers, to clear the water of impurities, and have no teeth, but feed by means of a soft leather-like mouth, capable of protrusion and contraction, on putrid vegetables and animal substances at the bottom; hence they have constant occasion to keep their bodies in the same inclined position as the extinct fossil Fishes, whose feeble brush-like teeth shew that they also fed on soft substances in similar situations.*

The Sharks employ their tail in another peculiar manner, to turn their body in order to bring the mouth, which is placed downwards beneath the head, into contact with their prey. We find an important provision in every animal to give a position of ease and activity to the head during the operation of feeding.†

* At the siege of Silistria, the Sturgeons of the Danube were observed to feed voraciously on the putrid bodies of the Turks and Russian soldiers that were cast into that river.

† This remarkable elongation of the superior lobe of the tail is found in every bony Fish of strata anterior to and including the Magnesian limestone; but in strata above this limestone the

Fishes of the Magnesian Limestone,
or Zechstein.

The Fishes of the Zechstein at Mansfeld and Eisleben have been long known, and are common in all collections; figures of many species are given by M. Agassiz. Examples of the Fishes of the Magnesian limestone of the north of England, are described and figured by Professor Sedgwick, in the Geol. Trans. of London, (2d Series, Vol. iii. p. 117, and Pl. 8, 9, 10). He states in this paper (p. 99), that the occurrence of certain Corals and Encrinites, and several species of Producta, Arca, Terebratula, Spirifier, &c. shews that the Magnesian limestone is more nearly allied in its zoological characters to the Carboniferous order, than to the calcareous formations which are superior to the New red sandstone. This conclusion accords with that which M. Agassiz has drawn from the character of its fossil Fishes.

tail is regular and symmetrical. In certain bony Fishes of the secondary period, the upper lobe of the tail is partly covered with scales, but without vertebræ. The bodies of all these Fishes also have an integument of rhomboidal bony scales, covered with enamel.

No species of Fish has been found common to the Carboniferous group, and to the Zechstein or Magnesian limestone; but certain genera occur in both, e. g. the genus Palæoniscus and Polypterus.

Fishes of the Muschelkalk, Lias, and Oolite
Formations.

The Fishes of the Muschelkalk are either peculiar to it, or similar to those of the Lias and Oolite. The figure engraved at Pl. 27ᶜ, is selected as an example of the character of a family of Fishes most abundant in the Jurassic or Oolite formation; it represents the genus Gyrodus in the family of Pycnodonts, or thick-toothed Fishes, which prevailed extensively during the middle ages of Geological History. Of this extinct family there are five genera. Their leading character consists in a peculiar armature of all parts of the mouth with a pavement of thick round and flat teeth, the remains of which, under the name of Bufonites, occur most abundantly throughout the Oolite formation.* The use of this peculiar apparatus was to crush small shells, and small Crustacea, and to comminute putres-cent sea-weeds. The habits of the family of Pycnodonts appear to have been omnivorous, and their power of progression slow.†

---

* Pl. 27ᶜ. Fig. 3. represents a five-fold series of these teeth on the palate of Pycnodus trigonus from Stonesfield; and Fig. 2, a series of similar teeth placed on the vomer in the palate of the Gyrodus Umbilicus from the great Oolite of Durrheim, in Baden.

† A similar apparatus occurs in a living family of the Order Cycloids, in the case of the modern omnivorous Sea Wolf, Anar-

Another family of these singular Fishes of the
ancient world, which was exceedingly abundant
in the Oolitic or Jurassic series, is that of the
Lepidoids, a family still more remarkable than
the Pycnodonts for their large rhomboidal bony
scales, of great thickness, and covered with beautiful enamel. The Dapedium of the lias (Pl. 1.
Fig. 54.) affords an example of these scales, well
known to geologists. They are usually furnished
on their upper margin with a large process or
hook, placed like the hook or peg near the upper
margin of a tile; this hook fits into a depression
on the lower margin of the scales placed next
above it. (See Pl. 27, Figs. 4, 5, and Pl. 15, Fig.
17.) All Ganoidian Fishes, of every formation,
prior to the Chalk, were enclosed in a similar
cuirass, composed of bony scales, covered with
enamel, and extending from the head to the rays
of the tail.*　One or two species only, having
this peculiar armature of enamelled bony scales,

rhicas Lupus, and other recent Fishes of different families. M.
Agassiz observes, that it is a common fact, in the class of Fishes,
to find nearly all the modifications which the teeth of these
animals present, recurring in several families, which in other
respects are very different.

* The Pycnodonts, as well as the fossil Sauroids, have enamelled scales, but it is in the Lepidoids that scales of this
kind are most highly developed. M. Agassiz has ascertained
nearly 200 fossil species that had this kind of armour. The
use of such an universal covering of thick bony and enamelled
scales, surrounding like a cuirass the entire bodies of so many
species of Fishes, in all formations anterior to the Cretaceous
deposits, may have been to defend their bodies against waters

have yet been discovered in the Cretaceous series; and three or four species in the Tertiary formations. Among living Fishes, scales of this kind occur only in the two genera, Lepidosteus and Polypterus.

Not a single genus of all that are found in the Oolitic series exists at the present time. The most abundant Fishes of the Wealden formation belong to genera that prevailed through the Oolitic period.*

Fishes of the Chalk Formation.

The next and most remarkable of all changes in the character of Fishes, takes place at the commencement of the Cretaceous formations. Genera of the first and second orders (Placoidean and Ganoidian), which had prevailed exclusively in all formations till the termination of the Oolitic series, ceased suddenly, and were replaced by genera of new orders (Ctenoidean and Cycloidean), then for the first time introduced. Nearly two-thirds of the latter also are now extinct; but these approach nearer to Fishes of the tertiary series, than to those which had preceded the formation of the Chalk.

that were warmer, or subject to more sudden changes of temperature, than could be endured by Fishes whose skin was protected only by such thin, and often disconnected coverings, as the membranous and horny scales of most modern Fishes.

* The most remarkable of these are the genus Lepidotus, Pholidophorus, Pycnodus, and Hybodus.

Comparing the Fishes of the Chalk with those of the elder Tertiary formation of Monte Bolca, we find not one species, and but few genera, that are common to both.*

Fishes of the Tertiary Formations.

As soon as we enter on the Tertiary strata, another change takes place in the character of fossil Fishes, not less striking than that in fossil Shells.

The fishes of Monte Bolca are of the Eocene period, and are well known by the figures engraved in the Ittiolitologia Veronese, of Volta; and in Knorr. About one-half of these fishes

---

* It has been already stated, that the remarkable deposit of fossil Fishes at Engi, in the Canton of Glaris, are referred by M. Agassiz to the lower portion of the Cretaceous system.

Many genera of these are identical with, and others closely approximate to, the fishes of the Inferior chalk (Pläner kalk) of Bohemia, and of the Chalk of Westphalia (see Leonhard and Bronn. Neues Jahrbuch, 1834). Although the mineral character of the slate of Glaris presents, as we have before stated, an appearance of high antiquity, its age is probably the same as that of the Gault, or Speeton clay of England. This alteration of character is consistent with the changes that have given an air of higher antiquity than belongs to them, to most of the Secondary and Tertiary formations in the Alps.

The Fishes of the Upper chalk are best known by the numerous and splendid examples discovered at Lewes by Mr. Mantell, and figured in his works. These Fishes are in an unexampled state of perfection; in the abdominal cavities of one species (Macropoma) the stomach, and coprolites are preserved entire, in their natural place.

belong to extinct genera, and not one is identical with any existing species; they are all marine, and the greater number approach most nearly to forms now living within the tropics.*

To this first period of the Tertiary formations belong also the Fishes of the London clay; many of the species found in Sheppy, though not identical with those of Monte Bolca, are closely allied to them. The Fishes of Libanus also are of this era. The Fishes in the gypsum of Mont Martre are referred to the same period by M. Agassiz, who differs from Cuvier, in attributing them all to extinct genera.

The Fishes of Oeningen have, by all writers, been referred to a very recent local lacustrine deposit. M. Agassiz assigns them to the second period of the Tertiary formations, coeval with the Molasse of Switzerland and the sandstone of Fontainbleau. Of seventeen extinct species, one only is of an extra-European genus, and all belong to existing genera.

The gypsum of Aix contains some species referrible to one of the extinct genera of Mont Martre, but the greatest part are of existing genera. M. Agassiz considers the age of this for-

---

* M. Agassiz has re-arranged these fishes under 127 Species, all extinct, and 77 Genera. Of these Genera 38 are extinct, and 39 still living; the latter present 81 fossil species at Monte Bolca, and the former 46 species. These 39 living Genera appear for the first time in this formation.

mation as nearly coinciding with that of the Oeningen deposits.

The Fishes of the Crag of Norfolk, and the superior Sub-apennine formation, as far as they are yet known, appear for the most part related to genera now common in tropical seas, but are all of extinct species.

Family of Sharks.

As the family of Sharks is one of the most universally diffused and most voracious among modern Fishes, so there is no period in geological history in which many of its forms did not prevail.* Geologists are familiar with the occurrence of various kinds of large, and beautifully enamelled teeth, some of them resembling the external form of a contracted leech, (Pl. 27ᵉ, and 27ᶠ): these are commonly described by the name of Palate bones, or Palates. As these teeth are usually insulated, there is little evidence to indicate from what animals they have been derived.

In the same strata with them are found large bony Spines, armed on one side with prickles, resembling hooked teeth, (see Pl. 27ᵈ. C. 3. a.) These were long considered to be jaws, and true teeth ; more recently they have been ascer-

---

* M. Agassiz has ascertained the existence of more than one hundred and fifty extinct species of fossil Fishes allied to this family.

tained to be dorsal spines of Fishes, and from their supposed defensive office, like those of the genus Balistes and Silurus, have been named Ichthyodorulites.

M. Agassiz has at length referred all these bodies to extinct genera in the great family of Sharks, a family which he separates into three sub-families, each containing forms peculiar to certain geological epochs, and which change simultaneously with the other great changes in fossil remains.

The first and oldest sub-family, *Cestra-cionts*, beginning with the Transition strata, appears in every subsequent formation, till the commencement of the Tertiary, and has only one living representative, viz. the Cestracion Philippi, or Port Jackson Shark. (Pl. 1. Fig. 18.) The second family, *Hybodonts*, beginning with the Muschel-kalk, and perhaps with the Coal formation, prevails throughout the Oolite series, and ceases at the commencement of the Chalk. The third family of "*Squaloids*," or true Sharks, commences with the Cretaceous formation, and extends through the Tertiary strata into the actual creation.*

* The character of the Cestracionts is marked by the presence of large polygonal obtuse enamelled teeth, covering the interior of the mouth with a kind of tesselated pavement. (Pl. 27^d. A. 1, 3, 4, and Pl. 27^d, B. 1, 2, 3, 4, 5.) In some species not less than sixty of these teeth occupied each jaw. They are rarely found connected together in a fossil state, in consequence of the

*Fossil Spines, or Ichthyodorulites.**

The bony spines of the dorsal fins of the Port Jackson Shark (Pl. 1. Fig. 18.) throw important light on the history of fossil Spines ; and enable

perishable nature of the cartilaginous bones to which they were attached ; hence the spines and teeth usually afford the only evidence of the former existence of these extinct fossil species. They are dispersed abundantly throughout all strata, from the Carboniferous series to the most recent Chalk.

In plate 27ᵉ, Figs. 1, 2, represent a series of teeth of the genus Acrodus, in the family of Cestracionts, from the lias of Somersetshire ; and Pl. 27ᶠ, a series of teeth of the genus Ptychodus, in the same family, a genus which occurs abundantly and exclusively in the Chalk formation.

In the section Pl. 1, Fig. 19 represents a tooth of Psammodus, and Fig. 19′, a tooth of Orodus, from the Carboniferous limestone ; and Fig. 18′, a recent tooth of the Cestracion Philippi. The Cestracion·Philippi, (Pl. 1, Fig. 18, and Pl. 27ᵈ, A.) is the only living species in the family of Sharks that has flat tesselated teeth, and enables us to refer numerous fossil teeth of similar construction to the same family. As the small anterior cutting teeth (Pl. 27ᵈ, A. Figs. 1. 2. 5.) in this species, present a character of true Sharks, which has not been found in any of the fossil Cestracionts, we have in this dentition of a living species, the only known link that connects the nearly extinct family of Cestracionts with the true Sharks or Squaloids.

The second division of the family of Sharks, Hybodonts, commencing probably with the Coal formation, prevailed during the deposition of all the Secondary strata beneath the Chalk ; the teeth of this division possess intermediate characters between the blunt polygonal crushing teeth of the sub-family Cestracion, and the smooth and sharp-edged cutting teeth of the Squaloids, or true Sharks, which commenced with the Cretaceous formations. They

* See Pl. 27ᵈ. C. 3.

us to refer those very common, but little understood fossils, which have been called Ichthyodorulites, to extinct genera and species of the sub-
family of Cestracionts. (See page 286). Several
living species of the great family of Sharks have

are distinguished from those of true Sharks by being plicated, both
on the external and internal surface of the enamel. (See Plate
27^d. B. Figs. 8, 9, 10). Plate 27^d C. 1.re represents a rare
example of a series of teeth of Hybodus reticulatus, still adhering
to the cartilaginous jaw bones, from the Lias of Lyme Regis.
Striated teeth of this family abound in the Stonesfield slate and
in the Wealden formation.

Another genus in the sub-family of Hybodonts, is the Onchus,
found in the Lias at Lyme Regis ; the teeth of this genus are represented, Pl. 27^d. B. 6, 7.

In the third, or Squaloid division of fossils of this family, we
have the character of true Sharks ; these appear for the first
time in the Cretaceous formations, and extend through all the
Tertiary deposits to the present era. (Pl. 27^d. B. 11, 12, 13.)
In this division the surface of the teeth is always smooth on the
outer side, and sometimes plicated on the inner side, as it is also
in certain living species ; the teeth are often flat and lancet-
shaped, with a sharp cutting border, which, in many species, is
serrated with minute teeth. Species of this Squaloid family
alone, abound in all strata of the Tertiary formation.

The greater strength, and flattened condition of the teeth of
the families of Sharks (Cestracionts and Hybodonts), that prevailed in the Transition and Secondary formations beneath the
Chalk, had relation, most probably, to their office of crushing
the hard coverings of the Crustacea, and of the bony enamelled
scales of the Fishes, which formed their food. As soon as Fishes
of the Cretaceous and Tertiary formations assumed the softer
scales of modern Fishes, the teeth of the Squaloid sub-family
assumed the sharp and cutting edges that characterise the teeth
of living Sharks. Not one species of the blunt-toothed Cestraciont family has yet been discovered in any Tertiary formation.

G.			U

smooth *horny* spines connected with the dorsal fin. In the Cestracion Philippi alone, (Pl. 1, Fig. 18), we find a *bony* spine armed on its concave side with tooth-like hooks, or prickles, similar to those that occur in fossil Ichthyodorulites : these hooks act as points of suspension and attachment, whereby the dorsal fin is connected with this bony spine, and its movements regulated by the elevation or depression of the spine, during the peculiar rotatory action of the body of Sharks. This action of the spine in raising and depressing the fin resembles that of a moveable mast, raising and lowering backwards the sail of a barge.

The common Dog-Fish, or Spine Shark, (Spinax Acanthias, Cuv.), and the Centrina Vulgaris, have a horny elevator spine on each of their dorsal fins, but without teeth or hooks; similar small toothless horny spines have been found by Mr. Mantell in the chalk of Lewes. These dorsal spines had probably a further use as offensive and defensive weapons against voracious fishes, or against larger and stronger individuals of their own species.*

The variety we find of fossil spines, from the Greywacke series to the Chalk inclusive, indi-

---

* Colonel Smith saw a captain of a vessel in Jamaica who received many severe cuts in the body from the spines of a Shark in Montego Bay. (See Griffith's Cuvier).

The Spines of Balistes and Silurus have not their base, like that of the spines of Sharks, simply imbedded in the flesh, and

cates the number of extinct genera and species of the family of Sharks, that occupied the waters throughout these early periods of time. Not less varied are the forms of palate bones and teeth, in the same formations that contain these spines; but as the cartilaginous skeletons to which they belonged have usually perished, and the teeth and spines are generally dispersed, it is chiefly by the aid of anatomical analogies, or from occasional juxtaposition in the same stratum, that their respective species can be ascertained.

Fossil Rays.

The Rays form the fourth family in the order Placoidians. Genera of this family abound among living fishes; but they have not been found fossil in any stratum older than the Lias; they occur also in the Jurassic limestone.

Throughout the tertiary formation they are very abundant; of one genus, Myliobatis, there are seven known species; from these have been derived the palates that are so frequent in the London clay and crag. (See Pl. 27ᵈ, B. Fig. 14.) The genus Trygon, and Torpedo, occur also in the Tertiary formations.

attached to strong muscles; but articulate with a bone beneath them. The Spine of Balistes also is kept erect by a second spine behind its base, acting like a bolt or wedge, which is simultaneously inserted, or withdrawn, by the same muscular motion that raises or depresses the spine.

Conclusion.

In the facts before us, we have an uninterrupted series of evidence, derived from the family
of Fishes, by which both bony and cartilaginous forms of this family, are shewn to have
prevailed during every period, from the first
commencement of submarine life, unto the present hour. The similarity of the teeth, and
scales, and bones, of the earliest Sauroid
Fishes of the coal formation (Megalichthys),
to those of the living Lepidosteus, and the correspondence of the teeth and bony spines of
the only living Cestraciont in the family of
Sharks, with the numerous extinct forms of that
sub-family, which abound throughout the Carboniferous and Secondary formations, conneet
extreme points of this grand verbetrated division
of the animal kingdom, by one unbroken chain,
more uniform and continuous than has hitherto
been discovered in the entire range of geological
researches.

It results from the review here taken of the
history of fossil Fishes, that this important class
of vertebrated animals presented its actual gradations of structure amongst the earliest inhabitants of our planet; and has ever performed
the same important functions in the general
economy of nature, as those discharged by their
living representatives in our modern seas, and

lakes, and rivers. The great purpose of their
existence seems at all times to have been, to fill
the waters with the largest possible amount of
animal enjoyment.

The sterility and solitude which have sometimes been attributed to the depths of the ocean,
exist only in the fictions of poetic fancy. The
great mass of the water that covers nearly three-
fourths of the globe is crowded with life, perhaps more abundantly than the air, and the
surface of the earth; and the bottom of the
sea, within a certain depth accessible to light,
swarms with countless hosts of worms, and creeping things, which represent the kindred families
of low degree which crawl upon the land.

The common object of creation seems ever
to have been, the infinite multiplication of life.
As the basis of animal nutrition is laid in the
vegetable kingdom, the bed of the ocean is not
less beautifully clothed with submarine vegetation, than the surface of the dry land with
verdant herbs and stately forests. In both
cases, the undue increase of herbivorous tribes
is controlled by the restraining influence of those
which are carnivorous; and the common result
is, and ever has been, the greatest possible
amount of animal enjoyment to the greatest
number of individuals.

From no kingdom of nature does the doctrine
of gradual Development and Transmutation of

species derive less support, than from the progression we have been tracing in the class of Fishes. The Sauroid Fishes occupy a higher place in the scale of organization, than the ordinary forms of bony Fishes; yet we find examples of Sauroids of the greatest magnitude, and in abundant numbers in the Carboniferous and Secondary formations, whilst they almost disappear and are replaced by less perfect forms in the Tertiary strata, and present only two genera among existing Fishes.

In this, as in many other cases, a kind of *retrograde* development, from complex to simple forms, may be said to have taken place. As some of the more early Fishes united in a single species, points of organization which, at a later period, are found distinct in separate families, these changes would seem to indicate in the class of Fishes a process of Division, and of Subtraction from more perfect, rather than of Addition to less perfect forms.

Among living Fishes, many parts in the organization of the Cartilaginous tribes, (e. g. the brain, the pancreas, and organs subservient to generation,) are of a higher order than the corresponding parts in the Bony tribes; yet we find the cartilaginous family of Squaloids co-existing with bony fishes in the Transition strata, and extending with them through all geological formations, unto the present time.

In no kingdom of nature, therefore, does it seem less possible to explain the successive changes of organization, disclosed by geology, without the direct interposition of repeated acts of Creation.

CHAPTER XV.

*Proofs of Design in the Fossil Remains of Mollusks.**

SECTION I.

FOSSIL UNIVALVE AND BIVALVE SHELLS.

WE are much limited in our means of obtaining information as to the anatomical structure of those numerous tribes of extinct animals which are comprehended under Cuvier's great division of Mollusks. Their soft and perishable bodies have almost wholly disappeared, and their external shells, and, in a few cases, an internal apparatus of the nature of shell, form the only evidence of the former existence of the myriads of these creatures that occupied the ancient waters.

The enduring nature of the calcareous cover-

* See note, p. 62.

ings which these animals had the power of
secreting, has placed our knowledge of Fossil
Shells almost on a footing with that of recent
Conchology.   But the plan of our present
enquiry forbids us here to take more than a
general review of the history, and economy of
the creatures by which they were constructed.

We find many and various forms, both of Univalve and Bivalve shells, mixed with numerous
remains of Articulated and Radiated animals, in
the most ancient strata of the Transition period
that contain any traces of organic life.   Many
of these shells agree so closely with existing
species, that we may infer their functions to
have been the same ; and that they were inhabited by animals of form and habits similar to
those which fabricate the living shells most
nearly resembling them.*

All Turbinated and simple shells are constructed by Mollusks of a higher Order than the
Conchifers, which construct Bivalves ; the former
have heads and eyes ; the Conchifers, or con-
structors of bivalves, are without either of these
important parts, and possess but a low degree of
any other sense than touch, and taste.   Thus
the Mollusk, which occupies a Whelk, or a
Limpet shell, is an animal of a higher Order

---

* See Mr. Broderip's Introduction to his Paper on some new
species of Brachiopoda, Zool. Trans., vol. 1, p. 141.

than the Conchifer enclosed between the two valves of a Muscle or an Oyster shell.

Lamarck has divided his Order of Trachelipods* into two great sections, viz. *herbivorous* and *carnivorous;* the carnivorous are also divisible into two families of different office, the one attacking and destroying *living* bodies, the other eating *dead* bodies that have perished in the course of nature, or from accidental causes; after the manner of those species of predaceous beasts and birds, e. g. the Hyænas and Vultures, which, by preference, live on carrion. The same principle of economy in nature, which causes the dead carcases of the hosts of terrestrial herbivorous animals to be accelerated in their decomposition, by forming the food of numerous carnivora, appears also to have been applied to the submarine inhabitants of the most ancient, as well as of the existing seas; thus converting the death of one tribe into the nutriment and support of life in others.

It is stated by Mr. Dillwyn, in a paper read before the Royal Society, June 1823, that Pliny has remarked, that the animal which was sup-

---

* This name is derived from the position of the foot, or locomotive apparatus, on the lower surface of the neck, or of the anterior part of the body. By means of this organ Trachelipods crawl like the common garden snail (Helix aspersa). This Helix offers also a familiar example of the manner in which they have the principal viscera packed within the spiral shell.

posed to yield the Tyrian dye, obtained its food by boring into other shells by means of an elongated tongue ; and Lamarck says, that all those Mollusks whose shells have a *notch* or canal at the base of their aperture, are furnished with a similar power of boring, by means of a retractile proboscis.* In his arrangement of invertebrate animals, they form a section of the Trachelipods, which he calls carnivorous. (*Zoophages*). In the other section of Trachelipods, which he calls herbivorous (*Phytiphages*) the aperture of the shell is *entire*, and the animals have jaws formed for feeding on vegetables.

Mr. Dillwyn further asserts, that every fossil Turbinated Univalve of the older beds, from the

---

* The proboscis, by means of which these animals are enabled to drill holes through shells, is armed with a number of minute teeth, set like the teeth of a file, upon a retractile membrane, which the animal is enabled to fix in a position adapted for boring or filing a hole from without, through the substance of shells, and through this hole to extract and feed upon the juices of the body within them. A familiar example of this organ may be seen in the retractile proboscis of Buccinum Lapillus, and Buccinum Undatum, the common whelks of our own shores. A valuable Paper on this subject has recently been published by Mr. Osler (Phil. Trans., 1832, Part 2, P. 497), in which he gives an engraved figure of the tongue of the Buccinum Undatum, covered with its rasp, whereby it perforates the shells of animals destined to become its prey. Mr. Osler modifies the rule or the distinction between the shells of carnivora and herbivora, by shewing that, although it is true that all beaked shells indicate their molluscous inhabitant to have been carnivorous, an entire aperture does not always indicate an herbivorous character.

Transition lime to the Lias, belongs to the *herbi-vorous* genera; and that the herbivorous class extends through every stratum in the entire series of geological formations, and still retains its place among the inhabitants of our existing seas. On the other hand, the shells of marine *carnivorous* Univalves are very abundant in the Tertiary strata above the Chalk, but are extremely rare in the Secondary strata, from the Chalk downwards to the Inferior oolite; beneath which no trace of them has yet been found.

Most collectors have seen upon the sea shore numbers of dead shells, in which small circular holes have been bored by the predaceous tribes, for the purpose of feeding upon the bodies of the animals contained within them; similar holes occur in many fossil shells of the Tertiary strata, wherein the shells of carnivorous Trachelipods also abound; but perforations of this kind are extremely rare in the fossil shells of any older formation. In the Green-sand and Oolite, they have been noticed only in those few cases where they are accompanied by the shells of equally rare carnivorous Mollusks; and in the Lias, and strata below it, there are neither perforations, nor any shells having the notched mouth peculiar to perforating carnivorous species.

It should seem, from these facts, that in the economy of submarine life, the great family of carnivorous Trachelipods, performed the same necessary office during the Tertiary period, which is allotted to them in the present ocean. We have further evidence to shew, that in times anterior to, and during the deposition of the Chalk, the same important functions were consigned to other carnivorous Mollusks, viz. the Testaceous Cephalopods ;* these are of comparatively rare occurrence in the Tertiary strata, and in our modern seas; but, throughout the Secondary and Transition formations, where carnivorous Trachelipods are either wholly wanting, or extremely scarce, we find abundant remains of carnivorous Cephalopods, consisting of the chambered shells of Nautili and Ammonites, and many kindred extinct genera of polythalamous shells of extraordinary beauty.   The Molluscous inhabitants of all these chambered shells, probably possessed the voracious habits of the modern Cuttle Fish, and by feeding like them upon young Testacea and Crustacea, restricted the excessive increase of animal life at the bottom of the more ancient seas.   Their sudden and nearly total disappearance at the commencement of the Tertiary era, would have caused a blank in the " police of nature," allowing the herbivorous tribes to increase to an excess, that would ultimately have been destructive of marine vegetation, as well as of themselves, had they not been replaced by a

---

* See explanation of the term Cephalopod, in note at p. 303.

different order of carnivorous creatures, destined to perform in another manner, the office which the inhabitants of Ammonites and various extinct genera of chambered shells then ceased to discharge. From that time onwards, we have evidence of the abundance of carnivorous Trachelipods, and we see good reason to adopt the conclusion of Mr. Dillwyn, that " in the formations above the Chalk, the vast and sudden decrease of one predaceous tribe has been provided for by the creation of many new genera, and species, possessed of similar appetencies, and yet formed for obtaining their prey by habits entirely different from those of the Cephalopods."*

The design of the Creator seems at all times to have been, to fill the waters of the seas, and cover the surface of the earth with the greatest possible amount of organized beings enjoying life ; and the same expedient of adapting the vegetable kingdom to become the basis of the life of animals, and of multiplying largely the amount of animal existence by the addition of Carnivora to the Herbivora, appears to have prevailed from the first commencement of organic life unto the present hour.*

* Mr. Dillwyn observes further, that all the herbivorous marine Trachelipods of the Transition and Secondary strata were furnished with an operculum, as if to protect them against the carnivorous Cephalopods which then prevailed abundantly; but that in the Tertiary formations, numerous herbivorous genera appear, which are not furnished with opercula, as if no longer requiring the protection of such a shield, after the extinction

Mr. De la Beche has recently published a list of the specific gravities of living shells of different genera, from which he shews that their weight and strength are varied in accommodation to the habits and habitation of the animals by which they are respectively constructed; and points out evidence of design, such as we discover, in all carefully conducted investigations of the works of nature, whether among the existing or extinct forms of the animal creation.*

of the Ammonites and of many cognate genera of carnivorous Trachelipods, at the termination of the Secondary period, i. e. after the deposition of the Chalk formation.

* " It can scarcely escape the observation of the reader, that, while the specific gravities of the *land* shells enumerated is generally greatest, the densities of the *floating marine* shells are much the smallest. The design of the difference is obvious: The land shells have to contend with all changes of climate, and to resist the action of the atmosphere, while, at the same time, they are thin for the purpose of easy transport, their density is therefore greatest. The Argonaut, Nautilus, and creatures of the like habits require as light shells as may be consistent with the requisite strength; the relative specific gravity of such shells is consequently small. The greatest observed density was that of a Helix, the smallest, that of an Argonaut. The shell of the Ianthina, a floating Molluscous creature, is among the smallest densities. The specific gravity of all the land shells examined was greater than that of Carara marble; in general more approaching to Arragonite. The freshwater and marine shells, with the exception of the Argonaut, Nautilus, Ianthina, Lithodomus, Haliotis, and great radiated crystalline Teredo from the East Indies, exceeded Carara marble in density. This marble and the Haliotis are of equal specific gravities."---De la Beche's Geological Researches, 1834, p. 76.

SECTION II.

FOSSIL REMAINS OF NAKED MOLLUSKS, PENS, AND INK-BAGS OF LOLIGO.

It is well known that the common Cuttle Fish, and other living species of Cephalopods,* which have no external shell, are protected from their enemies by a peculiar internal provision, consisting of a bladder-shaped sac, containing a black and viscid ink, the ejection of which defends them, by rendering opaque the water in which they thus become concealed. The most familiar examples of this contrivance are found in the Sepia vulgaris, and Loligo of our own seas. (See Pl. 28, Fig. 1.)

It was hardly to be expected that we should find, amid the petrified remains of animals of the

---

* The figure of the common Calmar, or Squid (Loligo Vulgaris Lam.---Sepia loligo of Linnæus), see Pl. 28, Fig. 1, illustrates the origin of the term Cephalopod, a term applied to a large family of molluscous animals, from the fact of their feet being placed around their heads. The feet are lined internally with ranges of horny cups, or suckers, by which the animal seizes on its prey, and adheres to extraneous bodies. The mouth, in form and substance resembles a Parrot's beak, and is surrounded by the feet. By means of these feet and suckers the Sepia octopus, or common Poulpe (the Polypus of the ancients), crawls with its head downwards, along the bottom of the sea.

ancient world, (remains which have been buried for countless centuries in the deep foundations of the earth,) traces of so delicate a fluid as the ink which was contained within the bodies of extinct species of Cephalopods, that perished at periods so incalculably remote; yet the preservation of this substance is established beyond the possibility of doubt, by the recent discovery of numerous specimens in the Lias of Lyme Regis,* in which the ink-bags are *preserved* in a fossil state, still distended, as when they formed parts of the organization of living bodies, and retaining the same juxta-position to a horny pen, which the ink-bag of the existing Loligo bears to the pen within the body of that animal. (Pl. 28, Fig. 1.)

Having before us the fact of the preservation of this fossil ink, we find a ready explanation of it, in the indestructible nature of the carbon of which it was chiefly composed. Cuvier describes the ink of the recent Cuttle Fish, as being a dense fluid of the consistence of pap, " bouillie," suspended in the cells of a thin net-work that pervades the interior of the ink-bag; it very much resembles common printers' ink. A substance of this nature would readily be trans-

---

* We owe this discovery to the industry and skill of Miss Mary Anning, to whom the scientific world is largely indebted, for having brought to light so many interesting remains of fossil Reptiles from the Lias at Lyme Regis.

ferred to a fossil state, without much diminution of its bulk.*

Pl. 28, Fig. 5, represents an ink-bag of a recent Cuttle Fish, in which the ink is preserved in a desiccated state, being not much diminished from its original volume. Its form is similar to that of many fossil ink-bags (Pl. 29, Figs. 3—10), and the indurated ink within it differs only from the fossil ink, inasmuch as the latter is impregnated with carbonate of lime.

In a communication to the Geological Society, February 1829, I announced that these fossil ink-bags had been discovered in the Lias at Lyme Regis, in connexion with horny bodies, resembling the pen of a recent Loligo.

These fossil pens are without any trace of nacre, and are composed of a thin, laminated,

---

* So completely are the character and qualities of the ink retained in its fossil state, that when, in 1826, I submitted a portion of it to my friend Sir Francis Chantrey, requesting him to try its power as a pigment, and he had prepared a drawing with a triturated portion of this fossil substance; the drawing was shewn to a celebrated painter, without any information as to its origin, and he immediately pronounced it to be tinted with sepia of excellent quality, and begged to be informed by what colourman it was prepared. The common sepia used in drawing is from the ink-bag of an oriental species of cuttle-fish. The ink of the cuttle-fishes, in its natural state, is said to be soluble only in water, through which it diffuses itself instantaneously; being thus remarkably adapted to its peculiar service in the only fluid wherein it is naturally employed.

semi-transparent substance, resembling horn.
Their state of preservation is such as to admit of
a minute comparison of their internal structure
with that of the pen of the recent Loligo; and
leads to the same result which we have collected
from the examination of so many other examples
of fossil organic remains; namely, that although
fossil species usually differ from their living representatives, still the same principles of construction have prevailed through every cognate genus,
and often also through the entire families under
which these genera are comprehended.

The petrified remains of fossil Loligo, therefore, add another link to the chain of argument
which we are pursuing, and aid us in connecting
successive systems of creation which have followed each other upon our Planet, as parts of
one grand and uniform Design. Thus the union
of a bag of ink with an organ resembling a pen
in the recent Loligo, is a peculiar and striking
association of contrivances, affording compensation for the deficiency of an external shell, to an
animal much exposed to destruction from its
fellow-tenants of the deep; we find a similar
association of the same organs in the petrified
remains of extinct species of the same family,
that are preserved in the ancient marl and limestone strata of the Lias. Cuvier drew his figures
of the recent Sepia with ink extracted from its
own body. I have drawings of the remains of

extinct species prepared also with their own ink;
with this fossil ink I might record the fact, and
explain the causes of its wonderful preservation.
I might register the proofs of instantaneous death
detected in these ink-bags, for they contain the
fluid which the living sepia emits in the moment
of alarm; and might detail further evidence of
their immediate burial, in the retension of the
forms of these distended membranes (Pl. 29.
Figs. 3—10.); since they would speedily have
decayed, and have spilt their ink, had they been
exposed but a few hours to decomposition in the
water.   The animals must therefore have died
suddenly, and been *quickly* buried in the sediment that formed the strata, in which their
petrified ink and ink-bags are thus preserved.
The preservation also of so fragile a substance
as the pen of a Loligo, retaining traces even of
its minutest fibres of growth, is not much less
remarkable than the fossil condition of the ink-
bags, and leads to similar conclusions.*

* We have elsewhere applied this line of argument to prove
the sudden destruction and burial of the Saurians, whose skeletons we find entire in the same Lias that contains the pens and
ink-bags of Loligo.   On the other hand, we have proofs of intervals between the depositions of the component strata of the Lias,
in the fact, that many beds of this formation have become the
repository of Coprolites, dispersed singly and irregularly at intervals far distant from one another, and at a distance from any
entire skeletons of the Saurians, from which they were derived:
and in the further fact, that those surfaces *only* of the Coprolites,
which lay *uppermost* at the bottom of the sea, have often

We learn from a recent German publication (Zeiten's Versteinerungen Württembergs. Stuttgart, 1832, Pl. 25 and Pl. 37,) that similar remains of pens and ink bags are of frequent occurrence in the Lias shale of Aalen and Boll.* Hence it is clear that the same causes which produced these effects during the deposition of the Lias at Lyme Regis, produced similar and nearly contemporaneous effects, in that part of Germany which presents such identity in the character and circumstances of these delicate organic remains.†

suffered partial destruction from the action of water before they were covered and protected by the muddy sediment that has afterwards permanently enveloped them. Further proof of the duration of time, during the intervals of the deposition of the Lias, is found in the innumerable multitudes of the shells of various Mollusks and Conchifers which had time to arrive at maturity, at the bottom of the sea, during the quiescent periods which intervened between the muddy invasions that destroyed, and buried suddenly the creatures inhabiting the waters, at the time and place of their arrival.

* As far as we can judge from the delineations and lines of structure in Zeiten's plate, our species from Lyme Regis is the same with that which he has designated by the name of Loligo Aalensis; but I have yet seen no structure in English specimens like that of his Loligo Bollensis.

† Although the resemblance between the pens of the Loligo and a feather (as might be expected from the very different uses to which they are applied) does not extend to their internal structure, we may still, for convenience of description, consider them as composed of the three following parts, which, in all our figures, will be designated by the same letters, A. B. C. First, the external filaments of the plume, (Pl. 28, 29, 30,

Paley has beautifully, and with his usual felicity, described the Unity and Universality of Providential care, as extending from the construction of a ring of two hundred thousand miles diameter, to surround the body of Saturn, and be suspended, like a magnificent arch, above the heads of his inhabitants, to the concerting

A.) analogous to those of a common feather. These filaments terminate inwards on a straight line, or base, where they usually form an acute angle with the outer edges of the marginal bands. Secondly, two marginal bands, B. B., dividing the base of the filaments from the body of the shaft; the surface of these bands, B., usually exhibits angular lines of growth in the smaller fossil pens (Pl. 28, Fig. 6, and Pl. 29, Fig. 2,) which become obtuse and vanish into broad curves, in larger specimens, Pl. 29, Fig. 1, and Pl. 30. Thirdly, the broad shaft, which forms the middle of the pen, is divided longitudinally into two equal parts by a straight line, or axis, C. : it is made up of a number of thin plates, of a horn-like substance, laid on each other, like thin sheets of paper in pasteboard; these thin plates are composed alternately, of longitudinal, and transverse fibres; the former (Pl. 28, Fig. 7, f. f.) straight, and nearly parallel to the axis of the shaft, the latter (Pl. 28, Fig. 7, e. e.) crossing the shaft transversely in a succession of symmetrical and undulating curves. These transverse fibres do not interlace the others, as the woof interlaces the weaver's warp, but are simply laid over, and adhering to them, as in the alternate laminæ of paper made from slices of papyrus; the strength of such paper much exceeds that made from flax or cotton, in which the fibres are disposed irregularly in all directions. The fibres of both kinds are also collected at intervals into fluted fasciculi, Pl. 30, f, and e, forming a succession of grooves and ridges fitted one into another, whereby the entire surface of each plate is locked into the surface of the adjacent plate, in a manner admirably calculated to combine elasticity with strength.

and providing an appropriate mechanism for the clasping and reclasping of the filaments in the feather of the Humming-bird. The geologist descries a no less striking assemblage of curious provisions, and delicate mechanisms, extending from the entire circumference of the crust of our planet, to the minutest curl of the smallest fibre in each component lamina of the pen of the fossil Loligo. He finds these pens uniformly associated with the same peculiar defensive provision of an internal ink-bag, which is similarly associated with the pen of the living Loligo in our actual seas; and hence he concludes, that such a union of contrivances, so nicely adjusted to the wants and weaknesses of the creatures in which they occur, could never have resulted from the blindness of chance, but could only have originated in the will and intention of the Creator.

SECTION III.

Proofs of Design in the Mechanism of Fossil Chambered Shells.

NAUTILUS.

I SHALL select from the family of Multilocular, or Chambered shells, the few examples which I shall introduce from mineral conchology, with a view of illustrating certain points that have relation to the object of the present Treatise.

I select these, first, because they afford proofs of mechanical contrivances, more obviously adapted to a definite purpose, than can be found in shells of simpler character. Secondly, because the use of many of their parts can be explained, by reference to the economy and organization of the existing animals, most nearly allied to the extinct fossil genera and species with which we are concerned. And, thirdly, because many of these chambered shells can be shewn, not merely to have performed the office of ordinary shells, as a defence for the body of their inhabitants; but also to have been hydraulic instruments of nice operation, and delicate adjustment, constructed to act in subordination to those universal and unchanging Laws, which appear to have ever regulated the movement of fluids.

The history of Chambered shells illustrates also some of those phenomena of fossil conchology, which relate to the limitation of species to particular geological Formations;* and affords striking proofs of the curious fact, that many

---

* Thus, the Nautilus multicarinatus is limited to strata of the Transition formation; the N. bidorsatus to the Muschelkalk; N. obesus, and N. lineatus, to the Oolite Formation; N. elegans, and N. undulatus, to the Chalk. The divisions of the Tertiary formations have also species of Nautili peculiar to themselves.

genera, and even whole families, have been called into existence, and again totally annihilated, at various and successive periods, during the progress of the construction of the crust of our globe.

The history of Chambered Shells tends further to throw light upon a point of importance in physiology, and shows that it is not always by a regular gradation from lower to higher degrees of organization, that the progress of life has advanced, during the early epochs of which geology takes cognizance. We find that many of the more simple forms have maintained their primeval simplicity, through all the varied changes the surface of the earth has undergone; whilst, in other cases, organizations of a higher order preceded many of the lower forms of animal life; some of the latter appearing, for the first time, after the total annihilation of many species and genera of a more complex character.*

---

* The introduction, in the Tertiary periods, of a class of animals of lower organization, viz. the carnivorous Trachelipods, (See Chap. XV. Section 1,) to fill the place which, during the Secondary periods, had been occupied by a higher order, namely, the carnivorous Cephalopods, affords an example of *Retrocession* which seems fatal to that doctrine of *regular Progression*, which is most insisted on by those who are unwilling to admit the repeated interferences of Creative power, in adjusting the successive changes that animal life has undergone.

It will appear, on examination of the shells of fossil Nautili, that they have retained, through strata of all ages, their abori-

The prodigious number, variety, and beauty, of extinct Chambered shells, which prevail throughout the Transition and Secondary strata, render it imperative that we should seek for evidence in living nature, of the character and habits of the creatures by which they were formed, and of the office they held in the ancient economy of the animal world. Such evidence we may expect to find in those inhabitants of the present sea, whose shells most nearly resemble the extinct fossils under consideration, namely, in the existing Nautilus Pompilius, (See Pl. 31, Fig. 1), and Spirula, (Pl. 44, Figs. 1, 2).*

ginal simplicity of structure; a structure which remains fundamentally the same in the Naütilus Pompilius of our existing seas, as it was in the earliest fossil species that we find in the Transition strata. Meantime the cognate family of Ammonites, whose shells were more elaborately constructed than those of Nautili, commenced their existence at the same early period with them in the Transition strata, and became extinct at the termination of the Secondary formations. Other examples of later creations of genera and species, followed by their periodical and total extinction, before, or at the same time with the cessation of the Ammonites, are afforded by those cognate Multilocular shells, namely, the Hamite, Turrilite, Scaphite, Baculite, and Belemnite, respecting each of which I shall presently notice a few particulars.

* I omit to mention the more familiar shell of the Argonauta or Paper Nautilus, because, not being a chambered species, it does not apply so directly to my present subject; and also, because doubts still exist whether the Sepia found within this shell be really the constructor of it, or a parasitic intruder into a shell formed by some other animal not yet discovered. Mr. Broderip, Mr. Gray, and Mr. G. Sowerby, are of opinion, that this shell is constructed by an animal allied to Carinaria.

I must enter at some length into the natural history of these shells, because the conclusions to which I have been led, by a long and careful investigation of fossil species, are at variance with those of Cuvier and Lamarck, as to the fact of Ammonites being external shells, and also with the prevailing opinions as to the action of the siphon and air chambers, both in Ammonites and Nautili.

Mechanical Contrivances in the Nautilus.

The Nautilus not only exists at present in our tropical seas, but is one of those genera which occur in a fossil state in formations of every age; and the molluscous inhabitants of these shells, having been among the earliest occupants of the ancient deep, have maintained their place through all the changes that the tenants of the ocean have undergone.

The recent publication of Mr. R. Owen's excellent Memoir on the Pearly Nautilus, (Nautilus Pompilius Lin.) 1832, affords the first scientific description ever given of the animal by which this long-known shell is constructed.*   This

---

* It is a curious fact, that although the shells of the Nautilus have been familiar to naturalists, from the days of Aristotle, and abound in every collection, the only authentic account of the animals inhabiting them, is that by Rumphius, in his history of Amboyna, accompanied by an engraving, which, though tole-

Memoir is therefore of high importance, in its relation to geology; for it enables us to assert, with a confidence we could not otherwise have assumed, that the animals by which all fossil Nautili were constructed, belonged to the existing family of Cephalopodous Mollusks, allied to the common Cuttle Fish.   It leads us further to infer, that the infinitely more numerous species of the family of Ammonites, and other cognate genera of Multilocular shells, were also constructed by animals, in whose economy they held an office analogous to that of the existing shell of the Nautilus Pompilius. We therefore entirely concur with Mr. Owen, that not only is the acquisition of this species peculiarly acceptable, from its relation to the Cephalopods of the present creation ; but that it is, at the same time, the living type of a vast tribe of organized beings, whose fossilized remains testify their existence at a remote period, and in another order of things.*

rably correct, as far as it goes, is yet so deficient in detail that it is impossible to learn any thing from it respecting the internal organization of the animal.

I rejoice in the present opportunity of bearing testimony to the value of Mr. Owen's highly philosophical and most admirable memoir upon this subject; a work not less creditable to the author, than honourable to the Royal College of Surgeons, under whose auspices this publication has been so handsomely conducted.

* A further important light is thrown upon those species of fossil Multilocular shells, e. g. Orthoceratites, Baculites, Ha-

By the help of this living example, we are prepared to investigate the question of the uses, to
which all fossil Chambered shells may have been
subservient, and to show the existence of design
and order in the mechanism, whereby they were
appropriated to a peculiar and important function, in the economy of millions of creatures long
since swept from the face of the living world.
From the similarity of these mechanisms to those
still employed in animals of the existing creation, we see that all such contrivances and
adaptations, however remotely separated by time
or space, indicate a common origin in the will
and design of one and the same Intelligence.

We enter then upon our examination of the
structure and uses of fossil Chambered shells,

mites, Scaphites, Belemnites, &c. (See Pl. 44), in which the
last, or external chamber, seems to have been too small to
contain the entire body of the animals that formed them, by
Peron's discovery of the well-known chambered shell, the
Spirula, partially enclosed within the posterior extremity of
the body of a Sepia (Pl. 44, Figs. 1, 2). Although some doubts
have existed respecting the authenticity of this specimen, in
consequence of a discrepance between two drawings professedly
taken from it (the one published in the Encyclopédie Méthodique, the other in Peron's Voyage), and from the loss of the
specimen itself before any anatomical examination of it had been
made, the subsequent discovery by Captain King of the same
shell, attached to a portion of the mutilated body of some
undescribed Cephalopod allied to the Sepia, leaves little doubt
of the fact that the Spirula was an internal shell, having its
dorsal margin *only* exposed, after the manner represented in both
the drawings from the specimen of Peron. (See Pl. 44, Fig. 1.)

with a preliminary knowledge of the facts, that the recent shells, both of N. Pompilius and Spirula, are formed by existing Cephalopods; and we hope, through them, to be enabled to illustrate the history of the countless myriads of similarly constructed fossil shells whose use and office has never yet been satisfactorily explained.

We may divide these fossils into two distinct classes; the one comprising external shells, whose inhabitants resided, like the inhabitant of the N. Pompilius, in the capacious cavity of their first or external chamber (Pl. 31, Fig. 1); the other, comprising shells, that were wholly or partially included within the body of a Cephalopod, like the recent Spirula, (Pl. 44, Figs. 1, 2). In both these classes, the chambers of the shell appear to have performed the office of air vessels, or *floats*, by means of which the animal was enabled either to raise itself and float near the surface of the sea, or sink to the bottom.

It will be seen by reference to Pl. 31, Fig 1,* that in the recent Nautilus Pompilius, the only organ connecting the air chambers, with the body of the animal, is a pipe, or *siphuncle*, which descends through an aperture and short project-

---

* The animal is copied from Pl. 1 of Mr. Owen's Memoir; the shell from a specimen in the splendid and unique collection of my friend W. J. Broderip, Esq., by whose unreserved communications of his accurate and extensive knowledge in Natural History, I have been long and largely benefited.

ing tube (y) in each successive transverse plate, till it terminates in the smallest chamber at the inner extremity of the shell.   By means of this pipe, the animal has power to increase or diminish its specific gravity, as Fishes do, by distending their membranous air bladder with air, or by causing it to collapse.   When the pipe of the Nautilus is filled with any fluid, the weight of that fluid, being added to the body and shell, renders the mass specifically heavier than water, and the animal sinks.   When it is inclined to rise, it withdraws the fluid from the pipe, and thus again, becoming specifically lighter, rises upwards to the surface.

The motion of the Nautilus, when floating, with its arms expanded, is retrograde, like that of the naked Cuttle Fish, being produced by the reaction of water, violently ejected from the funnel (k).   The fingers and tentacula (p, p,) are here represented as closed around the beak, which is consequently invisible ; when the animal is in action, they are probably spread forth like the expanded rays of the sea Anemone.

The horny beak of this recent Nautilus (See Pl. 31, Fig. 2, 3) resembles the bill of a Parrot. Each mandible is armed in front, with a hard and indented calcareous point, adapted to the office of crushing shells and crustaceous animals, of which latter, many fragments were found in the stomach of the individual here represented.

As these belonged to species of hairy brachy-
urous crustacea, that live exclusively at the bottom
of the sea, they shew that this Nautilus, though
occasionally foraging at the surface, obtains part
of its food from the bottom.   As it also had a
gizzard, much resembling that of a fowl, we see
in this organ, further evidence that the existing
Nautilus has the power of digesting hard shells.*

A similar apparatus is shewn to have existed
in the beaks of the inhabitants of many species
of fossil Nautili, and Ammonites, by the abundance of fossil bodies called Rhyncholites, or
beak-stones, in many strata that contain these
fossil shells, e. g. in the Oolite of Stonesfield, in
the Lias at Lyme Regis and Bath, and in the
Muschel-kalk at Luneville.

As we are warranted in drawing conclusions
from the structure of the teeth in quadrupeds,
and of the beak in birds, as to the nature of the

---

* In Pl. 31, Fig. 3 represents the lower mandible, armed in
front like Fig. 2. with a hard and calcareous margin; and Fig. 4
represents the anterior calcareous part of the palate of the upper
Mandible Fig. 2. formed of the same hard calcareous substance
as its point; this substance is of the nature of shell.

These calcareous extremities of both mandibles are of sufficient
strength to break through the coverings of Crustacea and shells,
and as they are placed at the extremity of a beak composed of
thin and tough horn, the power of this organ is thereby materially
increased.

In examining the contents of the stomach of the Sepia vulgaris, and Loligo, I have found them to contain numerous shells
of small Conchifera.

food on which they are respectively destined to feed, so we may conclude, from the resemblance of the fossil beaks, or Rhyncholites, (Pl. 31, Fig. 5—11), to the calcareous portions of the beak of the Cephalopod, inhabiting the N. Pompilius, that many of these Rhyncholites were the beaks of the cephalopodous inhabitants of the fossil shells with which they are associated; and that these Cephalopods performed the same office in restraining excessive increase among the Crustaceous and Testaceous inhabitants of the bottom of the Transition and Secondary seas, that is now discharged by the living Nautili, in conjunction with the carnivorous Trachelipods.*

Assuming, therefore, on the evidence of these analogies, that the inhabitants of the shells of the fossil Nautili and Ammonites were Cephalopods, of similar habits to those of the animal which constructs the shell of the N. Pompilius, we shall next endeavour to illustrate, by the organization and habits of the living Nautilus, the manner in which these fossil shells were adapted to the use of creatures, that sometimes moved and fed at the bottom of deep seas, and at other times rose and floated upon the surface.

The Nautili (see Pl. 31. Fig. 1. and Pl. 32. Figs. 1. 2.) constitute a natural genus of spiral discoidal shells, divided internally into a series

* See p. 250.

of chambers that are separated from each other by transverse plates; these plates are perforated by a tube or siphon, passing through the transverse plates, either at their *centre*, or towards their *internal* margin. (Pl. 1. Fig. 31. Pl. 32. Fig. 2. and Pl. 33.)

The *external* open chamber is very large, and forms the receptacle of the body of the animal. The *internal* close chambers are void, and have no communication with the outer chamber, excepting for the passage of a membranous tube, which descends through an aperture in each plate to the innermost extremity of the shell, (Pl. 31, y. y. a. b. c. d. e. and Pl. 32, a. b. d. e. f.). These air chambers are destined to counterbalance the weight of the increasing body and shell of the animal, and thereby to render the whole so nearly of the weight of water, that the difference arising from the membrane of the *siphuncle* being either empty, or filled with a fluid, may cause the mass to swim or sink.*

* The siphuncle represented in Pl. 31, Fig. 1, illustrates the structure and uses of that organ; in the smallest whorls, from d. inwards, it is enclosed by a thin calcareous covering, or *sheath*, of so soft a nature as to be readily scraped off by the point of a quill; this sheath may admit of expansion or contraction, together with the membraneous tube enclosed within it. In the fossil Nautili, a similar calcareous sheath is often preserved, as in Pl. 32, Figs. 2, 3, and Pl. 33, and forms a connected series of tubes of carbonate of lime, closely fitted to the collar of each transverse plate. In four chambers of the

As neither the siphuncle, nor the external shell have any kind of aperture through which a fluid could pass into the close chambers,* it follows that these chambers contain nothing

recent shell (Pl. 31, Fig. 1, a. b. c. d.) this sheath is partially removed from the desiccated membranous pipe within it, which has assumed the condition of a black elastic substance, resembling the black continuous siphuncular pipe that is frequently preserved in a carbonaceous state in fossil Ammonites.

At that part of each transverse plate, which is perforated for the passage of the siphuncle, (Pl. 31, Fig. 1, y. y.), a portion of its shelly matter projects *inwards* to about one-fourth of the distance across each chamber, and forms a *collar*, around the membranous pipe, thus directing its passage through the transverse plates, and also affording to it, when distended with fluid, a strong support at each collar. A similar projecting collar is seen in the transverse plate of a fossil Nautilus. (Pl. 32, Fig. 2, e, and Fig. 3, e, i. and Pl. 33.) A succession of such supports placed at short intervals from one another, divides this long and thin membranaceous tube, when distended, into a series of short compartments, or small oval sacs, each sac communicating with the adjacent sacs by a contracted aperture or neck at both its ends, and being firmly supported around this neck by the collar of each transverse plate. (See Pl. 32, Figs. 2, 3, and Pl. 33.)

The strength of each sac is thus increased by the shortness of the distance between its two extremities, and the entire pipe, thus subdivided into thirty or forty distinct compartments, derives from every subdivision an accession of power to sustain the pressure of any fluid that may be introduced to its interior.

* We learn from Mr. Owen, that there was no possibility of the access of water to the air chambers between the exterior of this pipe and the siphonic apertures of the transverse plates; because the entire circumference of the mantle in which the siphuncle originates, is firmly attached to the shell by a horny girdle, impenetrable to any fluid.—*Memoir on Nautilus Pompilius, p.* 47.

more than air, and must consequently be exposed to great pressure when at the bottom of the sea. Several contrivances are therefore introduced to fortify them against this pressure.

First, the circumference of the external shell is constructed every way upon the principles of an *Arch*, (see Pl. 31, Fig. 1, and Pl. 32, Fig. 1.) so as to offer the greatest resistance to pressure tending to force it inwards in all directions.

Secondly, this arch is further fortified by the addition of numerous *minute Ribs*, which are beautifully marked in the fossil specimens represented at Pl. 32, Fig. 1. In this fossil the external shell exhibits fine wavy lines of growth, which, though individually small and feeble, are collectively of much avail as ribs to increase the aggregate amount of strength. (See Pl. 32, Fig. 1. a. to b.)

Thirdly, the arch is rendered still stronger by the position of the edges of the internal *Transverse plates*, nearly at right angles to the sides of the external shell, (see Pl. 32, Fig. 1, b. to c.) The course of the edges of these transverse plates beneath the ribs of the outer shell is so directed, that they act as cross braces, or spanners, to fortify the sides of the shell against the inward pressure of deep water. This contrivance is analogous to that adopted in fortifying a ship for voyages in the Arctic Seas, against the pressure of ice-bergs, by the introduction of an extraordinary number of transverse beams and bulk heads.*

We may next notice a fourth contrivance by which the apparatus that gives the shell its power of floating, is progressively maintained in due proportion to the increasing weight and bulk of the body of the animal, and of the external chamber in which it resides ; this is effected by successive additions of new transverse Plates across the bottom of the outer chamber, thus converting into *air chambers* that part, which had become too small for the body of the Sepia. This operation, repeated at intervals in due proportion to the successive stages of growth of the outer shell, maintains its efficacy as a *float,*

---

* The disposition of the curvatures of the transverse ribs, or lines of growth, in a different direction from the curvatures of the internal transverse plates, affords an example of further contrivance for producing strength in the shells both of recent and fossil Nautili.   As the internal transverse plates are convex *inwards,* (see Pl. 32, Fig. 1, b. to c.) whilst the ribs of the outer shell are in the greater part of their course convex *outwards,* these ribs intersect the curved edges of the transverse plates at many points, and thus divide them into a series of curvilinear parallelograms; the two shorter sides of each parallelogram being formed by the edges of transverse plates, whilst its two longer sides are formed by segments of the external ribs.   The same principle of construction here represented in our plate of Nautilus hexagonus, extends to other species of its family of Nautilus, in many of which the ribs are more minute; it is also applied in other families of fossil chambered shells; e. g. the Ammonites, Pl. 35, and Pl. 38. Scaphites, Pl. 44, Fig. 15. Hamites, Pl. 44, Fig. 8—13. Turrilites, Pl. 44, Fig. 14, and Baculites, Pl. 44, Fig. 5.

enlarging gradually and periodically until the animal has arrived at full maturity.*

A fifth consideration is had of mechanical advantage, in disposing the *Distance* at which these successive transverse Plates are set from one another. (See Pl. 31. Fig. 1. and Pl. 32, Fig. 1, 2). Had these distances increased in the same proportion as the area of the air chambers, the external shell would have been without due support beneath those sides of the largest chambers, where the pressure is greatest : for this a remedy is provided in the simple contrivance of placing the transverse plates proportionally nearer to one another, as the chambers, from becoming larger, require an increased degree of support.

Sixthly, The last contrivance, which I shall here notice, is that which regulates the ascent and descent of the animal by the mechanism of the *Siphuncle*. The use of this organ has never yet been satisfactorily made out; even Mr. Owen's most important Memoir leaves its manner of operation uncertain ; but the appearances which it occasionally presents in a fossil state, (See Pl. 32, Fig. 2, 3, and Pl. 33,)† supply

* In a young Nautilus Pompilius in the collection of Mr. Broderip, there are only seventeen Septa.   Dr. Hook says that he has found in some shells as many as forty. A cast, expressing the form of a single air chamber, of the Nautilus Hexagonus is represented in Pl. 42, Fig. 1.

† Pl. 32, Fig. 2, represents a fractured portion of the interior of a Nautilus Hexagonus, having the transverse plates (c. c′.)

evidence, which taken in conjunction with Mr. Owen's representation of its termination in a large sac surrounding the heart of the animal, (P. 34, p, p, a. a.) appears sufficient to decide this long disputed question. If we suppose this sac (p, p.) to contain a *pericardial fluid*, the place of which is alternately changed from the pericardium to the siphuncle, we shall find in this shifting fluid an hydraulic balance or adjusting power, causing the shell to sink when the pericardial fluid is forced into the siphuncle, and to

encrusted with calcareous spar; the Siphuncle also is similarly encrusted, and distended in a manner which illustrates the action of this organ. (Pl. 32, Fig. 2, a. a¹. a². a³. d. e. f, and Fig. 3, d. e. f). The fracture at Fig. 2, b. shews the diameter of the siphuncle, where it passes through a transverse plate, to be much smaller than it is midway between these Plates (at d. e. f). The transverse sections at Fig. 2 a. and b., and the longitudinal sections at Fig. 2, d. e. f. and Fig. 3, d. e. f., shew that the interior of the siphuncle is filled with stone, of the same nature with the stratum in which the shell was lodged. These earthy materials, having entered the orifice of the pipe at *a* in a soft and plastic state, have formed a cast which shews the interior of this pipe, when distended, to have resembled a string of oval beads, connected at their ends by a narrow neck, and enlarged at their centre to nearly double the diameter of this neck.

A similar distension of nearly the entire siphuncle by the stony material of the rock in which the shell was imbedded, is seen in the specimen of Nautilus striatus from the Lias of Whitby, represented at Pl. 33. The Lias which fills this pipe, must have entered it in the state of *liquid mud*, to the same extent that the *pericardial fluid* entered, during the hydraulic action of the siphuncle in the act of sinking; not one of the air-chambers has admitted the smallest particle of this mud; they are all filled

become buoyant, whenever this fluid returns to the pericardium. On this hypothesis also the chambers would be continually filled with air alone, the elasticity of which would readily admit of the alternate expansion and contraction of the siphuncle, in the act of admitting or rejecting the pericardial fluid.

The principle to which we thus refer the rising and sinking of the living Nautilus, is the same which regulates the ascent and descent of the Water Balloon: the application of external pres-

with calcareous spar, *subsequently* introduced by gradual infiltration, and at *successive periods* which are marked by changes in the colour of the spar. In both these fossil Nautili, the entire series of the earthy casts within the siphuncle represents the bulk of fluid which this pipe could hold.

The sections, Pl. 32, Fig. 3, d. e. f., shew the edges of the calcareous sheath surrounding the oval casts of three compartments of the expanded siphuncle. This calcareous sheath was probably flexible, like that surrounding the membranous pipe of the recent Nautilus Pompilius. (Pl. 31, Fig. 1, b. d. e.) The continuity of this sheath across the air chambers, (Pl. 32, Figs. 2, d. e. f. Fig. 3, d. e. f. and Pl. 33), shows that there was *no communication* for the passage of any fluid from the siphuncle into these chambers: had any such existed, some portion of the fine earthy matter, which in these two fossils forms the casts of the siphuncle, must have passed through it into these chambers. Nothing has entered them, but *pure crystallized* spar, introduced by infiltration through the pores of the shell, after it had undergone sufficient decomposition to be percolated by water, holding in solution carbonate of lime.

The same argument applies to the solid casts of pure crystallized carbonate of lime, which have entirely filled the chambers of the specimen Pl. 32, Fig. 1; and to all fossil Nautili and Ammonites, in which the air chambers are either wholly void, or

sure through a membrane that covers the column
of water in a tall glass, forces a portion of this
water into the cavity, or single air-chamber of
the balloon, which immediately begins to sink;
on the removal of this pressure, the elasticity
of the compressed air causing it to return to its
former volume, again expels the water, and the
balloon begins to rise.*

I shall conclude this attempt to illustrate the
structure and economy of fossil Nautili by
those of the living species, by shewing in what
manner the chambers of the pearly Nautilus,
supposing them to be permanently filled *only*
with air, and the action of the siphuncle, supposing it to be the receptacle *only* of a fluid
secretion, interchanging its place alternately
from the siphuncle to the pericardium, would
be subsidiary to the movements of the animal,
both at the surface, and bottom of the sea.

First, The animal was seen and captured by

partially, or entirely filled with pure crystallized carbonate of
lime. (See Pl. 42, Fig, 1, 2, 3, and Pl. 36). In all such cases,
it is clear that no communication existed, by which water could
pass from the interior of the siphon to the air chambers. When
the pipe was ruptured, or the external shell broken, the earthy
sediment, in which such broken shells were lodged, finding
through these fractures admission to the air chambers, has filled
them with clay, or sand or limestone.

* The substance of the siphuncle is a thin and strong membrane, with no appearance of muscular fibres, by which it could
contract or expand itself; its functions, therefore, must have
been entirely *passive*, in the process of admitting or ejecting any
fluid to or from its interior.—See Owen's Memoir, p. 10.

Mr. Bennett, floating at the surface, with the upper portion of the shell raised above the water, and kept in a vertical position by means of the included air (see Pl. 31. Fig. 1.); this position is best adapted to the retrograde motion, which a Sepia derives from the violent ejection of water through its funnel (k); thus far, the air-chambers, serve to maintain both the shell and body of the animal in a state of equilibrium *at the surface.*

Secondly, The next point to be considered is the mode of operation of the siphuncle and air-chambers, *in the act of sinking* suddenly from the surface to the bottom.   These are explained in the note subjoined.*

* It appears from the figure of the animal, Pl. 34, with which I have been favoured by Mr. Owen, that the upper extremity of the siphuncle marked by the insertion of the probe b., termi-nates in the cavity of the pericardium p, p.   As this cavity may contain a fluid, more dense than water, excreted by the glandular follicles d. d., and is apparently of such a size that its contents would suffice to fill the siphuncle, it is probable that this fluid forms *the circulating medium of adjustment,* and regulates the ascent or descent of the animal by its interchange of place from the pericardium to the siphuncle.

When the arms and body are expanded, the fluid remains in the pericardium, and the siphuncle is empty, and collapsed, and surrounded by the portions of air that are permanently confined within each air-chamber; in this state, the specific gravity of the body and shell together is such as to cause the animal to rise, and be sustained floating at the surface.

When, on any alarm, the arms and body are contracted, and drawn into the shell, the retraction of these parts, causing pres-sure on the exterior of the pericardium, forces its fluid contents downwards into the siphuncle; and the bulk of the body being

Thirdly, It remains to consider the effect of the air, supposing it to be retained continually within the chambers, *at the bottom* of the sea. Here, if the position of the moving animal be beneath the mouth of the shell, like that of a snail as it crawls along the ground, the air within the chambers would maintain the shell, buoyant, and floating over the body of the animal in a

thus diminished, without increasing the bulk of the shell, into whose cavities the fluid is withdrawn, the specific gravity of the whole mass is suddenly increased, and the animal begins to *sink*.

The air within each chamber remains under compression, as long as the siphuncle continues distended by the pericardial fluid; and returning, by its elasticity, to its former state, as soon as the pressure of the arms and body is withdrawn from the pericardium, forces the fluid back again into the cavity of this organ ; and thus the shell, diminished as to its specific gravity, has a tendency to *rise*.

The place of the pericardial fluid, therefore, will be always in the pericardium, excepting when it is forced into and retained in the siphuncle, by muscular pressure, during the contraction of the arms and body closed up within the shell. When these are expanded, either on the surface, or at the bottom of the sea, the water will have free access to the branchiæ, and the movements of the heart will proceed freely in the distended pericardium; which will be emptied of its fluid at those times *only*, when the body is closed, and the access of water to the branchiæ consequently impeded.

The following experiments shew that the weight of fluid requisite to be added to the shell of a Nautilus, in order to make it sink, is about half an ounce.

I took two perfect shells of a Nautilus Pompilius, each weighing about six ounces and a half in air, and measuring about seven inches across their largest diameter; and having stopped the Siphuncle with wax, I found that each shell, when placed in fresh water, required the weight of a few grains more than an ounce to

vertical position, with little or no muscular exertion, and leave the creature at ease to regulate the movements of its tentacula (p) in crawling and seizing its prey.*

Dr. Hook considered (Hook's Experiments, 8vo. 1726, page 308) that the air chambers were *filled alternately with air or water;* and Parkinson (Organic Remains, vol. iii. p. 102), admitting that these chambers were not accessible to water, thinks that the act of rising or sinking depends on the *alternate introduction of air or water into the siphuncle;* but he is at a loss to find the source from which this air could be obtained at the bottom of the sea, or to

---

make it sink. As the shell, when attached to the living animal, was probably a quarter of an ounce heavier than these dry dead shells, and the specific gravity of the body of the animal may have exceeded that of water to the amount of another quarter of an ounce, there remains about half an ounce, for the weight of fluid which being introduced into the siphuncle, would cause the shell to sink; and this quantity seems well proportioned to the capacity both of the pericardium, and of the distended siphuncle.

* If the chambers were filled with water, the shell could not be thus suspended without muscular exertion, and instead of being poised vertically over the body, in a position of ease and safety, would be continually tending to fall flat upon its side; thus exposing itself to injury by friction, and the animal to attacks from its enemies. Rumphius states, that at the bottom, He creeps with his boat above him, and with his head and barbs (tentacula) on the ground, making a tolerably quick progress. I have observed that a similar vertical position is maintained by the shell of the Planorbis corneus, whilst in the act of crawling at the bottom.

explain " in what manner the animal effected those modifications of the tube and its contained air, on which the variation of its buoyancy depended."* The theory which supposes *the chambers of the shell to be permanently filled with air alone, and the siphuncle to be the organ which regulates the rising or sinking of the animal, by changing the place of the pericardial fluid*, seems adequate to satisfy every hydraulic condition of a Problem that has hitherto received no satisfactory solution.

I have dwelt thus long upon this subject, on account of its importance, in explaining the complex structure, and hitherto imperfectly understood functions, of all the numerous and widely disseminated families of fossil chambered shells, that possessed siphunculi. If, in all these families, it can be shewn that the same principles of mechanism, under various modifications, have prevailed from the first commencement of organic life unto the present hour, we can hardly avoid the conclusion which would refer such unity of organizations to the will and agency of one and the same intelligent First Cause, and lead us to regard them all as " emanations of that Infinite Wisdom, that appears in the shape and structure of all other created beings."†

---

* The recent observations of Mr. Owen shew, that there is no gland connected with the siphuncle, similar to that which is supposed to secrete air in the air-bladder of fishes.

† Dr. Hook's Experiments, p. 306.

SECTION IV.

AMMONITES.

HAVING entered thus largely into the history of the Mechanism of the shells of Nautili, we have hereby prepared ourselves for the consideration of that of the kindred family of Ammonites, in which all the essential parts are so similar in principle to those of the shells of Nautili, as to leave no doubt that they were subservient to a like purpose in the economy of the numerous extinct species of Cephalopodous Mollusks, from which these Ammonites have been derived.

Geological Distribution of Ammonites.

The family of Ammonites extends through the entire series of the fossiliferous Formations, from the Transition strata to the Chalk inclusive. M. Brochant, in his Translation of De la Beche's Manual of Geology, enumerates 270 species; these species differ according to the age of the strata in which they are found,† and vary in

† Thus one of the first forms under which this family appeared, the Ammonites Henslowi, (Pl. 40, Fig. 1), ceased with the Transition formation; the A. Nodosus (Pl. 40, Figs. 4, 5.) began and terminated its period of existence with the Muschel-Kalk. Other genera and species of Ammonites, in like manner, begin and end with certain definite strata, in the Oolitic and Cretaceous formations; e. g. the A. Bucklandi (Pl. 37, Fig. 6.) is

size from a line to more than four feet in diameter.*

peculiar to the Lias; the A. Goodhalli to the Greensand; and the A. Rusticus to the Chalk. There are few, if any, species which extend through the whole of the Secondary periods, or which have passed into the Secondary, from the Transition period.

The following Tabular Arrangement of the distribution of Ammonites, in different geological formations, is given by Professor Phillips in his Guide to Geology, 1834, p. 77.

SUB-GENERA OF AMMONITES.

LIVING SPECIES.	Goniatites.	Ceratites.	Arietes.	Falciferi.	Amalthei.	Capricorni.	Planulati.	Dorsati.	Coronarii.	Macrocephali.	Armati.	Dentati.	Ornati.	Flexuosi.
In Tertiary strata....														
In Cretaceous system.				2	4					9	14	13	2	3
In Oolitic system....				22	27	12	26	5	11	11	11	4	5	3
In Saliferous system.		3	12											
In Carboniferous system.	7													
‡ In Primary strata..	17													

Total, 223 species.

" It is easy to see how important, in questions concerning the relative antiquity of stratified rocks, is a knowledge of Ammonites, since whole sections of them are characteristic of certain systems of rocks."—Phillips's Guide to Geology, 8vo. 1834, sec. 82.

---

‡ The strata here termed *primary* are those which, in the Section, (Pl. 1), I have included in the lower region of the *transition* series.

* Mr. Sowerby (Min. Conch. vol. iv. p. 79 and p. 81,) and Mr. Mantell speak of Ammonites in Chalk, having a diameter of three feet. Sir T. Harvey, and Mr. Keith Milnes, have recently measured Ammonites in the Chalk near Margate, which exceeded *four feet* in diameter; and this in cases where the diameter can have been in a very small degree enlarged by pressure.

It is needless here to speculate either on the physical, or final causes, which produced these curious changes of species, in this highest order of the Molluscous inhabitants of the seas, during some of the early and the middle ages of geological chronology; but the exquisite symmetry, beauty, and minute delicacy of structure, that pervade each variation of contrivance throughout several hundred species, leave no room to doubt the exercise of Design and Intelligence in their construction; although we cannot always point out the specific uses of each minute variation, in the arrangement of parts fundamentally the same.

The geographical distribution of Ammonites in the ancient world, seems to have partaken of that *universality*, we find so common in the animals and vegetables of a former condition of our globe, and which differs so remarkably from the *varied distribution* that prevails among existing forms of organic life. We find, the same genera, and, in a few cases, the same species of Ammonites, in strata, apparently of the same age, not only throughout Europe, but also in distant regions of Asia, and of North and South America.*

---

* Dr. Gerard has discovered at the elevation of sixteen thousand feet in the Himmalaya Mountains, species of Ammonites, e. g. A. Walcoti, and A. Communis, identical with those of the Lias at Whitby and Lyme Regis. He has also found in the same parts of the Himmalaya, several species of Belemnite, with Te-

Hence we infer that during the Secondary and Transition periods a *more general distribution of the same species*, than exists at present, prevailed in regions of the world most remotely distant from one another.

An Ammonite, like a Nautilus, is composed of three essential parts : 1st. An *external shell*, usually of a flat discoidal form, and having its surface strengthened and ornamented with ribs (see Pl. 35, and Pl. 37.) 2nd. A series of *internal air chambers* formed by transverse plates, intersecting the inner portion of the shell, (see Pl. 36 and 41). 3rd. A *siphuncle*, or pipe, commencing at the bottom of the outer chamber, and thence passing through the entire series of air chambers to the innermost extremity of the shell, (see Pl. 36, d. e. f. g. h. i.) In each of these parts, there are evidences of mechanism, and

rebratulæ, and other Bivalves, that occur in the English Oolite; thereby establishing the existence of the Lias, and Oolite formations in that elevated and distant region of the world. He has also collected in the same Mountains, shells of the genera Spirifer, Producta, and Terebratula, which occur in the Transition formations of Europe and America.

The Greensand of New Jersey also contains Ammonites mixed with Hamites and Scaphites, as in the Greensand of England. and Captain Beechey and Lieutenant Belcher found Ammonites on the coast of Chili, in Lat. 36. S. in the Cliffs near Conception; a fragment of one of these Ammonites is preserved in the Museum of Hasler Hospital at Gosport.

Mr. Sowerby possesses fossil shells from Brazil resembling those of the Inferior Oolite of England.

consequently of design, a few of which I shall endeavour briefly to point out.

External Shell.

The use and place of the shells of Ammonites has much perplexed geologists and conchologists. Cuvier and Lamarck, guided by the analogies afforded by the Spirula, supposed them to be internal shells.* There is, however, good reason to believe that they were entirely external, and that the position of the body of the animal within these shells was analogous to that of the inhabitant of the Nautilus Pompilius. (See Pl. 31, Fig. 1).

* The smallness of the outer chamber, or place of lodgment for the animal, is advanced by Cuvier in favour of his opinion that Ammonites, like the Spirula, were internal shells. This reason is probably founded on observations made upon imperfect specimens. The outer chamber of Ammonites is very seldom preserved in a perfect state, but when this happens, it is found to bear at least as large a proportion to the chambered part of the shell, as the outer cell of the N. Pompilius bears to the chambered interior of that shell. It often occupies more than half, (see Pl. 36. a. b. c. d.) and, in some cases, the whole circumference of the outer whorl. This open chamber is not thin and feeble, like the long anterior chamber of the Spirula, which is placed within the body of the animal producing this shell; but is nearly of equal thickness with the sides of the close chambers of the Ammonite.

Moreover, the margin of the mature Ammonite is in some species reflected in a kind of scroll, like the thickened margin of the shell of the garden snail, giving to this part a strength which would apparently be needless to an internal shell. (See Pl. 37. Fig. 3. d.)

The presence of spines also in certain species, (as in A.

G.   Z

Mr. De la Beche has shewn that the mineral condition of the outer chamber of many Ammonites, from the Lias at Lyme Regis, proves that the entire body was contained within it; and that these animals were suddenly destroyed and buried in the earthy sediment of which the lias is composed, before their bodies had either undergone decay, or been devoured by the crustaceous Carnivora with which the bottom of the sea then abounded.*

As all these shells served the double office of affording protection, and acting as floats, it was necessary that they should be thin, or they would have been too heavy to rise to the surface: it was not less necessary that they should be strong, to resist pressure at the bottom of the sea; and accordingly we find them fitted for this double function, by the disposition of their ma-

Armatus, A. Sowerbii,) affords a strong argument against the theory of their having been internal shells. These spines which have an obvious use for protection, if placed externally, would seem to have been useless, and perhaps noxious in an internal position, and are without example in any internal structure with which we are acquainted.

* In the Ammonites in question, the outer extremity of the first great chamber in which the body of the animal was contained, is filled with stone only to a small depth, (see Pl. 36, from a. to b.); the remainder of this chamber from b. to c., is occupied by brown calcareous spar, which has been ascertained by Dr. Prout to owe its colour to the presence of animal matter, whilst the internal air chambers and siphuncle are filled with pure white spar. The extent of the brown calcareous spar, therefore, in the outer chamber, represents the space which was

terials, in a manner calculated to combine lightness and buoyancy with strength.

First, The entire shell, (Pl. 35,) is one continuous *arch*, coiled spirally around itself in such
a manner, that the base of the outer whorls rests
upon the crown of the inner whorls, and thus
the keel or back is calculated to resist pressure,
in the same manner as the shell of a common
hen's egg resists great force, if applied in the
direction of its longitudinal diameter.

Secondly, besides this general arch-like form,
the shell is further strengthened by the insertion
of *ribs*, or transverse arches, which give to many
of the species their most characteristic feature,
and produce in all, that peculiar beauty which
invariably accompanies the symmetrical repetition of a series of spiral curves.   See Pl. 37,
Figs. 1—10.)

From the disposition of these ribs over the

occupied by the body of the animal after it had shrunk within
its shell, at the moment of its death, leaving void the outer
portion only of its chamber, from a. to b., to receive the muddy
sediment in which the shell was imbedded.

I have many specimens from the lias of Whitby, of the Ammonites Communis, in which the outer chamber thus filled with
spar, occupies nearly the entire last whorl of the shell, its largest
extremity only being filled with lias.   From specimens of this
kind we also learn, that the animal inhabiting the shell of an
Ammonite, had no ink bag; if such an organ existed, traces
of its colour must have been found within the cavity which contained the body of the animal at the moment of its death.   The
protection of a shell seems to have rendered the presence of an
ink bag superfluous.

surface of the external shell, there arise mechanical advantages for increasing its strength,
founded on a principle that is practically applied
in works of human art. The principle I allude
to, is that by which the strength and stiffness
of a thin metallic plate are much increased by
corrugating, or applying *flutings* to its surface.
A common pencil-case, if made of corrugated or
fluted metal, is stronger than if the same quantity of metal were disposed in a simple tube.
Culinary moulds of tin and copper are in the
same way strengthened, by folds or flutings
around their margin, or on their convex surfaces.
The recent application of thin plates of *corrugated* iron to the purpose of making self-sup-
porting roofs, in which the corrugations of the
iron supply the place, and combine the power of
beams and rafters, is founded on the same principle that strengthens the vaulted shells of Ammonites. In all these cases, the ribs or elevated
portions, add to the plates of shell, or metal,
the strength resulting from the convex form of
an arch, without materially increasing their
weight; whilst the intermediate depressed parts
between these arches, are suspended and supported by the tenacity and strength of the
material. (See Pl. 37, Figs. 1—10.*)

---

* The figures engraved at Pl. 37, afford examples of various
contrivances to give strength and beauty to the external shell.
The first and simplest mode, is that represented in Pl. 35 and

The general principle of dividing and sub-dividing the ribs, in order to multiply supports as the vault enlarges, is conducted nearly on the same plan, and for the same purpose, as the divisions and subdivisions of the ribs beneath the *groin work*, in the flat vaulted roofs of the florid Gothic Architecture.

Another source of strength is introduced in many species of Ammonites by the elevation of parts of the ribs into little dome-shaped *tuber-*

Pl. 37, Fig. 1 and 6. Here each rib is single, and extends over the whole surface, becoming gradually wider, as the space enlarges towards the outer margin, or back of the shell.

The next variation is that represented (Pl. 37, Figs. 2, 7, 9,) where the ribs, originating singly on the inner margin, soon bifurcate into two ribs that extend outwards, and terminate upon the dorsal keel.

In the third case, (Pl. 37. Fig 4), the ribs, originate simply, and bifurcating as the shell enlarges, extend this bifurcation entirely around its circular back. Between each pair of bifurcated ribs, a third or auxiliary short rib is interposed, to fill up the enlarged space on the dorsal portion where the shell is broadest.

In a fourth modification, (Pl. 37, Fig 3), the ribs, originating singly on the internal margin, soon become trifurcate, and expand outwards, around the circular back of the shell. A perfect mouth of this shell is represented at Pl. 37. Fig. 3, d.

A fifth case is that (Pl. 37, Fig. 5,) in which the simple rib becomes trifurcate as the space enlarges, and one or more auxiliary short ribs are also interposed, between each trifurcation. These subdivisions are not always maintained with numerical fidelity through every individual of the same species, nor over the whole surface of the same shell; their use, however, is always the same, viz. to cover and strengthen those spaces which the expansion of the shell towards its outer circumference, would have rendered weak without the aid of some such compensation.

cles, or bosses, thus superadding the strength of a dome to that of the simple arch, at each point where these bosses are inserted.*

The bosses thus often introduced at the origin, division, and termination of the ribs, resemble those applied by architects to the intersections of the ribs in Gothic roofs, and are much more efficient in producing strength.† These tubercules have the effect of little vaults or domes; and they are usually placed at those parts of the external shell, beneath which there is no immediate support from the internal transverse plates (see Pl. 37, Fig. 8. Pl. 42, Fig. 3. c. d. e. and Pl. 40, Fig. 5.)‡

* These places are usually either at the point of bifurcation, as in Pl. 37, Figs. 2, 7, 9, 10, or at the point of trifurcation, as in Fig. 3.

† The ribs and bosses in vaulted roofs project *beneath* the under surface of the arch; in the shells of Ammonites, they are raised *above* the convex surface.

‡ In Pl. 37, Fig. 9 (A. varians), the strength of the ribs and proportions of the tubercles are variable, but the general character exhibits a triple series of large tubercles, rising from the surface of the transverse ribs. Each of these ribs commences with a small tubercle near the inner margin of the shell. At a short distance outwards is a second and larger tubercle, from which the rib bifurcates, and terminates in a third tubercle, raised at the extremity of each fork upon the dorsal margin.

Many species of Ammonites have also a dorsal ridge or keel, (Pl. 37, Figs. 1. 2. 6.) passing along the back of the shell, immediately over the siphuncle, and apparently answering, in some cases, the further purpose of a cut-water, and keel (Pl. 37, Figs. 1, 2.). In certain species, e. g. in the A. lautus (Pl. 37,

Similar tubercles are introduced with the same advantage of adding Strength as well as Beauty in many other cognate genera of chambered shells. (Pl. 44, Fig. 9. 10. 14. 15.)

In all these cases, we recognize the exercise of Discretion and Economy in the midst of Abundance; distributing internal supports but sparingly, to parts which, from their external form, were already strong, and dispensing them abun-

Fig. 7, a. c.) there is a double keel, produced by a deep depression along the dorsal margin; and the keels are fortified by a line of tubercles placed at the extremity of the transverse ribs. In the A. varians (Pl. 37, 9. a. b. c.) which has a triple keel, the two external ones are fortified by tubercles, as in Fig. 7, and the central keel is a simple convex arch.

Pl. 37, Fig. 8, offers an example of domes, or bosses, compensating the weakness that, without them, would exist in the A. catena, from the minuteness of its ribs, and the flatness of the sides of the shell. These flat parts are all supported by an abundant distribution of the edges of the transverse plates directly beneath them, whilst those parts which are elevated into bosses, being sufficiently strong, are but slightly provided with any other support. The back of this shell also, being nearly flat, (Pl. 37, Fig. 8. b. c.) is strongly supported by ramifications of the transverse plates.

In Pl. 37, Fig. 6, which has a triple keel, (that in the centre passing over the siphuncle), this triple elevation affords compensation for the weakness that would otherwise arise from the great breadth and flatness of the dorsal portion of this species. Between these three keels, or ridges, are two depressions or dorsal furrows, and as these furrows form the weakest portion of the shell, a compensation is provided by conducting beneath them the denticulated edges of the transverse plates, so that they present long lines of resistance to external pressure.

dantly beneath those parts only, which without them, would have been weak.

ˑWe find an infinity of variations in the form and sculpture of the external shell, and a not less beautiful variety in the methods of internal fortification, all adapted, with architectural advantage, to produce a combination of Ornament with Utility. The ribs also are variously multiplied, as the increasing space demands increased support; and are variously adorned and armed with domes and bosses, wherever there is need of more than ordinary strength.

Transverse Plates, and Air Chambers.

The uses of the internal air chambers will best be understood by reference to our figures. Pl. 36 represents a longitudinal section of an Ammonite bisecting the transverse plates in the central line where their curvature is most simple. On each side of this line, the curvature of these plates become more complicated, until, at their termination in the external shell, they assume a beautifully sinuous, or foliated arrangement, resembling the edges of a parsley leaf, (Pl. 38), the uses of which, in resisting pressure, I shall further illustrate by the aid of graphic representations.

At Pl. 35, from d. to e. we see the edges of the same transverse plates which, in Pl. 36, are simple curves, becoming foliated at their junction with the outer shell, and thus distributing their support more equally beneath all its parts, than if these simple curves had been continued to the extremity of the transverse plates. In more than two hundred known species of Ammonites, the transverse plates present some beautifully varied modifications of this foliated expansion at their edges; the effect of which, in every case, is to increase the strength of the outer shell, by multiplying the subjacent points of resistance to external pressure. We know that the pressure of the sea, at no great depth, will force a cork into a bottle filled with air, or crush a hollow cylinder or sphere of thin copper; and as the air chambers of Ammonites were subject to similar pressure, whilst at the bottom of the sea, they required some peculiar provision to preserve them from destruction,* more especially as most zoologists

---

* Captain Smyth found, on two trials, that the cylindrical copper air tube, under the vane attached to Massey's patent log, collapsed, and was crushed quite flat under a pressure of about three hundred fathoms. A claret bottle, filled with air, and well corked, was burst before it had descended four hundred fathoms. He also found that a bottle filled with fresh water, and corked, had the cork forced at about a hundred and eighty fathoms below the surface; in such cases, the fluid sent down is replaced by salt water, and the cork which had been forced in, is sometimes inverted.

Captain Beaufort also informs me, that he has frequently sunk corked bottles in the sea more than a hundred fathoms deep,

agree that they existed at great depths, " dans les grandes profondeurs des mers."*

Here again we find the inventions of art anticipated in the works of nature, and the same principle applied to resist the inward pressure of the sea upon the shells of Ammonites, that an engineer makes use of in fixing transverse stays beneath the planks of the wooden centre on which he builds his arch of stone.

The disposition of these supports assumes throughout the family of Ammonites a different arrangement from the more simple curvature of the edges of the transverse plates within the shells of Nautili; and we find a probable cause for this variation, in the comparative thinness of the outer shells of many Ammonites; since this external weakness creates a need of more internal support under the pressure of deep water, than was requisite in the stronger and thicker shells of Nautili.

This support is effected by causing the edges of the transverse plates to deviate from a simple

some of them empty, and others containing a fluid. The empty bottles were sometimes crushed, at other times, the cork was forced in, and the bottle returned full of sea water. The cork of the bottles containing a fluid was uniformly forced in, and the fluid exchanged for sea water; the cork was always returned to the neck of the bottle, sometimes, but not always, in an inverted position.

* See Lamarck, who cites Bruguières with approbation on this point.—Animaux sans: Vert: vol. vii. p. 635.

curve, into a variety of attenuated ramifications and undulating sutures. (See Pl. 38. and Pl. 37, Figs. 6, 8). Nothing can be more beautiful than the sinuous windings of these sutures in many species, at their union with the exterior shell; adorning it with a succession of most graceful forms, resembling festoons of foliage, and elegant embroidery. When these thin septa are converted into iron pyrites, their edges appear like golden filigrane work, meandering amid the pellucid spar, that fills the chambers of the shell.*

* The A. Heterophyllus, Pl. (38), is so called from the apparent occurrence of two different forms of foliage; its laws of dentation are the same as in other Ammonites, but the ascending secondary saddles (Pl. 38. S. S.) which, in all Ammonites are round, are in this species longer than ordinary, and catch attention more than the descending points of the lobes, (Pl. 38. d. l.)

The figures of the edge of one transverse plate are repeated in each successive plate. The animal, as it enlarged its shell, thus leaving behind it a new chamber, more capacious than the last, so that the edges of the plates never interfere or become entangled.

Although the pattern on the surface of this Ammonite is apparently so complicated, the number of transverse plates is but sixteen in one revolution of the shell; in this, as in almost all other cases, the extreme beauty and elegance of the foliations result from the repetition, at regular intervals, of one symmetrical system of forms, viz. those presented by the external margin of a single transverse plate. No trace of these foliations is seen on the outer surface of the external shell. (See Pl. 38, c.)

The figures of A. obtusus, (Pl. 35 and Pl. 36), shew the relations between the external shell and the internal transverse par-

The shell of the Ammonites Heterophyllus (Pl. 38, and Pl. 39,) affords beautiful exemplifications of the manner in which the mechanical strength of each transverse plate is so disposed, as to vary its support in proportion to the different degrees of necessity that exist for it in different parts of the same shell.*

titions of an Ammonite. Pl. 35 represents the form of the external shell, wherein the body occupied the space extending from a. to c., and corresponding with the same letters in Pl. E. 36.

This species has a single series of strong ribs passing obliquely across the shell of the outer chamber, and also across the air-chambers. From c. to the inmost extremity of the shell, these ribs intersect, and rest on the sinuous edges of the transverse plates which form the air chambers. These edges are not seen where the outer shell is not removed. (Pl. 35, e.) A small portion of the shell is also preserved at Pl. 35, b.

From d. inwards, these sinuous lines mark the terminations of the transverse plates at their junction with the external shell; they are not coincident with the direction of the ribs, and therefore more effectually co-operate with them in adding strength to the shell, by affording it the support of a series of various props and buttresses, set nearly at right angles to its internal surface.

* Thus on the back or keel, Pl. 39, from V. to B., where the shell is narrow, and the strength of its arch greatest, the intervals between the septa are also greatest, and their sinuosities comparatively distant; but as soon as the flattened sides of the same shell, Pl. 38, assume a form that offers less resistance to external pressure, the foliations at the edges of the transverse plates approximate more closely; as in the flatter forms of a Gothic roof, the ribs are more numerous, and the tracery more complex, than in the stronger and more simple forms of the pointed arch.

The same principle of multiplying and extending the ramifications of the edges of the transverse plates, is applied to other species of Ammonites, in which the sides are flat, and require a

At Plate 41. we have a rare and most beautiful example of the preservation of the transverse plates of the Ammonites giganteus converted to chalcedony, without the introduction of any earthy matter into the area of the air-chambers.

This shell is so laid open as to shew the manner in which each transverse plate forms a tortuous partition between the successive air-chambers. By means of these winding plates, the external shell, being itself a continuous arch, is further fortified with a succession of compound arches, passing transversely across its internal cavity ; each arch being disposed in the form of a double tunnel, vaulted not only at the top, but having a corresponding series of inverted arches along the bottom.

We can scarcely imagine a more perfect instrument than this for affording universal resistance to external pressure, in which the greatest possible degree of lightness is combined with the greatest strength.

similar increase of support; whilst in those species to which the more circular form of the sides gives greater strength (as in A. obtusus, Pl. 35.) the sinuosities of the septa are proportionately few.

It seems probable that some improvement might be made, in fortifying the cylindrical air-tube of Massey's Patent Log for taking soundings at great depths, by the introduction of transverse plates, acting on the principle of the transverse plates of the chambered portion of the shells of Nautili and Ammonites, or rather of Orthoceratites, and Baculites, (see Pl. 44, Figs. 4. and 5.)

The form of the air-chambers in Ammonites is much more complex than in the Nautili, in consequence of the tortuous windings of the foliated margin of the transverse plates.*

Siphuncle.

It remains to consider the mechanism of the Siphuncle, that important organ of hydraulic adjustment, by means of which the specific gravity of the Ammonites was regulated. Its

* Pl. 42, Fig. 1, represents the cast of a single chamber of N. Hexagonus, convex inwards, and concave outwards, and bounded at its margin by lines of simple curvature. In a few species only of Nautilus the margin is undulated, (as in Pl. 43, Fig. 3, 4,) but it is never jagged, or denticulated like the margin of the casts of the chambers of Ammonites.

In Ammonites, the chambers have a double curvature, and are, at their centre, convex outwards (see Pl. 36. d. and Pl. 39. d. V.). Pl. 42, Fig. 2, represents the front view of the cast of a single chamber of A. excavatus; d, is the dorsal lobe enclosing the siphuncle, and e. f. the auxiliary ventral lobes, which open to receive the inner whorl of the shell. Pl. 42. Fig. 3. represents a cast of three chambers of A. catena, having two transverse plates still retained in their proper place between them. The foliated edges of these transverse plates have regulated the foliations of the calcareous casts, which, after the shell has perished, remain locked into one another, like the sutures of a skull.

The substance of the casts in all these cases is pure crystalline carbonate of lime, introduced by infiltration through the pores of the decaying shell. Each species of Ammonite has its peculiar form of air-chamber, depending on the specific form of its transverse plates. Analogous variations in the form of the air-chambers are co-extensive with the entire range of species in the family of Nautili.

mode of operation as a pipe, admitting or rejecting a fluid, seems to have been the same as that
we have already considered in the case of Nautili.*

The universal prevalence of such delicate
hydraulic contrivances in the Siphuncle, and of
such undeviating and systematic union of buoyancy and strength in the air-chambers, throughout the entire family of Ammonites and Nautili,
are among the most prominent instances of order
and method, that pervade these remains of former

* In the family of Ammonites, the place of the Siphuncle is
always upon the exterior, or *dorsal margin* of the transverse plates.
(See Pl. 36. d. e. f. g. h. i., and Pl. 42, Fig. 3. a, b.)   It is
conducted through them by a ring, or collar, projecting *outwards;*
this collar is seen, well preserved, at the margin of all the transverse plates in Pl. 36.   In Nautili, the collar projects uniformly
inwards, and its place is either at the *centre*, or near the *inner
margin* of the transverse plates. (See Pl. 31, Fig, 1. y. and
Pl. 42. 1.)

The Siphuncle represented at Pl. 36, is preserved in a black
carbonaceous state, and passes from the bottom of the external
chamber (d.) to the inner extremity of the shell.   At e. f. g. h. its
interior is exposed by section, and appears filled, like the adjacent
air-chambers, with a cast of pure calcareous spar.   At Pl. 42.
Fig. 3. b. a similar cast fills the tube of the Siphuncle, and also the
air-chambers.   Here again, as in Pl. 36, its diameter is contracted at its passage through the collar of each transverse plate,
with the same mechanical advantages as in the Nautilus.

The shell engraved at Pl. 42. Fig. 4. from a specimen found by
the Marquis of Northampton in the Green sand of Earl Stoke,
near Devizes, and of which Figs. 5. 6. are fragments, is remarkable for the preservation of its Siphuncle, distended and empty,
and still fixed in its place along the interior of the dorsal margin
of the shell.   This Siphuncle, and also the shell and transverse

races that inhabited the ancient seas; and strange indeed must be the construction of that mind, which can believe that all this order and method can have existed, without the direction and agency of some commanding and controlling Intelligence.

Theory of Von Buch.

Besides the uses we have attributed to the sinuous arrangement of the transverse septa of Ammonites, in giving strength to the shell to

plates, are converted into thin chalcedony, the pipe retaining in these empty chambers the exact form and position it held in the living shell.

The entire substance of the pipe, thus perfectly preserved in a state that rarely occurs, shews no kind of aperture through which any fluid could have passed to the interior of the air-chambers. The same continuity of the Siphuncle appears at Pl. 42, Fig. 3. and in Pl. 36, and in many other specimens. Hence we infer, that nothing could pass from its interior into the air-chambers, and that the office of the Siphuncle was to be more or less distended with a fluid, as in the Nautili, for the purpose of adjusting the specific gravity, so as to cause the animal to float or sink.

Dr. Prout has analyzed a portion of the black material of the Siphuncle, which is so frequently preserved in Ammonites, and finds it to consist of animal membrane penetrated by carbonate of lime. He explains the black colour of these pipes, by supposing that the process of decomposition, in which the oxygen and hydrogen of the animal membrane escaped, was favourable to the evolution of carbon, as happens when vegetables are converted into coal, under the process of mineralization. The lime has taken the place of the oxygen and hydrogen which existed in the pipe before decomposition.

resist the pressure of deep water, M. Von Buch has suggested a further use of the lobes thus formed around the base of the outer chamber, as affording points of attachment to the mantle of the animal, whereby it was enabled to fix itself more steadily within its shell. The arrangement of these lobes varies in every species of Ammonite, and he has proposed to found on these variations, the specific character of all the shells of this great family.*

* The most decided distinction between Ammonites and Nautili, is founded on the place of the siphon. In the Ammonite, this organ is always on the *back* of the shell, and never so in the Nautilus. Many other distinctions emanate from this leading difference; the animal of the Nautilus having its pipe usually fixed near the *middle*, (See Pl. 31, Fig. 1), or towards the *ventral margin* (Pl. 32, Fig. 2, and Pl. 42, Fig. 1.) of the transverse plates, is thereby attached to the bottom of the external chamber, which is generally concave, and without any jagged termination, or sinuous flexure, of its margin. As the siphon in Ammonites is comparatively small, and always placed on the *dorsal* margin (Pl. 36, d. and Pl. 39, d.), it would have less power than the siphuncle of Nautili to keep the mantle in its place at the bottom of the shell; another kind of support was therefore supplied by a number of depressions along the margin of the transverse plate, forming a series of lobes at the junction of this plate with the internal surface of the shell.

The innermost of these, or ventral lobe, is placed on the inner margin of the shell (Pl. 39, V.); opposite to this, and on the external margin, is placed the dorsal lobe (D), embracing the siphon and divided by it into two divergent arms. Beneath the dorsal lobe are placed the superior lateral lobes (L), one on each side of the shell; and still lower, the inferior lateral lobe (l), next above the ventral lobe.

The separations between these lobes form seats, or saddles,

The uses ascribed by Von Buch to the lobes of Ammonites in affording attachment to the base of the mantle around the margin of the transverse plates, would in no way interfere with the service we have assigned to the same lobes, in supporting the external shell against the pressure of deep water. The union of two beneficial results from one and the same mechanical expedient, confirms our opinion of the excellence of the workmanship, and increases our admiration of the Wisdom in which it was contrived.

upon which the mantle of the animal rested, at the bottom of the outer chamber; these saddles are distinguished in the same manner as the lobes—that between the dorsal and superior lateral lobe, forming the dorsal saddle (S. d.), that between the superior and inferior lateral lobes, forming the lateral saddle (S. L.), and that between the inferior lateral and ventral lobe, the ventral saddle (S. V.). This general disposition, variously modified, pervades all forms of Ammonites; but when, as in Pl. 39, the turn of the shell increases rapidly in width, so that the last whorl nearly, or entirely, covers the preceding whorls, the additional part is furnished with small auxiliary lobes, varying with the growth of the Ammonite to the number of three, four, or five pairs. (Pl. 39, a. 1, a. 2, a. 3, a. 4, a. 5.)

All the lobes, as they dip inwards, are subdivided by numerous dentations, which afford points of attachment to the mantle of the animal; thus each lobe is flanked by a series of accessory lobes, and these again are provided with further symmetrical dentations, the extremities of which produce all the beautiful appearances of complicated foliage, which prevail through the family of Ammonites, and of which we have a striking example on the surface of Pl. 38.

The extremities of the dentations are always sharp and pointed, inwards, towards the air chamber, (Pl. 38, d. l.); but

Conclusion.

On examining the proofs of Contrivance and Design that pervade the testaceous remains of the family of Ammonites, we find, in every species, abundant evidence of minute and peculiar mechanisms, adapting the shell to the double purpose of acting as a float, and of forming a protection to the body of its inhabitant.

As the animal increased in bulk, and advanced along the outer chamber of the shell, the spaces left behind it were successively converted into air chambers, simultaneously increasing the power of the float. This float, being regulated by a pipe, passing through the whole series of the chambers, formed an hydraulic instrument of extraordinary delicacy, by which the animal

are smooth and rounded upwards towards the body of the animal, (Pl. 38, S. S.), and thus the jagged terminations of these lobes may have afforded holdfasts whereby the base of the mantle could fix itself firmly, and as it were take root, around the bottom of the external chamber.

No such dentations exist in any species of Nautilus. In the N. Pompilius, Mr. Owen has shewn that the base of the mantle adheres to the outer shell, near its junction with the transverse plate by means of a strong horny girdle; a similar contrivance probably existed also in all the fossil species of Nautili. The sides of the mantle also of the N. Pompilius are fixed to the sides of the great external chamber by two strong broad lateral muscles, the impressions of which are visible in most specimens of this shell.

could, at pleasure, control its ascent to the surface, or descent to the bottom of the sea.

To creatures that sometimes floated, a thick and heavy shell would have been inapplicable; and as a thin shell, inclosing air, would be exposed to various, and often intense degrees of pressure at the bottom, we find a series of provisions to afford resistance to such pressure, in the mechanical construction both of the external shell, and of the internal transverse plates which formed the air chambers. First, the shell is made up of a tube, coiled round itself, and externally convex. Secondly, it is fortified by a series of ribs and vaultings disposed in the form of arches and domes on the convex surface of this tube, and still further adding to its strength. Thirdly, the transverse plates that form the air chambers, supply also a continuous succession of supports, extending their ramifications, with many mechanical advantages, beneath those portions of the shell which, being weakest, were most in need of them.

If the existence of contrivance proves the exercise of mind; and if higher degrees of perfection in mechanism are proof of more exalted degrees of intellect in the Author from whom they proceeded; the beautiful examples which we find in the petrified remains of these chambered shells, afford evidence coeval and co-extensive with the mountains wherein they are entombed, attesting the Wisdom in which such exquisite contrivances originated, and setting forth the Providence and Care of the Creator, in regulating the structure of every creature of his hand.

SECTION V.

NAUTILUS SYPHO, AND NAUTILUS ZIC ZAC.

THE name of Nautilus Sypho* has been applied to a very curious and beautiful chambered shell found in the Tertiary strata at Dax, near Bourdeaux; and that of Nautilus Zic Zac to a cognate shell from the London clay. (See Pl. 43, Figs. 1, 2, 3, 4.)

These fossil shells present certain deviations from the ordinary characters of the genus Nautilus, whereby they in some degree partake of the structure of an Ammonite.

These deviations involve a series of compensations and peculiar contrivances, in order to render the shell efficient in its double office of acting as a float, and also as a defence and chamber of residence to the animal by which it was constructed.

Some details of these contrivances, relating

---

* This shell has been variously described by the names of Ammonites Atus, Nautilus Sypho, and N. Zonarius. (See M. de Basterot. Mem. Geol. de Bourdeaux).

to the Nautilus Sypho will be found in the subjoined note.*

As the place of the siphon in this species is
upon the internal margin of the transverse plates
(Pl. 43, Fig. 2, b¹, b², b³,) it had less power than
the more central siphuncle of the Nautilus to
attach the mantle of the animal to the bottom
of the outer chamber.   For this defect we find a

---

* The transverse plates, (Pl. 43, Fig 1, a. a¹. a².), present a
peculiarity of structure in the prolongation of the *collar*, or
siphuncular aperture entirely across the area of the air chambers,
so that the whole series of transverse plates are connected in one
continuous spiral chain.   This union is effected by the enlargement and elongation of the collar for the passage of the
siphuncle into the form of a long and broad funnel, the point of
which b. fits closely into the neck of the funnel next beneath it,
c. whilst its inner margin, resting upon the arch of the subjacent
whorl of the shell, transfers to this arch a portion of the external
pressure upon the transverse plates, thereby adding to their
strength.

As this structure renders it impossible for the flexible siphuncle to expand itself into the area of the air chambers, as in other
Nautili and in Ammonites, the diameter of each funnel is made
large enough to allow space within it for the distension of the
siphuncle, by a sufficient quantity of fluid to cause the animal to
sink.

At each articulation of the funnels, the diameter of the siphuncle is contracted, as the siphuncles of Ammonites and Nautili are
contracted at their passage through the collars of their transverse
plates.

Another point in the organization of the siphuncle is illustrated
by this shell, namely, the existence of a soft calcareous sheath,
(Pl. 43, Fig. 1, b. c. d.), analogous to that of the N. Pompilius,
(Pl. 31, Fig. 1, a. b. c. d.), between each shelly funnel and the
membranous pipe or siphuncle enclosed within it.   At Pl. 43,
Fig. 1, b, we have a section of this sheath folding round the

compensation, resembling that which Von Buch
considers to have been afforded by the lobes of
Ammonites to the inhabitants of those shells.
This compensation will be illustrated by a comparison of the lobes in N. Sypho (Pl. 43, Fig. 2.),
with a similar provision in the Nautilus Zic Zac
(Pl. 43, Figs. 3. 4.)*

smaller extremity of the funnel a'. From c, to d, it lines the
inside of the subjacent funnel a^2; and from d, continues downwards to the termination of the funnel a^c, on the inside of e.
At e, and f, we see the upper termination of two perfect sheaths,
similar to that of which a section is represented at b. c. d. This
sheath, from its insertion between the point of the upper siphon
and mouth of the lower one, (Fig. 1, c.), must have acted as
a collar, intercepting all communication between the interior
of the shelly siphuncular tube and the air chambers. The area
of this shelly tube is sufficient, not only to have contained the
distended siphuncle, but also to allow it to be surrounded with a
volume of air, the elasticity of which would act in forcing back
the pericardial fluid from the siphuncle, in the same manner as
we have supposed the air to act within the chambers of the
N. Pompilius.

    * On each side of the transverse plate in both these species
there is an undulation, or sinus, producing lobes (Pl. 43, Fig. 2.
a^1, a^2, a^3, Fig. 3. a. and Fig. 4. a. b.) There is also a deep
backward curvature of the two ventral lobes, Fig. 4. c. c. All
these lobes may have acted conjointly with the siphuncle, to
give firm attachment to the mantle of the animal at the bottom
of the outer chamber. The shell Fig. 1. is broken in such a
manner, that no portion of any lateral lobe is visible on the side
here represented. At Fig. 2. a^1, we see the projection of the
lateral lobes, on each side of the convex internal surface of a
transverse plate; at a^2 we see the interior of the same lobes, on
the concave side of another transverse plate; and at a^3 the
points of a third pair of lobes attached to the sides of the largest
air-chamber that remains in this fragment.

A still more important use of the lobes formed
by the transverse plates both of N. Sypho and
N. Zic Zac, may be found in the strength which
they impart to the sides of the external shell (see
Pl. 43, Figs. 1, 2, 3, 4.), underpropping their flattest
and weakest part, so as to resist pressure more
effectually than if the transverse plates had been
curved simply, as in N. Pompilius.   One cause
which rendered some such compensation necessary, may be found in the breadth of the intervals between each transverse plate ; the weakness
resulting from this distance, being compensated
by the introduction of a single lobe, acting on
the same principle as the more numerous and
complex lobes in the genus Ammonite.

The N. Sypho and N. Zic Zac seem, therefore,
to form *Links* between the two great genera of
Nautilus and Ammonite, in which an intermediate
system of mechanical contrivances is borrowed,
as it were, from the mechanics of the Ammonite,
and applied to the Nautilus. The adoption of
lobes, analogous to the lobes of the Ammonite,
compensating the disadvantages, that would
otherwise have followed from the marginal position of the siphuncle in these two species, and
the distances of their transverse plates.*

---

* In some of the most early forms of Ammonites which we find
in the Transition strata, e. g. A. Henslowi, A. Striatus, and A.
Sphericus, Pl. 40, Figs. 1, 2, and 3,) the lobes were few, and
nearly of the same form as the single lobe of the Nautilus Sypho,

It is a curious fact, that contrivances, similar to those which existed in some of the most early forms of Ammonite, should have been again adopted in some of the most recent species of fossil Nautili, in order to afford similar compensation for weakness that would otherwise have been produced by aberrations from the normal structure of the genus Nautilus. All this seems inexplicable on any theory which would exclude the interference of controlling Intelligence.

SECTION VI.

CHAMBERED SHELLS ALLIED TO NAUTILUS AND AMMONITE.

WE have reason to infer, from the fact of the recent N. Pompilius being an external shell, that all fossil shells of the great and ancient family of Nautili, and of the still more numerous family of Ammonites, were also external shells, inclosing in their outer chamber the body

and of N. Zic zac; like them also the margin was simple and destitute of fringed edges. The A. nodosus (Pl. 40, Figs. 4 and 5.), which is peculiar to the early Secondary deposits of the Muschel-kalk, offers an example of an intermediate state, in which the fringed edge is partially introduced, on the descending or inward portions only, of the lobated edge of the transverse plates.

of a Cephalopod. We further learn, from Peron's discovery of the shell of a Spirula partially enclosed within the body of a Sepia,* (see Pl. 44, Fig. 1, 2), that many of those genera of fossil chambered shells, which, like the Spirula, do not terminate externally in a wide chamber, were probably internal, or partially enclosed shells, serving the office of a float, constructed on the same principles as the float of the Spirula. In the class of fossil shells thus illustrated by the discovery of the animal inclosing the Spirula, we may include the following extinct families, occurring in various positions from the earliest Transition strata to the most recent Secondary formations :—Orthoceratite, Lituite, Baculite, Hamite, Scaphite, Turrilite, Nummulite, Belemnite.†

* The uncertainty which has arisen respecting the animal which constructs the Spirula, from the circumstance of the specimen discovered by Peron having been lost, is in some degree removed by Captain King's discovery of another of these shells, attached to a fragment of the mantle of an animal of unknown species resembling a Sepia, which I have seen in the possession of Mr. Owen, at the Royal College of Surgeons, London.

† In the genus Lituite, Orthoceratite, and Belemnite, Pl. 44, *f.* 3, 4, 17, the simple curvature of the transverse plates resembles the character of the Nautilus. In the Baculite, Hamite, Scaphite, and Turrilite, Pl. 44, Fig. 5, 8, 12, 13, 14, 15, the sinuous foldings and foliated edges of the transverse plates resemble those of the Ammonites.

Orthoceratite, Pl. 44, Fig. 4.

The Orthoceratites (so called from their usual form,—that of a straight horn) began their existence at the same early period with the Nautili, in the seas which deposited the Transition strata; and are so nearly allied to them in structure, that we may conclude they performed a similar function as floats of Cephalopodous Mollusks. This genus contains many species, which abound in the strata of the Transition series, and is one of those which, having been called into existence amongst the earliest inhabitants of our planet, was at an early period also consigned to almost total destruction.*

An Orthoceratite (see Pl. 44, Fig 4) is, like the Nautilus, a multilocular shell, having its chambers separated by transverse plates, concave externally, and internally convex; and pierced, either at the centre or towards the margin, by a Siphuncle, (a.) This pipe varies in size, more

---

* See D'Orbigny's Tableau Méthodique des Céphalopodes.

There are, I believe, only two exceptions yet known to the general fact, that the genus Orthoceratite became extinct before the deposition of the Secondary strata had commenced. The most recent rocks in which they have been noticed, are a small and problematical species in the Lias at Lyme, and another species in Alpine Limestone of the Oolite formation, at Halstadt, in the Tyrol.

than that of any other multilocular shell, viz.
from one-tenth to one-half of the diameter of the
shell; and often assumes a tumid form, which
would admit of the distension of a membranous
siphon. The base of the shell beyond the last
plate presents a swelling cavity, wherein the
body of the animal seems to have been partly
contained.

The Orthoceratites are straight and conical,
and bear the same relation to the Nautili which
Baculites (see Pl. 44, Fig. 5) bear to Ammonites; the Orthoceratites, with their simple
transverse septa, resembling straight Nautili;
and the Baculites, with a sinuous septa, having
the appearance of straight Ammonites. They
vary considerably in external figure, and also in
size; some of the largest species exceeding a
yard in length, and half a foot in diameter. A
single specimen has been known to contain
nearly seventy air chambers. The body of the
animal which required so large a float to balance
it, must have greatly exceeded, in all its proportions, the most gigantic of our recent Cephalopods; and the vast number of Orthoceratites
that are occasionally crowded together in a single
block of stone, shows how abundantly they must
have swarmed in the waters of the early seas.
These shells are found in the greatest numbers
in blocks of marble, of a dark red colour, from
the transition Limestone of Oeland, which some

years ago was imported largely to various parts of Europe for architectural purposes.*

Lituite.

Together with the Orthoceratite, in the Transition Limestone of Oeland, there occurs a cognate genus of Chambered shells, called Lituites. (Pl. 44, Fig. 3.) These are partially coiled up into a spiral form at their smaller extremity, whilst their larger end is continued into a straight tube, of considerable length, separated by transverse plates, concave outwards, and perforated by a siphuncle (a). As these Lituites closely resemble the shell of the recent Spirula (Pl. 44, Fig. 2), their office may have been the same, in the economy of some extinct Cephalopod.

Baculite.

As in rocks of the Transition series, the form of a straight Nautilus is presented by the genus

* Part of the pavement in Hampton Court Palace, that of the hall of University College, Oxford, and several tombs of the kings of Poland in the cathedral at Cracow, are formed of this marble, in which many shells of Orthoceratites are discoverable. The largest known species are found in the Carboniferous limestone of Closeburn, near Edinburgh, being nearly of the size of a man's thigh. The presence of such gigantic Mollusks seems to indicate a highly exalted temperature, in the then existing climate of these northern regions of Europe. See Sowerby's Min. Con. Pl. 246.

Orthoceratite, so we find in the Cretaceous formation alone, the remains of a genus which may be considered as a straight Ammonite. (See Pl. 44, Fig. 5.)

The Baculite (so called from its resemblance to a straight staff) is a conical, elongated, and symmetrical shell, depressed laterally, and divided into numerous chambers by transverse plates, which, like those in the Ammonite, are sinuous, and terminated by foliated dentations at their junction with the external shell; being thus separated into dorsal, ventral, and lateral lobes and saddles, analogous to those of Ammonites.*

It is curious, that this straight modification of the form of Ammonites should not have appeared, until this Family had arrived at the last stage of the Secondary deposits, throughout which it had occupied so large an extent; and that, after a comparatively short duration, the Baculite should have become extinct, simultaneously with the last of the Ammonites, at the termination of the Chalk formation.

---

* The external chamber (a) is larger than the rest, and swelling; and capable of containing a considerable portion of the animal. The outer shell was thin, and strengthened, like the Ammonite, by oblique ribs. Near the posterior margin of the shell, the transverse plates are pierced by a Siphuncle (Pl. 44, 5^b, c,). This position of the Siphuncle, and the sinuous form and denticulated edges of the transverse plates, are characters which the Baculite possesses in common with the Ammonite.

Hamite.

If we imagine a Baculite to be bent round near its centre, until the smaller extremity became nearly parallel to its larger end, it would present the most simple form of that cognate genus of chambered shells, which, from their frequently assuming this hooked form, have been called Hamites. At Pl. 44, Fig. 9, 11, represent portions of Hamites which have this most simple curvature; other species of this genus have a more tortuous form, and are either closely coiled up, like the small extremity of a Spirula, (Pl. 44, Fig. 2,) or disposed in a more open spiral. (Pl. 44, Fig. 8.)*

It is probable that some of these Hamites

---

* Both these forms of Hamite bear the same relation to Ammonites that Lituites bear to Nautili; each being nearly such as shells of these genera would respectively present, if partially unrolled. See Phillips' Geol. Yorkshire, Pl. 1, Figs. 22, 29, 30.

Baculites and Hamites have two characters which connect them with Ammonites; first, the position of the Siphuncle, on the back, or outer margin of the shell, (Pl. 44, Figs. 5[b], c. 8[a], a. 10. 11, a. 12, a. 13, a.); secondly, the foliated character of the margin of the transverse plates, at their junction with the external shell. (Pl. 44, Fig. 5, 8, 12, 13.) The external shell of Hamites is also fortified by transverse folds or ribs, increasing the strength both of the outer chambers and of the air chambers, upon the same principles that we have pointed out in the case of Ammonites. (See Pl. 44, Fig. 8, 9, 11, 12, 13.)

In certain species of Hamites, as in certain Ammonites, the marginal Siphuncle has a keel-shaped pipe raised over it. Others have a series of spines on each side of the back. (Pl. 44, Fig. 9, 10.)

were partly internal, and partly external shells;
where the spines are present, the portion so armed
was probably external. Nine species of Hamites
occur in the single formation of Gault or Speeton
clay immediately below the chalk, near Scarborough. (See Phillip's Geology of Yorkshire.)
Some of the larger species equal a man's wrist
in diameter.*

Scaphite.

The Scaphites constitute a genus of Elliptical
chambered shells, (see Pl. 44, Fig. 15, 16,) of
remarkable beauty, which are almost peculiar
to the Chalk formation ; they are so rolled up at
each extremity, whilst their central part continues nearly in a horizontal plane, as to resemble
the ancient form of a boat; whence the name of
Scaphite has been applied to them.†

---

* The Hamites grandis, (Sowerby, M. C. 593,) from the Green
sand at Hythe, is of these large dimensions.

† The inner extremity of the Scaphite is coiled up like that of
an Ammonite, (Pl. 44, Fig. 15, c. and Fig. 16) in whorls embracing
one another ; the last and outer chamber (a) is larger than all
the rest together, and is sometimes (probably in the adult state)
folded back so as to touch the spire, and thereby materially to
contract the mouth, which is narrower than the last or outer
chamber. (Pl. 44, Fig. 15, b.) In this character of the external
chamber, the Scaphite differs from the Ammonite; in all other
respects it essentially agrees with it; its transverse plates being
numerous, and pierced by a marginal Siphuncle, at the back
of the shell (Fig.16, a.); and their edges being lobated, deeply
cut, and foliated. (Fig. 15, c.)

It is remarkable that those approximations to the structure of Ammonites which are presented by Scaphites and Hamites, should have appeared but very rarely, and this in the lias and inferior oolite,* until the period of the cretaceous formations, when the entire type of the ancient and long continued genus Ammonite was about to become extinct.

Turrilite.

The last genus I shall mention, allied to the family of Ammonites, is composed of spiral shells, of another form, coiled around themselves in the form of a winding tower, gradually diminishing towards the apex (Pl. 44, Fig. 14).†

The same essential characters and functions pervade the Turrilites, which we have been tracing in the Scaphites, Hamites, Baculites, and Ammonites. In each of these genera it is the exterior form of the shell that is principally

---

* The Scaphites bifurcatus occurs in the Lias of Wurtemburg, and Hamites annulatus in the Inferior oolite of France.

† The shells of the Turrilites are extremely thin, and their exterior is adorned and strengthened (like that of Ammonites), with ribs and tubercles. In all other respects also, except the manner in which they are coiled up, they resemble Ammonites; their interior being divided into numerous chambers by transverse plates, which are foliated at their edges, and pierced by a siphuncle, near the dorsal margin. (Pl. 44, Fig. 14, a, a.) The outer chamber is large.

G.   B B

varied, whilst the interior is similarly constructed
in all of them, to act as a float, subservient to
the movements of Cephalopodous Mollusks. We
have seen that the Ammonites, beginning with
the Transition strata, appear in all formations,
until the termination of the Chalk, whilst the
Hamites and Scaphites are very rare, and the
Turrilites and Baculites do not appear at all,
until the commencement of the Cretaceous formations. Having thus suddenly appeared, they
became as suddenly extinct at the same period
with the Ammonites, yielding up their place and
office in the economy of nature to a lower order
of Carnivorous mollusks in the Tertiary and existing seas.

In the review we have taken of genera in the
family of Chambered shells, allied to Nautilus,
and Ammonite, we have traced a connected series
of delicate and nicely adjusted instruments,
adapted to peculiar uses in the economy of
every animal to which they were attached.
These all attest undeviating Unity of design,
pervading many varied adaptations of the same
principle; and afford cumulative evidence, not
only of the exercise of Intelligence, but also of
the *same* Intelligence through every period of
time, in which these extinct races inhabited the
ancient deep.

SECTION VII.

Belemnite.

WE shall conclude our account of chambered shells with a brief notice of Belemnites. This extensive family occurs only in a fossil state, and its range is included within that series of rocks which in our section are called Secondary.* These singular bodies are connected with the other families of fossil chambered shells we have already considered ; but differ from them in having their chambers inclosed within a cone-shaped fibrous sheath, the form of which resembles the point of an arrow, and has given origin to the name they bear.

M. de Blainville, in his valuable memoir on Belemnites, (1827) has given a list of ninety-one authors, from Theophrastus downwards, who have written on this subject. The most intelligent among them agree in supposing these bodies to have been formed by Cephalopods allied to the modern Sepia. Voltz, Zeiten, Raspail, and Count Münster, have subsequently published important memoirs upon the same subject. The principal English notices on Belemnites are those of Miller, Geol. Trans. N. S.

* The lowest stratum in which Belemnites are said to have been found is the Muschel-kalk, and the highest the upper Chalk of Maestricht.

London, 1826, and that of Sowerby, in his Min. Conch. vol. vi. p. 169, et seq.

A Belemnite was a compound internal shell, made up of three essential parts, which are rarely found together in perfect preservation.

First, a fibro-calcareous cone-shaped shell, terminating at its larger end in a hollow cone (Pl. 44, Fig. 17. and Pl. 44′, Fig. 7, 9, 10, 11, 12.*

Secondly, a conical thin horny sheath, or cup, commencing from the base of the hollow cone of the fibro-calcareous sheath, and enlarging rapidly as it extends outwards to a considerable

---

* This part of the Belemnite is usually called the *sheath*, or *guard*: it is made up of a pile of cones, placed one within another, having a common axis, and the largest enclosing all the rest. (See Pl. 44, Fig. 17.) These cones are composed of crystalline carbonate of lime, disposed in fibres that radiate from an eccentric axis to the circumference of the Belemnite. The crystalline condition of this shell seems to result from calcareous infiltrations (subsequent to interment), into the intervals between the radiating calcareous fibres of which it was originally composed. The idea that the Belemnite was a heavy solid stony body, whilst it formed part of a living and floating sepia, would be contrary to all analogies afforded by the internal organs of living Cephalopods. The odour, resembling burnt horn, produced on burning this part of a Belemnite, arises from the remains of horny membranes interposed between each successive fibro-calcareous cone.

An argument in favour of the opinion that Belemnites were internal organs, arises from the fact of their surface being often covered with vascular impressions, derived from the mantle in which it was inclosed. In some species of Belemnites the back is granulated, like the back of the internal shell of Sepia officinalis.

distance. Pl. 44', Fig. 7, b, e, e', e". This horny cup formed the anterior chamber of the Belemnite, and contained the ink bag, (c), and some other viscera.*

Thirdly, a thin conical internal chambered shell, called the *Alveolus*, placed within the calcareous hollow cone above described. (Pl. 44, Fig. 17, a. and Pl. 44', Fig. 7, b, b'.)

This chambered portion of the shell is closely allied in form, and in the principles of its construction, both to the Nautilus and Orthoceratite. (See Pl. 44, Fig. 17, a, b. and Fig. 4.) It is divided by thin transverse plates into a series of narrow air-chambers, or *areolæ*, resembling a pile of watch-glasses, gradually diminishing towards the apex. The transverse plates are outwardly concave, inwardly convex; and are perforated by a continuous siphuncle, (Pl. 44, Fig. 17, b.), placed on the inferior, or ventral margin.

We have already (Ch. XV. Section II.) described the horny pens and ink-bags of the Loligo, found in the Lias at Lyme Regis. Similar ink-bags have recently been found in connection with Belemnites in the same Lias. Some of these ink-bags are nearly a foot in length, and show

---

* This laminated horny sheath is rarely preserved in connection with the fibro-calcareous shelly sheath; but in the Lias at Lyme Regis it is frequently found without the shell. Certain portions of it are often highly nacreous, whilst other parts of the same sheath retain their horny condition.

that the *Belemno-sepiæ*,† from which they were
derived, attained great size.

The fact of these animals having been provided with so large a reservoir of ink, affords an
à priori probability that they had no external
shell; the ink-bag, as far as we yet know, being
a provision confined to naked Cephalopods,

---

† In 1829, I communicated to the Geological Society of London a notice respecting the *probable* connection of Belemnites
with certain fossil ink-bags, surrounded by brilliant nacre, found
in the Lias at Lyme Regis. (See Phil. Mag. N. S. 1829, p. 388.)
At the same time I caused to be prepared the drawings of fossils,
engraved in Pl. 44″, which induced me to consider these ink-
bags as derived from Cephalopods connected with Belemnites.
I then withheld their publication, in the hope of discovering
certain demonstration, in some specimen that should present
these ink-bags in connection with the sheath or body of a Belemnite, and this demonstration has at length been furnished by
a discovery made by Professor Agassiz (October, 1834), in the
cabinet of Miss Philpotts, at Lyme Regis, of two important
specimens, which appear to be decisive of the question. (See
Pl. 44′, Figs. 7, 9.)

Each of these specimens contains an ink-bag within the anterior portion of the sheath of a perfect Belemnite; and we are
henceforth enabled with certainty to refer all species of Belemnites to a family in the class of Cephalopods, for which I would,
in concurrence with M. Agassiz, propose the name of *Belemno-
sepia.* Such ink-bags are occasionally found in contact with
traces of isolated alveoli of Belemnites: they are more frequently
surrounded only by a thin plate of brilliant nacre.

The specimen (Pl. 44″, Fig. 1), was procured by me from
Miss Mary Anning in 1829, who considered it as appertaining to
a Belemnite. Near its lower end we see the lines of growth of
the horny anterior sheath, but no traces of the posterior calcareous sheath; within this horny sheath is placed the ink-bag.
The conical form of this anterior chamber seems to have been

which have not that protection from an external
shell, which is afforded by the shell of the N.
Pompilius to its inhabitant, that has no ink-bag.
No ink, or ink-bags have been ever seen within
the shell of any fossil Nautilus or Ammonite:
had such a substance existed in the body of
the animals that occupied their outer chamber,

altered by pressure. It is composed of a thin laminated substance (see Pl. 44″, Fig. 1, d.), which in some parts is brilliantly
nacreous, whilst in other parts it presents simply the appearance of horn. The outer surface of this cup is marked transversely with gentle undulations, which probably indicate stages
of growth. Miss Baker has a Belemnite from the inferior Oolite
near Northampton, in which one half of the fibrous cup being
removed, the structure of the conical shell of the alveolus is seen
impressed on a cast of iron-stone, and exhibits undulating lines
of growth, like those on the exterior of the shell of N. Pompilius.

M. Blainville, although he had not seen a specimen of Belemnite in which the anterior horny conical chamber is preserved, has
argued from the analogy of other cognate chambered shells that
such an appendage was appertinent to this shell. The soundness of his reasoning is confirmed by the discovery of the specimen before us, containing this part in the form and place which
he had predicted. "Par analogie elle était donc évidemment
dorsale et terminale, et lorsqu'elle était complète, c'est-à-dire
pourvue d'une cavité, l'extremité postérieure des viscères de
l'animal (très-probablement l'organe sécréteur de la génération
et partie du foie) y était renfermée."—Blainville Mém. sur les
Bèlemnites. 1827. Page 28.

Count Munster (Mem. Geol. par A. Boue, 1832, V. 1, Pl. 4,
Figs. 1, 2, 3, 15) has published figures of very perfect Belemnites
from Solenhofen, in some of which the anterior horny sheath is
preserved, to a distance equal to the length of the solid calcareous
portion of the Belemnite (Pl. 44′, Figs. 10, 11, 12, 13), but in
neither of these are there any traces of an ink-bag.

some traces of it must have remained in those beds of lias at Lyme Regis, which are loaded with Nautili and Ammonites, and have preserved the ink of naked Cephalopods in so perfect a condition. The young Sepia officinalis, whilst included within the transparent egg, exhibits its ink-bag distended with ink, provided beforehand for use as soon as it is excluded; and this ink-bag is surrounded by a covering of brilliant nacreous matter, similar to that we find on certain internal membranes of many fishes.*

* I would here add a few words in explanation of the curious fact, that among the innumerable specimens of Belemnites which have so long attracted the attention of naturalists, not one has till now been found entire in all its parts, having the ink within its external chamber; either the fibro-calcareous sheath is found detached from the horny sheath and ink-bag, or the ink-bag is found apart from the Belemnite, and surrounded only by the nacreous horny membrane of its anterior chamber. We know from the condition of the compressed nacreous Ammonites in the Lias-shale at Watchet, that the nacreous lining *only* of these shells is here preserved, whilst the shell itself has perished. This fact seems to explain the absence of the calcareous sheath and shell in almost every specimen of ink-bags at Lyme Regis, which is surrounded with iridescent nacre, like that of the Ammonites of Watchet. The matrix in these cases may have had a capacity for preserving nacreous or horny substances, whilst it allowed the more soluble calcareous matter of shells to be removed, probably dissolved in some acid.

The greater difficulty is to explain the reason, why amidst the millions of Belemnites that are dispersed indiscriminately through almost all strata of the Secondary series, and sometimes form entire pavements in beds of shale connected with the Lias and Inferior oolite, it so rarely happens that either the horny sheath, or the ink-bag, have been preserved. We may, I think, explain

Comparing the shell of Belemnite, with that of Nautilus, we find the agreement of all their

the absence of the nacreous horny sheath, by supposing that a condition of the matrix favourable to the preservation of the calcareous sheath was unfavourable to the preservation of horny membrane; and we may also explain the absence of ink-bags, by supposing that the decomposition of the soft parts of the animal usually caused the ink to be dispersed, before the body was buried in the earthy sediment then going on.

At the base of Golden Cap hill, near Charmouth, the shore presents two strata of marl almost paved with Belemnites, and separated by about three feet only of comparatively barren marl. As great numbers of these Belemnites have Serpulæ, and other extraneous shells attached to them, we learn from this circumstance that the bodies and ink-bags had decomposed, and the Belemnites lain some time uncovered at the bottom. These facts are explained by supposing that the sea near this spot was much frequented by Belemno-sepiæ during the intervals of the deposition of the Lias. Similar conclusions follow, from the state of many Belemnites in the chalk of Antrim, which had been perforated by small boring animals, whilst they lay at the bottom of the sea, and these perforations filled with casts of chalk or flint, when the matter of the chalk strata was deposited upon them, in a soft and fluid state. (See Allan's Paper on Belemnite, Trans. Royal Soc. Edin., and Miller's Paper, Geol. Trans. Lond. 1826, p. 53.)

Thus of the millions of Belemnites which crowd the Secondary formations, only the fibro-calcareous sheath and chambered alveoli are usually preserved; whilst in certain shale beds this sheath and shell have sometimes entirely disappeared, and the horny or nacreous sheath and ink-bag alone remain. See Pl. 44″, Fig. 1, 2, 3, 4, 5, 6, 7, 8. In the rare case, Pl. 44′, Fig. 7, which has afforded the clue to this hitherto unexplained enigma, we have all the three essential parts of a Belemnite preserved in their respective places nearly entire. The ink-bag (c) is placed within the anterior horny cup (e, e′, e″); and the chambered alveolus (b b′) within the hollow cone of the posterior fibro-calcareous shell, or common Belemnite.

most important parts to be nearly complete ;*
and the same analogies might be traced through
the other genera of chambered shells.†

* The air chambers and siphuncle are, in both these families,
essentially the same.

In Belemnites, the anterior extremity of the fibro-calcareous
shell, which forms a hollow *straight* cone, surrounding the
transverse plates of the chambered alveolus, represents the hollow *coiled-up* cone containing all the transverse plates, which
make up the alveolus of the Nautilus.

The anterior horny cup, or outer chamber of the Belemnite,
surrounding the ink bag, and other viscera, represents the large
anterior shelly chamber which contains the body of the Nautilus.

The posterior portion of the Belemnite, which is elongated
backwards into a fibrous pointed shaft, is a modification of the
apex of the straight cone of this shell, to which there seems to
be no equivalent in the apex of the coiled-up cone of Nautilus.
The cause of this peculiar addition to the ordinary parts of
shells, seems to rest in the peculiar uses of the shaft of the Belemnite, as an internal shell, acting like the internal shell of the
Sepia Officinalis, to support the soft parts of the animals, within
the bodies of which they were respectively enclosed. The fibrous
structure of this shaft is such as is common to many shells, and
is most obvious in the Pinnæ.

† Comparing the Belemnite, or internal shell of Belemno-
sepia with the Sepiostaire, (*Blainville*), or internal shell of the
Sepia Officinalis, we have the following analogies. In the
Sepiostaire, (Pl. 44′, Fig. 2, a. e. and Figs. 4, 4′, 5), the small
conical apex (a) represents the apex of the long calcareous posterior sheath of the Belemnite, (Fig. 7, a.), and the calcareous
plates, alternating with horny plates, which form the shield and
shallow cup of the Sepiostaire, (Pl. 44′, Fig. 2, e. and Fig. 5. e.),
represent the hollow fibro-calcareous cone or cup of the Belemnite, surrounding its alveolus.

The margin of the horny plates, interposed between the calcareous plates of the shield and cup of the Sepiostaire, (Pl. 44′,
Fig. 4, e, e, e′, e′.), represents the horny marginal cavity of the

Eighty-eight species of Belemnites have already been discovered ;* and the vast numerical amount to which individuals of these species were extended, is proved by the myriads of their fossil remains that fill the Oolitic and Cretaceous formations. When we recollect that throughout both these great formations, the still more numerous extinct family of Ammonites is co-extensive with the Belemnites; and that each species of Ammonite exhibits also contrivances, more complex and perfect than those retained in the few

cone of the Belemnite, beyond the base of its hollow calcareous cone, (Pl. 44′, Fig. 7, e. e′. e″). This horny sheath of the Belemnite was probably formed by the prolongation of the horny laminæ which were interposed between its successive cones of fibro-calcerous matter.

The chambered alveolus of the Belemnite is represented by the congeries of thin transverse plates, (Pl. 44, Fig. 4, b.) which occupy the interior of the shallow cup of Sepiostaire, (e. e′.); these plates are composed of horny matter, penetrated with carbonate of lime.

The hollow spaces between them, (Fig. 5, b, b′,), which are nearly a hundred in number in the full grown animal, act as air chambers to make the entire shell permanently lighter than water ; but there is no siphuncle to vary the specific gravity of this shell; and the thin chambers between its transverse plates are studded with an infinity of minute columnar, and sinuous partitions, planted at right angles to the plates, and giving them support. (Fig. 6′, 6″, 6‴).

The absence of a siphuncle renders the Sepiostaire an organ of more simple structure, and of lower office, than the more compound shell of Belemnite.

* (See index to M. Brochant de Villiers' Translation of De la Beche's Manual of Geology).

existing cognate genera of Cephalopods; we cannot but infer that these extinct families filled a larger space, and performed more important functions among the inhabitants of the ancient seas, than are assigned to their few living representatives in our modern oceans.

Conclusion.

It results from the view we have taken of the zoological affinities between living and extinct species of chambered shells, that they are all connected by one plan of organization; each forming a link in the common chain, which unites existing species with those that prevailed among the earliest conditions of life upon our globe; and all attesting the Identity of the design, that has effected so many similar ends through such a variety of instruments, the principle of whose construction is, in every species, fundamentally the same.

Throughout the various living and extinct genera of Chambered shells, the use of the air chambers and siphon, to adjust the specific gravity of the animals in rising and sinking, appears to have been identical. The addition of a new transverse plate within the conical shell added a new air chamber, larger than the preceding one, to counterbalance the increase of weight that attended the growth of the shell and body of all these animals.

These beautiful arrangements are, and ever have been, subservient to a common object, viz. the construction of *hydraulic instruments* of essential importance in the economy of creatures destined to move sometimes at the bottom, and at other times upon or near the surface of the sea. The delicate adjustments whereby the same principle is extended through so many grades and modifications of a single type, show the uniform and constant agency of some controlling Intelligence ; and in searching for the origin of so much method and regularity amidst variety, the mind can only rest, when it has passed back, through the subordinate series of Second causes, to that great First Cause, which is found in the will and power of a common Creator.

SECTION VIII.

FORAMINATED POLYTHALAMOUS SHELLS.

Nummulites.

If the present were a fit occasion for such minute inquiries, the investigations of the various known species of Microscopic shells would unfold a series of contrivances having relation to the economy of the minute Cephalopods by which they were constructed, not less striking than those we have been examining in the shells of extinct

Genera and species of larger Cephalopods. M. D'Orbigny has noticed from 600 to 700 species of these shells, and has prepared magnified models of 100 species, comprehending all the Genera.[*]

The greater number of these shells are microscopic, and swarm in the Mediterranean and Adriatic. Their fossil species abound chiefly in the Tertiary formations, and have hitherto been noticed principally in Italy. (See Soldani, as quoted at page 117 of this volume.) They occur also in the Chalk of Meudon, in the Jura Limestone of the Charente inferieure, and the Oolite of Calne. They have been found by the Marquis of Northampton in Chalk flints from the neighbourhood of Brighton.

[*] M. D'Orbigny, in his Classification of the shells of Cephalopodous Mollusks, has established three orders. 1. Those that have but a single chamber, like the shell of the sepia and horny pen of the Loligo. 2. Polythalamous shells, which have a siphuncle passing through all the internal chambers, and which terminate in a large external chamber, beyond the last partition, such as Nautili, Ammonites, and Belemnites. 3. Polythalamous internal shells, which have no chamber beyond their last partition.

Shells of this Order have no siphuncle, but the chambers communicate with each other by means of one or many small foramina. On this distinction he has founded his Order *Foraminiferes*, containing five families and fifty-two genera.

It may be necessary to apprize the reader that doubts have been entertained as to the cephalopodous structure of some of these minute multilocular shells; and that there are not wanting those who attribute to them a different organization.

The Nummulite is the only Genus I shall select on the present occasion from this Order. It is included in M. D'Orbigny's Section Nautiloids.

Nummulites (Pl. 44, Fig. 6, 7,) are so called from their resemblance to a piece of money, they vary in size from that of a crown piece to microscopic littleness; and occupy an important place in the history of fossil shells, on account of the prodigious extent to which they are accumulated in the later members of the Secondary, and in many of the Tertiary strata. They are often piled on each other nearly in as close contact as the grains in a heap of corn. In this state they form a considerable portion of the entire bulk of many extensive mountains, e. g. in the Tertiary limestones of Verona and Monte Bolca, and in secondary strata of the Cretaceous formation in the Alps, Carpathians, and Pyrenees. Some of the pyramids, and the Sphinx, of Egypt are composed of limestone loaded with Nummulites.

It is impossible to see such mountain-masses of the remains of a single family of shells thus added to the solid materials of the globe, without recollecting that each individual shell once held an important place within the body of a living animal; and thus recalling our imagination to those distant epochs when the waters of the ocean which then covered Europe were filled with floating swarms of these extinct Mollusks, thick as the countless myriads of Beröe and Clio Borealis that now crowd the waters of the polar seas.*

The Nummulites, like the Nautilus and Ammonite, are divided into air chambers, which served the office of a float; but there is no enlargement of the last chamber which could have contained any part of the body of the animal. The chambers are very numerous, and minutely divided by transverse plates; but are without a

* We have an analogy to this supposed state of crowded population of Nummulites in the ancient sea, in the marvellous fecundity of the northern ocean at the present time. It is stated by Cuvier, in his memoir on the Clio Borealis, that in calm weather, the surface of the water in these seas swarms with such millions of these mollusks (rising for a moment to the air at the surface, and again instantly sinking towards the bottom), that the whales can scarce open their enormous mouths without gulping in thousands of these little gelatinous creatures, an inch long, which, together with Medusæ, and some smaller animals, constitute the chief articles of their food; and we have a farther analogy in the fact mentioned in Jameson's Journal, vol. ii. p. 12. "That the number of small Medusæ in some parts of the Greenland seas is so great, that in a cubic inch, taken up at random, there are no less than 64. In a cubic foot this will amount to 110,592; and in a cubic mile (and there can be no doubt of the water being charged with them to that extent), the number is such, that allowing one person to count a million in a week, it would have required 80,000 persons, from the creation of the world, to complete the enumeration."—See Dr. Kidd's admirable Introductory Lecture to a course of Comparative Anatomy, Oxford, 1824, p. 35.

siphuncle.* The form of the essential parts varies in each species of this genus, but their principles of construction, and manner of operation, appear in all to have been the same.

The remains of Nummulites are not the only animal bodies which have contributed to form the calcareous strata of the crust of the earth; other, and more minute species of Chambered shells have also produced great, and most surprising effects. Lamarck (Note, v. 7. p. 611), speaking of the *Miliola*, a small multilocular shell, no larger than a millet seed, with which the strata of many quarries in the neighbourhood of Paris are largely interspersed, notices the important influence which these minute bodies have exercised by reason of their numerical abundance. We scarcely condescend, says he, to examine microscopic shells, from their insignificant size; but we cease to think them insignificant, when we reflect that it is by means of the smallest objects, that Nature every where produces her most remarkable and astonishing phenomena. Whatever she may seem to lose in point of volume in the production of living bodies, is amply made up by the number of the individuals, which she multiplies with admirable

---

* In Pl. 44, Figs. 6, 7, sections of two species of Nummulite are copied from Parkinson. These show the manner in which the whorls are coiled up round each other, and divided by oblique septa.

G. C C

promptitude to infinity. The remains of such minute animals have added much more to the mass of materials which compose the exterior crust of the globe, than the bones of Elephants, Hippopotami, and Whales.

Chapter XVI.

Proofs of Design in the Structure of Fossil Articulated Animals.

THE third grand division in Cuvier's arrangement of the animal kingdom, viz. the articulated animals, comprehends four classes.

1. The Annelidans, or worms with red blood.

2. Crustaceans, most familiar to us under the form of Crabs and Lobsters.

3. Arachnidans, or Spiders.

4. Insects.

SECTION I.

First Class of Articulated Animals.

FOSSIL ANNELIDANS.

HOWEVER numerous may have been the ancient species of Annelidans without a shelly covering, naked worms of this class can have left but slight traces of their existence, except the holes they perforated, and the little accumulations of sand or mud cast up at the orifice of these perforations; in a preceding chapter* we have noticed examples of this kind. We have also abundant evidence of the early and continued prevalence of that order of Annelidans, which formed shelly calcareous tubes, in the occurrence of fossil Serpulæ in nearly all formations, from the Transition periods to the present time.

SECTION II.

Second Class of Articulated Animals.

FOSSIL CRUSTACEANS.

THE history of fossil Crustaceans has been hitherto almost untouched by Palæontologists, and their relations to the existing Genera of this great Class of the Animal Kingdom are too little known to admit of discussion in this place. We may judge of their extent in certain Formations, from the fact, that in the cabinet of Count Munster, there are nearly sixty species collected from a single stratum of the Jurassic Limestone of Solenhofen.

* See note at page 260.

A rich harvest, therefore, remains in store for the Naturalist who will trace this interesting subject through the entire series of Geological formations.

The analogies between existing species, and certain fossil remains of Crustaceans have been beautifully illustrated by the investigations of M. Desmarest. From him we learn, that all the inequalities of the external shell in the living species have constant relation to distinct compartments in their internal organization. By the application of these distinctions to fossil species, he has pointed out a method of comparing them with living Crustaceans in a new and unexpected manner, and has established satisfactory analogies between the extinct and existing members of this very numerous Class, in cases where the legs and other parts on which generic distributions have been founded, were entirely wanting.*

Referring my readers to these valuable com-

---

* H. Von Meyer has recently noticed five or six extinct genera of *Macrourous Decapods* in the Muschel-kalk of Germany. (Leonhardt and Bronn Jahrbuch, 1835.)

The subject of English fossil Astacids (*Crawfishes*) is at this time receiving important illustration in the able hands of Prof. Phillips.

In a recent communication to the Geological Society (June 10, 1835), Mr. Broderip describes some very interesting remains of Crustaceans from the Lias at Lyme Regis, in the collection of Viscount Cole. In one of these, the Lamellæ of the external Antennæ, the form and situation of the eyes, and other characters show that it was a *macrourous decapod* intermediate between Palinurus and the Shrimps.

A fragment of another *macrourous decapod* proves the exist-

mencements of the history of fossil Crustaceans,
I proceed to select one very remarkable family,
the *Trilobites*, and to devote to them that detailed
consideration, to which they seem peculiarly entitled, from their apparently anomalous structure, and from the obscurity in which their history
has been involved.

Trilobites.

The great extent to which Trilobites are distributed over the surface of the globe, and their
numerical abundance in the places where they
have been discovered, are remarkable features
in their history; they occur at most distant
points, both of the Northern and Southern Hemisphere. They have been found all over Northern
Europe, and in numerous localities in North
America; in the Southern Hemisphere they occur
in the Andes,* and at the Cape of Good Hope.

ence at this early period of a crustacean approaching to Palinurus, and as large as our common Sea Crawfish.

Two other specimens exhibit the breathing organs of another
delicate Crustacean, with the tips of the four larger and four
smaller branchiæ preserved, and pointing towards the region of
the heart, showing that these fossil Crustaceans belonged to the
highest division of the *Macroura*. They reminded Mr. Broderip
of the living Arctic forms of the macrourous decapods.

* I learn from Mr. Pentland that M. D'Orbigny has lately
found Trilobites accompanied by Strophomena and Producta in
the Greywacke slate formation of the Eastern Cordillera of the
Andes of Bolivia. Fresh water shells, Melania, Melanopsis, and
probably Anodon, occur also in the same rock; a fact which

No Trilobites have yet been found in any strata more recent than the Carboniferous series; and no other Crustaceans, except three forms which are also Entomostracous, have been noticed in strata coeval with any of those that contain the remains of Trilobites;* so that, during the long periods that intervened between the deposition of the earliest fossiliferous strata and the termination of the Coal formation,† the Trilobites appear to have been the chief representatives of a class which was largely multiplied into other orders and families, after these earliest forms of marine Crustaceans became extinct.

The fossil remains of this family have long

seems analogous to the recent discovery of similar fossils in the Transition rocks of Ireland, Germany, and the United States. The Fresh water fossils occurred near Potosi, at an elevation of 13,200 feet.

M. D'Orbigny's specimens also confirm Mr. Pentland's view, as to the analogies between the great Limestone formation of this district, and the Carboniferous limestones of England; and as to the great extent also of the Red Marl, and New red sandstone formations on the Continent of South America.

* In Scotland two genera of Entomostracous Crustaceans, the Eurypterus, and Cypris, occur in the Fresh water lime-stone beneath the Mid Lothian Coal Field; the Eurypterus at Kirkton, near Bathgate, and the Cypris at Burdiehouse, near Edinburgh. (*Trans. Royal Soc. Edin.* vol. xiii.) The third Genus, Limulus, has but recently been recognised in the Coal Formation, and will be described presently. The Entomostracans appear to have been the only representatives of the Class Crustaceans until after the Deposition of the Carboniferous Strata.

† Trilobites of a new species have lately been found in Iron-stone from the centre of the coal measures in Coalbrook-dale. Lond. and Edin. Phil. Mag. vol. 4. 1834, p. 376.

attracted attention, from their strange peculiarities of configuration. M. Brogniart, in his
valuable History of Trilobites, 1822, enumerated
five genera,* and seventeen species; other writers
(Dalman, Wahlenberg, Dekay, and Green,) have
added five more genera, and extended the number
of species to fifty-two ; examples of four of these
genera are given in Plate 46. Fossils of this
family were long confounded with Insects, under
the name of Entomolithus paradoxus; after many
disputes respecting their true nature, their place
has now been fixed' in a separate section of the
class Crustaceans, and although the entire family
appears to have been annihilated at so early a
period as the termination of the Carboniferous
strata, they nevertheless present analogies of
structure, which place them in near approximation to the inhabitants of the existing seas.†

The anterior segment of the Trilobites (Pl. 46,
a, *passim*,) is composed of a large semi-circular,
or crescent-shaped shield, succeeded by an abdomen, or body (c), composed of numerous segments folding over each other, like those in a
Lobster's tail, and generally divided by two

* The names of these Genera are Calymene, Asaphus, Ogyges,
Paradoxus, and Agnostus. Some of these terms are devised expressly to denote the obscure nature of the bodies to which they
are attached ; e. g. Asaphus, from ἀσαφὴς, obscure; Calymene,
from κεκάλυμμένη, concealed ; παράδοξος, wonderful ; ἀγνωστος,
unknown.

† See M. Audouin's Récherches sur les Rapports naturels qui
existent entre les Trilobites et les Animaux articulés.

longitudinal furrows into three ranges of lobes, from which they. have derived the name of Trilobites. Behind this body, in many species, is placed a triangular or semi-lunar tail or post-abdomen (d), less distinctly lobated than the body. One of these Genera, the Calymene, has the property of rolling itself up into a ball like a common Wood-Louse. (See Pl. 46, Figs. 1, 3, 4, 5.)

The nearest approach among living animals to the external form of Trilobites is that afforded by the genus Serolis in the class Crustacea. (See Pl. 45, Figs. 6, 7.)* The most striking difference between this animal, and the Trilo-

---

* The Genus Serolis was first established by Dr. Leach, on the authority of specimens collected by Sir Joseph Banks, in the Straits of Magellan (or rather of Magalhaens, the proper name of the navigator, according to Capt. King) during Sir Joseph's voyage with Captain Cook, and given by Sir Joseph to the Linnæan Society; and of another specimen of the same Genus from Senegal given by Mr. Dufresne to Dr. Leach. From these Dr. Leach described and named the species represented in our plate; his description of this Genus is published in the Dictionnaire des Sciences Naturelles, v. 12, p. 340. Captain King has lately collected many specimens of Serolis on the east coast of Patagonia, lat. 45. S. 30 miles from the shore, and brought up by dredging in 40 fathoms water; and also at Port Famine, in the Straits of Magalhaens, where it was thrown upon the beach by the tide; here Captain King saw the beach literally covered with them dead; he has observed them alive swimming close to the bottom among the sea-weed; their motions were slow and gradual, and not like those of a shrimp; he never saw them swimming near the surface; their legs seemed shaped for swimming and crawling on the bottom.

bites, consists in there being a fully developed series of crustaceous legs and antennæ in the Serolis (Pl. 45, Fig. 7.), whilst no traces of either of these organs have yet been discovered in connexion with any Trilobite. M. Brogniart explains the absence of these organs, by conceiving that the Trilobites hold precisely that place in the class Crustaceans (*Gymnobranchia*), in which the antennæ become very small, or altogether fail; and that the legs being transformed to soft and perishable paddles (*pattes*), bearing branchiæ, (or filamentous organs for breathing in water), were incapable of preservation.

A second approximation to the character of Trilobites occurs in the Limulus, or King crab (Lamarck, T. 5, p. 145.), a genus now most abundant in the seas of warm climates, chiefly in those of India, and of the coasts of America (see Pl. 45, Figs. 1. 2.) The history of this genus is important, on account of its relations, both to the existing and extinct forms of Crustaceans; it has been found fossil in the Coal formation of Staffordshire and Derbyshire; and in the Jurassic limestone of Aichstadt, near Pappenheim, together with many other marine Crustaceans of a higher Order.*

---

* In the genus Limulus (see Pl. 45, Figs. 1. 2.) there are but faint traces of antennæ, and the shield (a.), which covers the anterior portion of the body, is expanded entirely over a series of small crustaceous legs (Fig. 2. a.). Beneath the second, or abdominal portion of the shell (c.), is placed a series of thin

A third example of this disposition, in an
animal belonging to the same class of Crustaceans, whereby the legs are reduced to soft
paddles, and combine the functions of respiration with those of locomotion, is afforded by
the Branchipus stagnalis, (Cancer stagnalis,
Lin.), of our English ponds, (see Pl. 45, Figs.
3, c. 4, e. 5, e.)

In the comparison here made between four
different families of Crustaceans, for the purpose
of illustrating the history of the long extinct
Trilobites, by the analogies we find in the Serolis,
Limulus, and Branchipus; we have a beautiful
example, taken from the extreme points of time

horny transverse plates (Fig. 2, e. 2, e'. and 2, e''.) supporting
the fibres of the branchiæ, and at the same time acting as paddles
for swimming. The same disposition of laminated branchiæ is
found also in the Serolis, Fig. 7. e. Fig. 8, is a magnified
representation of these laminated branchiæ, very similar to those
at Figs. 3, e. and 5, e.

Thus while the Serolis (Fig. 7) presents an union of antennæ
and crustaceous legs with soft paddles bearing the Branchiæ, we
have in the Limulus (Fig. 2), a similar disposition of legs and
paddles, and only slight traces of antennæ; in the Branchipus,
(Figs. 3 and 5,) we find antennæ, but no crustaceous legs; while
the Trilobite, being without antennæ, and having *all its legs*
represented by soft paddles, as in Branchipus, is by the latter
condition placed near Branchipus among the Entomostracous
Crustaceans, in the Order of Branchiopods, whose feet are represented by ciliated paddles, combining the functions of respiration
and natation. At Pl. 45, Fig. 3. e, Fig. 4. e, Fig. 5. e, represent
the soft branchiæ of Branchipus, performing the double office
of feet and lungs.

of which Geology takes cognizance, of that systematic and uniform arrangement of the Animal Kingdom, under which every family is nearly connected with adjacent and cognate families. Three of the families under consideration are among the present inhabitants of the water, while the fourth has been long extinct, and occurs only in a fossil state. When we see the most ancient Trilobites thus placed in immediate contact with our living Crustaceans, we cannot but recognize them as forming part and parcel of one great system of Creation, connected through its whole extent by perfect unity of design, and sustained in its minutest parts by uninterrupted harmonies of organization.

We have in the Trilobites an example of that peculiar, and, as it is sometimes called, rudimentary development of the organs of locomotion in the Class Crustaceans, whereby the legs are made subservient to the double functions of paddles and lungs. The advocate for the theory of the derivation of existing more perfect species, by successive changes from more simple ancient forms, might imagine that he sees in the Trilobite the extinct parent stock from which, by a series of developments, consecutive forms of more perfect Crustaceans may, during the lapse of ages, have been derived; but according to this hypothesis, we ought no longer to find the same simple condition as that of the Trilobite still retained in

the living Branchipus, nor should the primeval
form of Limulus have possessed such an intermediate character, or have remained unadvanced
in the scale of organization, from its first appearance in the Carboniferous Series,* through the
midway periods of the secondary formations, unto
the present hour.

Eyes of Trilobites.

Besides the above analogies between the Trilobites and certain forms of living Crustaceans,
there remains a still more important point of
resemblance in the structure of their eyes. This
point deserves peculiar consideration, as it affords the most ancient, and almost the only

---

* The very rare fossil engraved in Martin's Petrifacata Derbi-
ensia (Tab. 45, Fig. 4,) by the name of Entomolithus Monocu-
lites (*Lunatus*) appears to be a Limulus. It was found in Iron
Stone of the Coal formation on the borders of Derbyshire.

A similar fossil in the collection of Mr. Anstice, of Madely, is
engraved in our Plate 46″, Fig. 3.

In the Secondary period, during the deposition of the Jurassic
limestone, the Limulus abounded in the seas which then covered
central Germany; and it still maintains its primeval intermediate form in the King Crab of the present ocean.

My friend Mr. Stokes has discovered, on the under side of
a fossil Trilobite from Lake Huron (Pl. 45, Fig. 12.), a crus-
aceous plate (f.) forming the entrance into the stomach, the
shape and structure of which resemble those of the analogous
parts in some recent Crabs. This organ forms another link of
connexion between the Trilobite and living Crustaceans.—Geol.
Trans. N. S. vol. i. p. 208, Pl. 27.

example yet found in the fossil world, of the preservation of parts so delicate as the visual organs of animals that ceased to live many thousands, and perhaps millions of years ago. We must regard these organs with feelings of no ordinary kind, when we recollect that we have before us the identical instruments of vision, through which the light of heaven was admitted to the sensorium of some of the first created inhabitants of our planet.

The discovery of such instruments in so perfect a state of preservation, after having been buried for incalculable ages in the early strata of the Transition formation, is one of the most marvellous facts yet disclosed by geological researches ; and the structure of these eyes supplies an argument, of high importance in connecting together the extreme points of the animal creation. An identity of mechanical arrangements, adapted to the construction of an optical instrument, precisely similar to that which forms the eyes of existing insects and Crustaceans, affords an example of agreement that seems utterly inexplicable without reference to the exercise of one and the same Intelligent Creative power.

Professor Müller and Mr. Straus* have ably and amply illustrated the arrangements, by which the eyes of Insects and Crustaceans are adapted

* See Lib. Ent. Knowledge, v. 12.; and Dr. Roget's Bridgewater Treatise, vol. ii. p. 486 et seq. and Fig. 422—428.

to produce distinct vision, through the medium of a number of minute facets, or lenses, placed at the extremity of an equal number of conical tubes, or microscopes; these amount sometimes, as in the Butterfly, to the number of 35,000 facets in the two eyes, and in the Dragon-fly to 14,000.

It appears that in eyes constructed on this principle, the image will be more distinct in proportion as the cones in a given portion of the eye are more numerous and long; that, as compound eyes see only those objects which present themselves in the axes of the individual cones, the limit of their field of vision is greater or smaller as the exterior of the eye is more or less hemispherical.

If we examine the eyes of Trilobites with a view to their principles of construction, we find both in their form, and in the disposition of the facets, obvious examples of optical adaptation.

In the Asaphus caudatus (see Pl. 45, Figs. 9 and 10.), each eye contains at least 400 nearly spherical lenses fixed in separate compartments on the surface of the cornea.*   The form of the

---

* As the Crystalline lens in the eyes of Fishes is spherical, and those in the Eye of Trilobites are nearly so, there seems to be in this form an adaptation to the medium of Water, which would lead us to expect to find a similar form of lens in the compound Eyes of all marine Crustacea, and probably a different form in the compound Eyes of Insects that live in Air.

general cornea is peculiarly adapted to the uses
of an animal destined to live at the bottom of the
water : to look downwards was as much impossible as it was unnecessary to a creature living at
the bottom; but for horizontal vision in every
direction the contrivance is complete.† The
form of each eye is nearly that of the frustum
of a cone (see Pl. 45, Figs. 9 and 10.), incom-
plete on that side only which is directly opposite
to the corresponding side of the other eye, and
in which if facets were present, their chief range
would be towards each other across the head,
where no vision was required. The exterior of
each eye, like a circular bastion, ranges nearly
round three-fourths of a circle, each commanding
so much of the horizon, that where the distinct
vision of one eye ceases, that of the other eye
begins, so that in the horizontal direction the
combined range of both eyes was panoramic.

If we compare this disposition of the eyes with
that in the three cognate Crustaceans, by which
we have been illustrating the general structure of
the Trilobites, we shall find the same mechanism
pervading them all, modified by peculiar adaptations to the state and habits of each ; thus in
the Branchipus (Pl. 45, Fig. 3, b, b′), which
moves with rapidity in all directions through the

† The facetted eyes of Bees are disposed most favourably for
horizontal vision, and for looking downwards.—Lib. Ent. Knowl.
v. xii. p. 130.

water, and requires universal vision, each eye is nearly hemispherical, and placed on a peduncle, by which it is projected to the distance requisite to effect this purpose. (See Pl. 45, Fig. 3, b, and b'.)

In the Serolis (Pl. 45, Fig. 6. b'.), the disposition of the eye, and its range of vision, are similar to those in the Trilobite; but the summit of the eye is less elevated; as the flat back of this animal presents little obstruction to the rays of light from surrounding objects.*

In the Limulus (Pl. 45, Fig. 1.), where the side eyes (b, b'.) are sessile, and do not command the space immediately before the head, two other simple eyes (b″) are fixed in front, compensating for the want of range in the compound eyes over objects in that direction.†

In the above comparison of the eyes of Trilobites, with those of the Limulus, Serolis, and Branchipus, we have placed side by side, examples of the construction of that most delicate and

---

* Fig. 1. b'. Fig. 3. b'. and Fig. 6. b'. are magnified representations of the eyes to which these figures are respectively adjacent. Figs. 10. and 11. are differently magnified forms of the eye of Asaphus caudatus, which in Fig. 9. is represented of its natural size. A few of these lenses are semi-transparent; they are still set in their original rims, or frame-work of the cornea, the whole being converted into calcareous spar.

† These eyes are placed so close together, that, having been mistaken for a single eye, they caused the name of Monoculus Polyphemus to be applied to this animal by Linnæus.

complex organ the eye, selected from each extreme, and from a midway place in the progressive series of animal creations. We find in Trilobites of the Transition rocks, which were among the most ancient forms of animal life, the same modifications of this organ which are at the present time adapted to similar functions in the living Serolis. The same kind of instrument was also employed in those middle periods of geological chronology when the Secondary strata were deposited at the bottom of a warm sea, inhabited by Limuli, in the regions of Europe which now form the elevated plains of central Germany.

The results arising from these facts are not confined to animal Physiology; they give information also regarding the condition of the ancient Sea and ancient Atmosphere, and the relations of both these media to Light, at that remote period when the earliest marine animals were furnished with instruments of vision, in which the minute optical adaptations were the same that impart the perception of light to Crustaceans now living at the bottom of the sea.

With respect to the waters wherein the Trilobites maintained their existence throughout the entire period of the Transition formation, we conclude that they could not have been that imaginary turbid and compound Chaotic fluid, from the precipitates of which some Geologists

G. D D

have supposed the materials of the surface of the earth to be derived ; because the structure of the eyes of these animals is such, that any kind of fluid in which they could have been efficient at the bottom, must have been pure and transparent enough to allow the passage of light to organs of vision, the nature of which is so fully disclosed by the state of perfection in which they are preserved.

With regard to the Atmosphere also we infer, that had it differed materially from its actual condition, it might have so far affected the rays of Light, that a corresponding difference from the eyes of existing Crustaceans would have been found in the organs on which the impressions of such rays were then received.

Regarding Light itself also, we learn, from the resemblance of these most ancient organizations to existing eyes, that the mutual relations of Light to the Eye, and of the Eye to Light, were the same at the time when Crustaceans endowed with the faculty of vision were first placed at the bottom of the primeval seas, as at the present moment.

Thus we find among the earliest organic remains, an Optical instrument of most curious construction, adapted to produce vision of a peculiar kind, in the then existing representatives of one great Class in the Articulated division of the Animal Kingdom. We do not find this instrument passing onwards, as it were, through a series of experimental changes, from more simple into more complex forms ; it was created at the very first, in the fulness of perfect adaptation to the uses and condition of the class of creatures, to which this kind of eye has ever been, and is still appropriate.

If we should discover a microscope, or telescope, in the hand of an Egyptian Mummy, or beneath the ruins of Herculaneum, it would be impossible to deny that a knowledge of the principles of Optics existed in the mind by which such an instrument had been contrived. The same inference follows, but with cumulative force, when we see nearly four hundred microscopic lenses set side by side, in the compound eye of a fossil Trilobite ; and the weight of the argument is multiplied a thousand fold, when we look to the infinite variety of adaptations by which similar instruments have been modified, through endless genera and species, from the long-lost Trilobites, of the Transition strata, through the extinct Crustaceans of the Secondary and Tertiary formations, and thence onwards throughout existing Crustaceans, and the countless hosts of living Insects.

It appears impossible to resist the conclusions as to Unity of Design in a common Author, which are thus attested by such cumulative evidences of Creative Intelligence and Power ;

both, as infinitely surpassing the most exalted faculties of the human mind, as the mechanisms of the natural world, when magnified by the highest microscopes, are found to transcend the most perfect productions of human art.

SECTION III.

Third Class of Articulated Animals.

FOSSIL ARACHNIDANS.

UNDER the relations that now subsist between the animal and vegetable kingdoms, the connection of terrestrial Plants with Insects is so direct and universal, that each species of plant is considered to afford nutriment to three or four species of insects. The general principle which we have traced throughout the Secondary and Tertiary formations, ever operating to maintain on the surface of the earth the greatest possible amount of life, affords a strong antecedent probability that so large a mass of terrestrial vegetables as that which is preserved in the Carboniferous strata of the Transition series, held the same relation, as the basis of nutriment to Insect families of this early date, that modern vegetables do to this most numerous class of existing terrestrial animals.

Still further, the actual provisions for restraining this Insect class within due bounds, by the controlling agency of the carnivorous Arachnidans would lead us to expect that Spiders and Scorpions were employed in similar service during the successive geological epochs, in which we have evidence of the abundant growth of terrestrial vegetables.

Some recent discoveries confirm the argument from these analogies, by the test of actual observation. The two great families in the higher order of living Arachnidans (Pulmonariæ) are Spiders and Scorpions; and we have evidence to shew that fossil remains of both these families exist in strata of very high antiquity.

Fossil Spiders.

Although no Spiders have been yet discovered in any rocks so ancient as the Carboniferous series, the presence of Insects in this series, and also of Scorpions, renders it highly probable that the cognate family of Spiders was co-ordinate with Scorpions, in restraining the Insect tribes of this early epoch, and that it will ere long be recognized among its fossil remains.*

* The animal found by Mr. W. Anstice in the Iron stone of Coalbrook Dale, and noticed by Mr. Prestwich as " apparently a Spider" (Phil. Mag. May, 1834, v. iv. p. 376), has been subsequently laid open by me, and shewn to be an Insect, belonging to the family of Curculionidæ. (Pl. 46″, Fig. 1.) At the time when it was figured, and supposed to be a Spider, its head and

The existence of Spiders in the Jurassic portion of the Secondary formations has been established, by Count Munster's discovery of two species in the lithographic limestone of Solenhofen. M. Marcel de Serres and Mr. Murchison have discovered fossil Spiders in Freshwater Tertiary strata near Aix in Provence. (See Pl. 46″, Fig. 12.)

Fossil Scorpions.

The address of my friend Count Sternberg to the members of the National Museum of Bohemia (Prague, 1835), contains an account of his discovery of a fossil Scorpion in the ancient Coal formation at the village of Chomle, near Radnitz, on the S. W. of Prague. This most instructive fossil (the first of its kind yet noticed) was found

tail were covered by Iron stone, and its appearance much resembled an animal of this kind. Mr. Prestwich announces also the discovery, in the same formation, of a Coleopterous Insect, which will be further described in our next section, as referrible also to the Curculionidæ.

It is scarcely possible to ascertain the precise nature of the animals, rudely figured as Spiders and Insects on Coal slate by Lhwyd, (Ichnograph. Tab. 4,) and copied by Parkinson, (Org. Rem. V. iii. Pl. 17, Figs. 3, 4, 5, 6); but his opinion of them is rendered highly probable by the recent discoveries in Coalbrook Dale: " Scripsi olim suspicari me Araneorum quorundam icones, unà cum Lithophytis in Schisto Carbonariâ observasse: hoc jam ulteriore experientiâ edoctus apertè assero. Alias icones habeo, quæ ad Scarabæorum genus quàm proximè accedunt. In posterum ergo non tantùm Lithophyta, sed et quædam Insecta in hoc lapide investigare conabimur." Lhwyd Epist. iii. ad fin.

in July, 1834, in a stone-quarry, on the outcrop of the Coal measures, near a spot where coal has been wrought since the sixteenth century. In the same quarry were found four erect trunks of trees, and numerous vegetable remains, of the same species that occur in the great Coal formation of England.

A series of drawings of this Scorpion was submitted to a select committee at the meeting of Naturalists and Physicians of Germany, in Stutgard, 1834 ; and from their report the subjoined particulars are taken. All our Figures, (Pl.46'.) are copied from those attached to this Report, in the Transactions of the Museum of Bohemia, April, 1835.*

* This fossil Scorpion differs from existing species, less in general structure than in the position of the eyes. In the latter respect, it approaches nearest to the genus *Androctonus*, which, like it, has twelve eyes, but differently disposed from those of the fossil species. From the nearly circular arrangement of these organs in the latter animal, it has been ranged under a new genus, *Cyclophthalmus*.

The sockets of all these twelve eyes are perfectly preserved, (Pl. 46'. fig. 3.) One of the small eyes, and the left large eye, still retain their form, with the cornea preserved in a wrinkled state, and their interior filled with earth.

The jaws also are very distinct, but in a reversed position. (Pl. 46' fig. 2. a.) Both these jaws have three projecting teeth, and one of them (Pl. 46', Figs. 4. 5.) exhibits, when magnified, the hairs with which its horny integument was covered.

The rings of the thorax, (apparently eight) and of the tail, are too much dislocated for their number to be accurately distinguished, but they differ from all known species. The view of the

As far as we can argue from the analogy of living species, the presence of large Scorpions is a certain index of the warmth of the climate in which they lived; and this indication is in perfect harmony with those afforded by the tropical aspect of the vegetables with which the Scorpion, found in the Bohemian coal-field, is associated.

back (Pl. 46', Fig. 1.) has been obtained by cutting into the stone from behind.

The under surface of the animal is well exposed in Fig 2, with its characteristic pincers on the right claw. Between this claw and the tail lies a fossil carbonized Seed, of a species common in the Coal formation.

The horny covering of this Scorpion is in a most extraordinary state of preservation, being neither decomposed nor carbonized. The peculiar substance *(Chitine or Elytrine)* of which, like the elytra of Beetles, it is probably composed, has resisted decomposition and mineralization. It can readily be stripped off, is elastic, translucent, and horny. It consists of two layers, both retaining their texture. The uppermost of these (Pl. 46' Fig. 6. a.) is harsh, almost opaque, of a dark-brown colour, and flexible; the under skin (Pl. 46', Fig. 6. b.) is tender, yellow, less elastic, and organized like the upper. The structure of both exhibits, under the microscope, hexagonal cells, divided by strong partitions. Both are penetrated at intervals by pores, which are still open, each having a sunk areola, with a minute opening at its centre for the orifices of the trachea. Fig. 7. represents impressions of the muscular fibres connected with the movement of the legs.

SECTION IV.

Fourth Class of Articulated Animals.

FOSSIL INSECTS.[*]

ALTHOUGH the numerical amount of living Insects forms so vast a majority of the inhabitants
of the present land, few traces of this large class
of Articulated animals have yet been discovered
in a fossil state. This may probably result from
the circumstance, that the greatest portion of
fossil animal remains are derived from the inhabitants of *salt water*, a medium in which only
one or two species of Insects are now supposed
to live.

Had no indications of Insects been discovered
in a fossil state, the presence in any strata, of
Scorpions or Spiders, both belonging to families
constructed to feed on Insects, would have afforded a strong *à priori* argument, in favour of
the probability, of the contemporaneous existence of that very numerous class of animals,
which now forms the prey of the Arachnidans.
This probability has been recently confirmed by
the discovery of two Coleoptera of the family
Curculionidæ in the Iron-stone of Coalbrook

---

[*] See Pl. 46″. Figs. 1. 2. & 4.—11.

Dale,* and also of the wing of a Corydalis, which
will be noticed in our description of Pl. 46".

It is very interesting and important, to have
discovered in the Coal formation fossil remains,
which establish the existence of the great Insectivorous Class Arachnidans, at this early
period. It is no less important to have found also
in the same formation the remains of Insects,
which may have formed their prey. Had neither
of these discoveries been made, the abundance
of Land plants would have implied the probable
abundance of Insects, and this probability would
have involved also that of the contemporaneous
existence of Arachnidans, to control their undue
increase. All these probabilities are now reduced
to certainty, and we are thus enabled to fill up
what has hitherto appeared a blank in the history
of animal life, from those very distant times when
the Carboniferous strata were deposited.

The Estuary, or Fresh-water formation of those
strata of the Carboniferous series which contain
shells of Unio, in Coalbrook Dale, and in other
Coal basins, renders the presence of Insects and
Arachnidans in such strata, easy of explanation ;
they may have been drifted from adjacent lands,
by the same torrents that transported the ter-

* Our figures (Pl. 46". Figs. 1. 2.) represent these fossils of
their natural size. See description of this Plate for further details respecting them.

restrial vegetables which have produced the beds of Coal.

The existence of the wing-covers of Insects in the Secondary Series, in the Oolitic slate of Stonesfield, has been long known ; these are all Coleopterous, and in the opinion of Mr. Curtis many of them approach most nearly to the Buprestis, a genus now most abundant in warm latitudes. (See Pl. 46". Figs. 4. 5. 6. 7. 8. 9. 10.)*

Count Munster has in his collection twenty-five species of fossil Insects, found in the Jurassic Limestone of Solenhofen ; among these are five species of the existing Family of Libellula, (See

---

* M. Aug. Odier has ascertained, that the Elytra and other parts of the horny covering of insects, contain the peculiar substance *Chitine, or Elytrine*, which approaches nearly to the vegetable principle *Lignine;* these parts of Insects burn without fusion, or swelling, like horn, and without the smell of animal matter; they also leave a Coal which retains their form.

M. Odier found that even the hairs of a *Scarabæus nasicornis* retained their form after burning, and therefore concludes that they are different from the hairs of vertebral animals. This circumstance explains the preservation of the hairs on the horny cover of the Bohemian Scorpion.

He ascertained also that the Sinews *(Nervures)* of Scarabæi, are composed of Chitine, and that the soft flexible laminæ of the shell of a crab, which remain after the separation of the Lime, also contain Chitine.

Cuvier observes, that the Integuments of Entomostracans, are rather horny than calcareous, and that in this respect they approximate to the nature of Insects and Arachnidans. See Zoological Journal. London, 1825, vol. i. p. 101.

Pl. 1, Fig. 49), a large Ranatra, and several Coleoptera.

Numerous fossil Insects have recently been discovered in the Tertiary Gypsum of Fresh-water formation at Aix, in Provence. M. Marcel de Serres speaks of sixty-two Genera, chiefly of the Orders Diptera, Hemiptera, and Coleoptera; and Mr. Curtis refers all the specimens he has seen from Aix to European forms, and most of them to existing Genera.* Insects occur also in the tertiary Brown coal of Orsberg on the Rhine.

General Conclusions.

We have seen from the examples cited in the last four sections, that all of the four existing great Classes of the grand Division of Articulated animals, viz. Annelidans, Crustaceans, Arachnidans, and Insects, and many of their Orders, had entered on their respective functions in the natural world, at the early Epoch of the Transition Formations. We find evidences of change in the Families of these Orders, at several periods of the Secondary and Tertiary series, very distant from one another; and we further find each Family variously represented during different intervals by Genera, some of which are known

---

* See Edinburgh New Phil. Journ. Oct. 1829.

only in a fossil state, whilst others (and these chiefly in the lower Classes,) have extended through all geological Eras unto the present time.

On these facts we may found conclusions which are of great importance in the investigation of the physical history of the Earth. If the existing Classes, Orders, and Families of marine and terrestrial Articulated animals have thus pervaded various geological epochs, since life began upon our planet, we may infer that the state of the Land and Waters, and also of the Atmosphere, during all these Epochs, was not so widely different from their actual condition as many geologists have supposed. We also learn that throughout all these epochs and stages of change, the correlative Functions of the successive Representatives of the animal and vegetable kingdoms have ever been the same as at the present moment; and thus we connect the entire series of past and present forms of organized beings, as parts of one stupendously grand, and consistent, and harmonious *Whole.*

Chapter XVII.

Proofs of Design in the Structure of Fossil Radiated Animals, or Zoophytes.

THE same difficulties which we have felt in selecting from other grand Divisions of the animal
kingdom, subjects of comparison between the extinct and living forms of their respective Classes,
Orders, and Families, embarrass our choice also
from the last Division that remains for consideration. Volumes might be filled with descriptions of fossil species of those beautiful genera
of Radiated Animals, whose living representatives crowd the waters of our present seas.

The result of all comparisons between the living and fossil species of these families would be,
that the latter differ almost always in species,
and often in genus, from those which actually
exist; but that all are so similarly constructed on
one and the same general Type, and show such
perfect Unity of Design throughout the infinitely
varied modifications, under which they now perform, and ever have performed the functions
allotted to them, that we can find no explanation
of such otherwise mysterious Uniformity, than
by referring it to the agency of one and the same
Creative Intelligence.

SECTION I.

FOSSIL ECHINODERMS.

THE animals that compose this highest Class in the grand division of Radiated animals, viz. Echinidans, Stelleridans, and Crinoïdeans, have, till lately, been considered as made up of many *similar* parts disposed like Rays around a common centre.

Mr. Agassiz has recently shewn, (London and Edin. Phil. Mag. Nov. 1834, p. 369), that they do not partake of this character, from which the division of radiated animals is named; but that their rays are *dissimilar*, and not always connected with an uniform centre; and that a *bilateral* symmetry, analogous to that of the more perfect classes of animals, exists throughout the families of Echini, Asteriæ, and Crinoidea.

ECHINIDANS AND STELLERIDANS.

The History of the fossil species of Echinidans and Stelleridans has been most beautifully illustrated, in the plates of the *Petrefacten* of Prof. Goldfuss. Though derived from Strata of various degrees of high antiquity, they are for the most part referred by him to existing Genera.

The family of Echinidans appears to have extended through all Formations, from the Epoch of the Transition series to the present time.*

No fossil Stelleridans have yet been noticed in strata more ancient than the Muschel-kalk.

As the structure of the fossil species of both these families is so nearly identical with that of existing Echini, and Star-fishes, I shall confine my observations respecting fossil animals in the class of Echinoderms to a family which is of rare occurrence, excepting in a fossil state, and which seems to have prevailed most abundantly in the most ancient fossiliferous formations.

CRINOIDEANS.

Among the fossil families of the Radiated division of animals, the Geologist discovers one whose living analogues are seldom seen, and whose vast numerical extent and extraordinary beauty entitle it to peculiar consideration.

Successions of strata, each many feet in thickness, and many miles in extent, are often half made up of the calcareous skeletons of Encrinites. The Entrochal Marble of Derbyshire, and the Black rock in the cliffs of Carboniferous

---

* I found many years ago fossil Echinidans in the Carboniferous limestone of Ireland, near Donegal, they are however rare in the Transition formation, become more frequent in the Muschel-kalk and Lias, and abound throughout the Oolitic and Cretaceous formations.

limestone near Bristol, are well known examples of strata thus composed; and shew how largely the bodies of Animals have occasionally contributed by their remains, to swell the Volume of materials that now compose the mineral world.

The fossil remains of this order have been long known by the name of Stone Lilies, or *Encrinites*, and have lately been classed under a separate order by the name of Crinoïdea.

This order comprehends many Genera and numerous Species, and is ranged by Cuvier after the Asteriæ, in the division Zoophytes. Nearly all these species appear to have been attached to the bottom of the Sea, or to floating extraneous bodies.*

The two most remarkable Genera of this family have been long known to Naturalists by the

* These animals form the subject of an elaborate and excellent work, by Mr. Miller, entitled a Natural History of the Crinoïdea, or Lily-shaped Animals. The representations at Pl. 48, and Pl. 49, Fig. 1. of one of the most characteristic species of this family, being that to which the name of *stone-lily* was first applied; and the figures of two other species at Pl. 47, Fig. 1, 2, 5, will exemplify the following definition given of them by Mr. Miller. " An Animal with a round, oval, or angular column, composed of numerous articulating joints, supporting at its summit, a series of plates, or joints, which form a cup-like body, containing the viscera, from whose upper rim proceed five articulated arms, dividing into tentaculated fingers, more or less numerous, surrounding the aperture of the mouth, (Pl. 47. Figs. 6, x. 7, x) situated in the centre of a plated integument, which extends over the abdominal cavity, and is capable of being contracted into a conical or proboscal shape."

G.   E E

names of *Encrinite* and *Pentacrinite;* the former
(see Pl. 49, Fig. 1, and Pl. 47, Figs. 1. 2. 5.) most
nearly resembling the external form of a Lily,
placed on a circular stem ; the latter (see Pl. 51,
and Pl. 52, Fig. 1, 3.) retaining the general
analogies of structure presented by the Encrinite, but, from the pentagonal form of its stem,
denominated Pentacrinite. A third Genus, called
Apiocrinites, or Pear Encrinite, (Pl. 47. Figs. 1,
2.) exhibits, on a large scale, the component parts
of bodies of this family; and has been placed by
Mr. Miller at the head of his valuable work on
the Crinoïdea, from which many of the following
descriptions and illustrations will be collected.

Two existing species of recent animals throw
much light on the nature of these fossil remains ;
viz. the Pentacrinus Caput Medusæ from the
West Indies, represented at Pl. 52, Fig. 1, and
the Comatula fimbriata,* figured in the first
plate of Miller's Crinoïdea.

We will proceed to consider the mechanical
provisions in the structure of two or three of the
most important fossil species of this family,
viewed in relation to their office as Zoophytes,

---

* The Comatula presents a conformity of structure with that
of the Pentacrinite, almost perfect in every essential part, excepting that the column is either wanting, or at least reduced to a
single plate. Peron states that the Comatula suspends itself by
its side arms from fuci, and Polyparies, and in this position
watches for its prey, and attains it by its spreading arms and
fingers. Miller, p. 182.

destined to find their nourishment by spreading
their nets and moving their bodies through a
limited space, from a fixed position at the bottom
of the sea; or by employing the same instruments, either when floating singly through the
water, or attached, like the modern Pentelasmis
anatifera, to floating pièces of wood.

Although the representatives of Crinoïdeans
in our modern seas are of rare occurrence,
this family was of vast numerical importance
among the earliest inhabitants of the ancient
deep.* The extensive range which it formerly
occupied among the earliest inhabitants of our
Planet, may be estimated from the fact, that the
Crinoïdeans already discovered have been arranged in four divisions, comprising nine genera,
most of them containing several species, and each
individual exhibiting, in every one of its many
thousand component little bones,† a mechanism
which shows them all to have formed parts of a
well-contrived and delicate mechanical instrument; every part acting in due connection with

---

* The monograph of Mr. Miller, exhibiting the minute details
of every variation in the structure of each component part in the
several Genera of the family of Crinoïdea, affords an admirable
exemplification of the regularity, with which the same fundamental
type is rigidly maintained through all the varied modifications
that constitute its numerous extinct genera and species.

† These so-called Ossicula are not true bones, but partake of
the nature of the shelly Plates of Echini, and the calcareous
joints of Star-fishes.

the rest, and all adjusted to each other with a view to the perfect performance of some peculiar function in the economy of each individual.

The joints, or little bones, of which the skeletons of all these animals were composed, resemble those of the star-fish : their use, like that of the bony skeleton in vertebral animals, was to constitute the solid support of the whole body, to protect the viscera, and to form the foundation of a system of contractile fibres pervading the gelatinous integument with which all parts of the animal were invested.*

The bony portions formed the great bulk of the animal, as they do in star-fishes. The calcareous matter of these little bones was probably secreted by a Periosteum, which in cases of accident, to which bodies so delicately constructed must have been much exposed in an element so stormy as the sea, seems to have had the power of depositing fresh matter to repair casual injuries. Mr. Miller's work abounds with examples of reparations of this kind in various fossil species of Crinoïdeans. Our Pl. 47, Fig. 2ª, represents a reparation near the upper portion of the stem of Apiocrinites Rotundus.

---

* As the contractile fibres of radiated animals are not set together in the same complex manner as the true muscles of the higher orders of animals, the term Muscle, in its strict acceptation, cannot with accuracy be applied to Crinoïdeans ; but, as most writers have designated by this term the more simple contractile fibres which move their little bones, it will be convenient to retain it in our descriptions of these animals.

In the recent Pentacrinus (Pl. 52, Fig. 1), one of the arms is under the process of being reproduced, as Crabs and Lobsters reproduce their lost claws and legs, and many lizards their tails and feet. The arms of star-fishes also, when broken off, are in the same manner reproduced.

From these examples we see that the power of reproduction has been always strongest in the lowest orders of animals, and that the application of remedial forces has ever been duly proportioned to the liability to injury, resulting from the habits and condition of the creatures in which these forces are most powerfully developed.

Encrinites Moniliformis.

As the best mode of explaining the general economy of the Crinoïdea, will be to examine in some detail the anatomy of a single species, I shall select, for this purpose, that which has formed the type of the order, viz. the Encrinites moniliformis (see Pl. 48, 49, 50). Minute and full descriptions are given by Parkinson and Miller of this fossil, shewing it to combine in its various organs an union of mechanical contrivances, which adapt each part to its peculiar functions in a manner infinitely surpassing the most perfect contrivances of human mechanism.

Mr. Parkinson* states that after a careful ex-

* Organic Remains, vol. ii. p. 180.

amination he has ascertained that, independently of the number of pieces which may be contained in the vertebral column, and which, from its probable length, may be very numerous, the fossil skeleton of the superior part of the Lily Encrinite (Encrinites Moniliformis) consists of at least 26,000 pieces. See Pl. 50, Figs. 1, 2, 3, 4, &c.*

Mr. Miller observes that this number would increase most surprisingly, were we to take into account the minute calcareous plates that are interwoven in the integument covering the abdominal cavity and inner surface of the fingers and tentacula.†

* Bones of the Pelvis	5
Ribs	5
Clavicles	5
Scapulæ	5
Arms.　Six bones in each of the ten arms	60
Hands.　Each hand being formed of two fingers, and each finger consisting of at least 40 ossicula, these in 20 fingers make	800
Tentacula.　30 proceeding from each of the 6 bones in each of the ten arms, make........	1800
30 proceeding, on the average, from each of the 800 bones of the fingers make	24,000
Total	26,680

† Although the names here used are borrowed from the skeleton of vertebrated animals, and are not strictly applicable to radiated Echinoderms, it will be convenient to retain them until the comparative anatomy of this order of animals has been arranged in some other more appropriate manner.

We will first examine the contrivances in the joints, of the vertebral column, which adapted it for flexure in every direction, and then proceed to consider the arrangement of other parts of the body.

These joints are piled on each other like the masonry of a slender Gothic shaft, but, as a certain degree of flexibility was requisite at every articulation, and the amount of this flexure varied in different parts of the column, being least at the base and greatest at the summit, we find proportionate variations both in the external and internal form and dimensions of each part.*   The

* The body (Pl. 49, Fig. 1) is supported by a long vertebral column attached to the ground by an enlargement of its base (Pl. 49, Fig. 2). It is composed of many cylindrical thick joints, articulating firmly with each other, and having a central aperture, like the spinal canal in the vertebra of a quadruped, through which a small alimentary cavity descends from the stomach to the base of the column, Pl. 49, Fig. 4, 6, 8, 10. The form of the column nearest the base is the strongest possible, viz. cylindrical. This column is interrupted, at intervals, which become more frequent as it advances upwards, by joints of wider diameter and of a globular depressed form (Pl. 49, Fig. 1. and Figs. 3, 4, a, a, a, a.) Near the summit of the column, (Pl. 49, Figs. 3, 4,) a series of thin joints, c, c, c, is placed next above and below each largest joint, and between these two thin joints, there is introduced a third series, b, b, b, of an intermediate size. The use of these variations in the size of the interpolated joints was to give increased flexibility to that part of the column, which being nearest to its summit required the greatest power of flexion.

At Plate 49, Figs. 6, 8, 10, are vertical sections of the columnar joints 5, 7, 9, taken near the base; and show the internal cavity of the column, to be arranged in a series of double

varieties of form and contrivance which occur in the column of a single species of Encrinite, may serve as an example of analogous arrangements in the columns of other species of the family of Crinoïdeans, (see Pl. 47. Figs. 1, 2, 5, and Pl. 49. Fig. 4 to Fig. 17).*

The name of Entrochi, or wheel stones, has with much propriety been applied to these insulated vertebræ. The perforations in the centre of these joints affording a facility for stringing them as beads, has caused them, in ancient times, to be used as rosaries. In the northern

hollow cones, like the intervertebral cavities in the back of a fish, and to be, like thèm, subsidiary to the flexion of the column ; they probably also formed a reservoir for containing the nutritious fluids of the animals.

The various kinds of Screw stone so frequent in the chert of Derbyshire, and generally in the Transition Limestone, are casts of the internal cavities of the columns of other species of Encrinites, in which the cones are usually more compressed than in the column of the E. moniliformis.

* At Pl. 49, Fig. 4 is a vertical section of Fig. 3, being a portion taken from near the summit of the column, where the greatest strength and flexure were required, and where also the risk of injury and dislocation was the greatest; the arrangement of these vertebræ is therefore more complex than it is towards the base, and is disposed in the following manner (see Fig. 4). The vertebræ, a. b. c. are alternately wider and narrower; the edges of the latter, c. are received into, and included within, the perpendicularly lengthened margin of the wider, a. b. ; the outer crenulated edge of the narrower included vertebræ, articulate with the inner crenulated edge of the wider vertebræ, which thus surround them with a collar, that admits of more oblique flexion

parts of England they still retain the appellation
of St. Cuthbert's beads.

On a rock by Lindisfarn
Saint Cuthbert sits, and toils to frame
The sea-born beads, that bear his name.

MARMION.

Each of these presents a similar series of articulations, varying as we ascend upwards through
the body of the animal, every joint being exactly
adjusted, to give the requisite amount of flexibility and strength. From one extremity of the
vertebral column to the other, and throughout

than the plane crenulated surfaces near the base of the column,
Figs. 9, 10, and at the same time renders dislocation almost impossible.

To these is superadded a third contrivance, which still further
increases the flexibility and strength of this portion of the column,
viz. that of making the alternate larger joints, b. b. considerably
thinner than the largest collar joints, a. a.

The Figures numbered from 11 to 26 inclusive, represent
single vertebræ taken from various portions of the column of
Encrinites moniliformis. The joints at Figs. 11, 13, 15, 17, 19,
21, 23, 25, are of their natural size and in their natural horizontal position, and show, at the margin of each, a crenated edge,
every tooth of which articulated with a corresponding depression
near the margin of the adjacent joint. The stellated figures
(12, 14, 16, 18, 20, 22, 24, 26,) placed beneath the horizontal
joints to which they respectively belong, are magnified representations of the various internal patterns presented by their
articulating surfaces, variously covered with an alternate series
of ridges and grooves, that like the cogs of two wheels, articulate
with corresponding depressions and elevations on the surfaces of
the adjacent vertebræ.

the hands and fingers (see Pl. 47, figs. 1, 2, 3. and Pl. 50, figs. 1, 2, 3.), the surface of each bone articulates with that adjacent to it, with the most perfect regularity and nicety of adjustment. So exact, and methodical is this arrangement, even to the extremity of its minutest tentacula, that it is just as improbable, that the metals which compose the wheels of a chronometer should for themselves have calculated and arranged the form and number of the teeth of each respective wheel, and that these wheels should have placed themselves in the precise position, fitted to attain the end resulting from the combined action of them all, as for the successive hundreds and thousands of little bones that compose an Encrinite, to have arranged themselves, in a position subordinate to the end produced by the combined effect of their united Mechanism; each acting its peculiar part in harmonious subordination to the rest, and all conjointly producing a result which no single series of them acting separately, could possibly have effected.

In Pl. 50 I have selected from Goldfuss, Parkinson, and Miller, details of the structure of the body and upper extremities of Encrinites Moniliformis, or Lily Encrinite, in which the component parts are indicated by letters, explained in the annexed note; and I must refer my readers to these authors for minute descriptions of the individual forms and uses of each successive series of plates.*

From the subjoined analysis of the component portions of the body of the E. Moniliformis, we see that it may be resolved into four series of plates each composed of five pieces, and bearing a distant analogy to those parts in the organization of superior animals from which they have been denominated. A similar system of plates,

* " On the summit of the vertebral column are placed successive series of little bones, see Pl. 50, Fig. 4, which from their position and uses may be termed the Pelvis E, Scapula H, Costal F, forming (with the pectoral and capital plates) a kind of sub-globular body (see Pl. 48. Pl. 49. Fig. 1. Pl. 50, Figs. 1, 2), having the mouth in its centre, and containing the viscera and stomach of the animal, from which the nourishing fluids were admitted to an alimentary cavity within the column, and also carried to the arms and tentaculated fingers." From the scapula (H) proceeded the five arms, (Pl. 50, Fig. 1, K) which, as they advanced, subdivided into hands (M) and fingers (N) terminating in minute tentacula (Pl. 50. Figs. 2, 3), the number of which extended to many thousands. These hands and fingers are. represented as closed, or nearly closed, in Pl. 48. and Pl. 49, Fig. 1. and Pl. 50, Figs. 1, 2. In Mr. Miller's restoration of the Pear Encrinite (Pl. 47, Fig. 1) they are represented as expanded in search of food. These tentaculated fingers, when thus expanded, would form a delicate net, admirably adapted to detain Acalephans, and other minute molluscous animals that might be floating in the sea, and which probably formed part of the food of the Crinoïdea. In the centre of these arms was placed the mouth (Pl. 47, Fig. 1), capable of elongation into a proboscis. Pl. 47. 6, x. 7, x. represent the bodies of Crinoïdea from which the arms have been removed.

In Pl. 50, Fig. 1 represents the superior portion of the animal, with its twenty fingers closed like the petals of a closed lily. Fig. 2 represents the same partially uncovered, with the tentacula

varying in number and holding the same place
between the column and the arms of the animal,
may be traced through each species of the family
of Crinoïdeans.   The details of all these specific
variations are beautifully illustrated by Mr.
Miller, to whose excellent work I must again
refer those who are inclined to follow him,
through his highly philosophical analysis of the
structure of this curious family of fossil animals.*

still folded up.   Fig. 3 is a side view of one of the fingers with
its tentacula.   Fig. 4 represents the interior of the body which
contained the viscera.   Fig. 5 represents the exterior of the
same body, and the surface by which the base articulates with
the first joint of the vertebral column.   Figs. 6, 7, 8, 9, represent a dissection of the four series of plates that compose the
body, forming successively the scapulæ, upper and lower costal
plates, and pelvis of the animal.   Fig. 10 is the upper extremity
of the vertebral column.   Fig. 11 represents the upper surfaces
of the five scapulæ, showing their articulations with the inferior
surfaces of the first bones of the arms.   Fig. 12 is the inferior
surface of the same series of scapular plates, showing their articulations with the superior surfaces of the upper or second series
of costal plates, Fig. 13.   Fig. 14 is the inferior surface of Fig.
13, and articulates with the first or lower series of costal plates,
Fig. 15.   Fig. 16 is the lower surface of Fig. 15, and articulates
with the upper surface of the bones of the pelvis, Fig. 17.   Fig.
18 is the inferior surface of the pelvis, Fig. 17, and articulates
with the first or uppermost joint of the vertebral column,
Fig. 10.

   * Our Pl. 47 gives Mr. Miller's restoration of two other
genera; fig. 1, the Apiocrinites rotundus, or Pear Ecrinite, with
its root or base of attachment, and its arms expanded.   Fig 2
is the same with its arms contracted.   Two young individuals
and the broken stumps of two other small specimens, are seen
fixed by their base to the root of the larger specimens, shewing

From the details I have thus selected from the best authorities, with a view to illustrate the most important parts that enter into the organization of the family of Encrinites, it is obvious that similar investigations might be carried to

the manner in which these roots are found attached to the upper surface of the great oolite at Bradford near Bath. When living, their roots were confluent, and formed a thin pavement at this place over the bottom of the sea, from which their stems and branches rose into a thick submarine forest, composed of these beautiful Zoophytes. The stems and bodies are occasionally found united, as in their living state : the arms and fingers have almost always been separated, but their dislocated fragments still remain, covering the pavement of roots that overspreads the surface of the subjacent Oolitic limestone rock.

This bed of beautiful remains has been buried by a thick stratum of clay. Fig. 3 represents the exterior of the body, and the upper columnar joints of this animal, about two-thirds of the natural size. Fig. 4 is a longitudinal section of the same, shewing the cavity for the viscera, and also the large open spaces for the reception of nourishment between the uppermost enlarged joints of the column.

At fig. 5 we have the Actinocrinites 30-dactylus, from the carboniferous limestone near Bristol. D. represents the auxiliary side arms which are attached to the column of this species, and B its base and fibres of attachment. Fig. 6 represents its body, from which the fingers are removed, shewing the pectoral plates, Q, and capital plates, R, which form an integument over the abdominal cavity of the body, and terminate in a mouth (x), capable of being protruded into an elongated proboscis by the contraction of its plated integument. Fig. 7 represents the body of an Encrinite in the British Museum, figured by Parkinson, vol. 2, fol. 17, fig. 3, by the name of Nave Encrinite. The mouth of this specimen also is seen at X, and between the mouth and the bases of the arms, the series of plates which form the upper and exterior integuments of the stomach.

an almost endless extent by examining the peculiarities of each part throughout their numerous
species. We may judge of the degree, to which
the individuals of these species multiplied among
the first inhabitants of the sea, from the countless
myriads of their petrified remains which fill so
many Limestone beds of the Transition Formations, and compose vast strata of Entrochal
marble, extending over large tracts of country in
Northern Europe and North America. The substance of this marble is often almost as entirely
made up of the petrified bones of Encrinites, as
a corn-rick is composed of straws. Man applies it to construct his palace and adorn his
sepulchre, but there are few who know, and fewer
still who duly appreciate the surprising fact,
that much of this marble is composed of the
skeletons of millions of organized beings, once
endowed with life, and susceptible of enjoyment, which after performing the part that was
for a while assigned to them in living nature,
have contributed their remains towards the composition of the mountain masses of the earth.*

Of more than thirty species of Crinoïdeans
that prevailed to such enormous extent in the
Transition period, nearly all became extinct before the deposition of the Lias, and only one

---

* Fragments of Encrinites are also dispersed irregularly
throughout all the depositions of this period, intermixed with the
remains of other cotemporary marine animals.

presents the angular column of the Pentacrinite ; with this one exception, pentangular columns first began to abound among the Crinoïdeans at the commencement of the Lias, and have from thence extended onwards into our present seas. Their several species and even genera are also limited in their extent; e. g, the great Lily Encrinite (E. moniliformis) is peculiar to the Muschel Kalk, and the Pear Encrinite to the middle region of the Oolitic formation.

The Physiological history of the family of Encrinites is very important; their species were numerous among the most ancient orders of created beings, and in this early state their construction exhibits at least an equal if not a higher degree of perfection than is retained in the existing Pentacrinites; and although the place, which, as Zoophytes, they occupied in the animal kingdom, was low, yet they were constructed with a perfect adaptation to that low estate, and in this primeval perfection they afford another example at variance with the doctrine of the progression of animal life from simple rudiments through a series of gradually improving and more perfect forms, to its fullest development in existing species. Thus, a comparison of one of the early forms of the Genus Pentacrinite, viz. the Briarean Pentacrinite of the Lias, (Pl. 51 and Pl. 52, Fig. 2, and Pl. 53) with the fossil species of more recent formations, and with the existing

Pentacrinus Caput Medusæ from the Caribbean Sea, Pl. 52, Fig. 1, shews in the organization of this very ancient species an equal degree of per- fection, and a more elaborate combination of analogous organs, than occurs in any other fossil species of more recent date, or in its living re- presentative.

Pentacrinite.

The history of these fossil bodies, that abound in the lower strata of the Oolite formation, and especially in the Lias, has been much illustrated by the discovery of two living forms of the same Genus, viz. the Pentacrinus Caput Me- dusæ,* (Pl.52, Fig.1,) and Pentacrinus Europæus, Pl. 52, Figs. 2. 2′. Of the first of these a few specimens only have been brought up from the bottom of deep seas in the West Indies; having their lower extremities broken, as if torn from a firm attachment to the bottom. The Pentacrinus Europæus† (see Pl. 52, Figs. 2. 2′,) is found at- tached to various kinds of Sertularia and Flus- tracea in the Cove of Cork,-and other parts of the coast of Ireland.

It appears that Pentacrinites are allied to

* See Miller's Crinoïdea, p. 45.

† See Memoir on Pentacrinus Europæus by T. V. Thompson, Esq. Cork, 1827. He has subsequently ascertained that this animal is the young of the Comatula.

the existing family of star-fishes, and approach most nearly to the Comatula; (see Miller's Crinoïdea, Pl. 1, and p. 127): the bony skeleton constitutes by far the largest portion of these animals. In the living species this bony framework is invested with a gelatinous membrane, accompanied by a muscular system, regulating the movements of every bone. Although, in the fossil species, these softer parts have perished, yet an apparatus for muscular attachment exists on each individual bone.*

The calcareous joints which compose the fingers of the P. Europæus, together with their tentacula, are capable of contraction and expansion in every direction; at one time spreading outwards, like the Petals of an open flower (Pl. 52, Fig. 2′), and at another rolled inwards over the mouth, like an unexpanded bud; the office of these organs is to seize and convey to the mouth its destined food. Thus the habits of living animals illustrate the movements and manner of life of the numerous extinct fossil members of this great family, and afford an example of the validity of the mode of argument, to which we are obliged to have recourse in the consideration of extinct species of organic remains. In this process we argue backwards, and from the mechanical arrangements that pervade the solid portions

* See the tubercles and corrugations on the surfaces of the bones engraved at Pl. 52, Figs. 7, 9, 11, 13, 14, 15, 16, 17.

G.　　　　　　　　F F

of fossil skeletons, infer the nature and functions of the muscles by which motion was imparted to each bone.

I shall select from the many fossil species of the Genus Pentacrinite, that, which from the extraordinary number of auxiliary side arms, placed along its column, has been called the *Briarean Pentacrinite*, and of which our figures (Pl. 51. Figs. 1, 2. ; Pl. 52. Fig. 3. ; and Pl. 53.) will give a more accurate idea than can be conveyed by verbal descriptions.*

* Pl. 51 represents a single specimen of Briarean Pentacrinite, which stands in high relief upon the surface of a slab of Lias, from Lyme Regis, almost entirely made up of a mass of other individuals of the same species.   The arms and fingers are considerably expanded towards the position they would assume in searching for food.   The side arms remain attached to the upper portion only of the vertebral column.

At Pl. 53. Fig. 1 and 2 represent two other specimens of the same species, rising in beautiful relief from a slab, which is composed of a congeries of fragments of similar individuals.   The columns of these specimens, Fig. 2, a, shew the side arms rising in their natural position from the grooves between the angular projections of the Pentagonal stem.   At Pl. 52. Fig. 1. $\overset{a}{\text{F}}$. $\overset{b}{\text{F}}$. are seen the costal plates surrounding the cavity of the body ; at H, the Scapulæ, with the arms and fingers proceeding from them to the extremities of the tentacula.

At Pl. 53. Fig. 3. exhibits the side arms rising from the lower part of a vertebral column, and entirely covering it.   Fig. 4. is another column, on which, the side arms being removed, we see the grooves wherein they articulated with the alternate vertebræ. Fig. 5. exhibits a portion of another column slightly contorted.

Vertebral Column.

The upper part of the vertebral column of Pentacrinites is constructed on principles analogous to those already described in the upper part of the column of the Encrinite.*

All the joints of the column, when seen transversely, present various modifications of pentagonal star-like forms ; hence their name of Asteriæ, or star-stones.

These transverse surfaces are variously covered with a succession of teeth, set at minute intervals from one another, and locking into the interstices between corresponding teeth on the surface of the next vertebræ, they are so disposed as to admit of flexure in all directions, without risk of dislocation. †

As the base or root of Pentacrinites was usu-

---

* The columnar joints of the Briarean Pentacrinite are disposed in pieces alternately thicker and thinner, with a third and still thinner joint interposed between every one of them. Pl. 53. Figs. 8, and 8ᵃ, a. b. c. The edges of this thinnest joint appear externally only at the angles of the column ; internally they enlarge themselves into a kind of intervertebral collar, c. c. c.

A similar alternation in the joints of the Pentacrinite's subangularis is represented in Pl. 52. Figs. 4 and 5.

† The ranges of tubercles upon the exterior surface of each joint in the fragments of columns, Pl. 52. Figs. 7. 9. 11. mark the origin and insertion of muscular fibres, by which the movement of every joint was regulated. At every articulation of the vertebræ, we see also the mode in which the crenated edges lock into one another, combining strength with flexibility. In Pl. 52,

ally fixed to the bottom of the sea, or to some
extraneous floating body, the flexibility of the
jointed column, which forms the stem, was subservient to the double office, first, of varying, in
every direction, the position of the body and arms
in search of food, and secondly, of yielding, with
facility, to the course of the current, or fury of
the storm, swinging, like a vessel held by her
cable, with equal ease in all directions around
her moorings.

The Root of the Briarean Pentacrinite was
probably slight, and capable of being withdrawn
from its attachment.*   The absence of any large

Figs. 11. and 13, the Vertebræ (d.) present five lateral surfaces
of articulation, whereby their side arms were attached to the
vertebral column at distant intervals, as in the Pentacrinus
Caput Medusæ, Pl. 52. Fig. 1.

The double series of crenated surfaces, which pass from the
centre to the points of each of the five radii of these star-shaped
vertebræ, Pl. 52. Figs. 6. to 17.; and Pl. 53. Figs. 9. to 13,
present a beautiful variety of arrangements, not only in each
species, but in different parts of the column of the same species,
according to the degree of flexion which each individual part
required.

* Mr. Miller describes a recent specimen of Pentacrinus Caput
Medusæ, as having the joints next the base partially consolidated, and admitting but little motion, where little is required ;
but higher up, the joints become thinner, and are disposed
alternately, a smaller and thinner joint succeeding a larger and
thicker, to allow a greater freedom of motion, till near the apex
this change is so conspicuous, that the small ones resemble thin
leather-like interpositions.   He also observed traces of the action
of contractile muscular fibres on the internal surfaces of each
vertebra.

solid Secretions, like those of the Pear Encrinite, by which this Pentacrinite could have been fixed permanently to the bottom, and the further fact of its being frequently found in contact with masses of drifted wood converted into jet (Pl. 52, Fig. 3.), leads us to infer that the Briarean Pentacrinite was a locomotive animal, having the power of attaching itself temporarily either to extraneous floating bodies, or to rocks at the bottom of the sea, either by its side arms, or by a moveable articulated small root.*

* The specimen of Briarean Pentacrinite at Pl. 52, Fig. 3. from the Lias at Lyme Regis, adheres laterally to a portion of imperfect jet, which forms part of a thin bed of Lignite, in the Lias marl, between Lyme and Charmouth.

Throughout nearly its whole extent, Miss Anning has constantly observed in this Lignite the following curious appearances : The lower surface *only* is covered by a stratum, entirely composed of Pentacrinites, and varying from one to three inches in thickness; they lie nearly in a horizontal position, with the foot stalks uppermost, next to the lignite. The greater number of these Pentacrinites are preserved in such high perfection, that they must have been buried in the clay that now invests them before decomposition of their bodies had taken place. It is not uncommon to find large slabs several feet long, whose *lower* surface only presents the arms and fingers of these fossil animals, expanded like plants in a Hortus Siccus; whilst the *upper* surface exhibits only a congeries of stems in contact with the under surface of the lignite. The greater number of these stems are usually parallel to one another, as if drifted in the same direction by the current in which they last floated.

The mode in which these animal remains are thus collected immediately *beneath* the Lignite, and never on its *upper* surface, seems to shew that the creatures had attached themselves, in large

Side Arms.

The Side Arms become gradually smaller towards the upper extremity of the column. In the P. Briareus (Pl. 52, Fig. 3. and Pl. 53, Fig. 1. and 3.) these amount to nearly a thousand in number.* The numerous side arms of the Briarean Pentacrinite, when expanded, would act as auxiliary nets to retain the prey of the animal, and also serve as hold-fasts to assist it in adhering to the bottom, or to extraneous bodies. In agitated water they would close and fold themselves along the column, in a position which would expose the least possible surface to the

groups, (like modern barnacles), to the masses of floating wood, which, together with them, were suddenly buried in the mud, whose accumulation gave origin to the marl, wherein this curious compound stratum of animal and vegetable remains is imbedded. Fragments of petrified wood occur also in the Lias, having large groups of Mytili, in the position that is usually assumed by recent mytili, attached to floating wood.

* If we suppose the lower portion of the specimen, Pl. 53, Fig. 2. a. to be united to the upper portion of the fractured stem, Fig. 3, we shall form a correct idea of the manner in which the column of this animal was surrounded with its thousand side-arms, each having from fifty to a hundred joints, Pl. 53, Fig. 14. The number of joints in the side-arms gradually diminishes towards the top of the vertebral column; but as one of the lowest and largest (Pl. 53, Fig. 14.) contains more than a hundred, we shall be much below the reality in reckoning fifty as their average number.

Each of these joints articulates with the adjacent joint, by processes resembling a mortice and tennon; and the form both of

element, and, together with the column and arms, would yield to the direction of the current.

Stomach.

The abdominal cavity, or stomach, of the Pentacrinite, (Pl. 51. Fig. 2.), is rarely preserved in a fossil state ; it formed a funnel-shaped pouch, of considerable size, composed of a contractile membrane, covered externally with many hundred minute calcareous angular plates. At the apex of this funnel was a small aperture, forming the mouth, susceptible of elongation into a proboscis for taking in food.* The place of this organ is in the centre of the body, surrounded by the arms.

the articulating surfaces and of the bone itself, varies so as to give more universal motion as they advance towards the small extremity of the arm. See Pl. 53, Fig. 13. a. b.

In all this delicate mechanism which pervades every individual side-arm, we see provision for the double purpose of attaching itself to extraneous bodies, and apprehending its prey. Five of these arms are set off from each of the largest joints of the vertebral column. At Pl. 53. Fig. 7. a. we see the bases, or first joints of these side-arms articulating with the larger vertebræ, and inclined alternately to the right and left, for the purpose of occupying their position most advantageously for motion, without interfering with each other, or with the flexure of the vertebral column.

In the recent Pentacrinus Caput Medusæ (Pl. 52, Fig. 1.) the side-arms (D.) are dispersed at distant intervals along the column.

* This unique specimen forms part of the splendid collection of James Johnson, Esq. of Bristol.

Body, Arms, and Fingers.

The body of the Pentacrinite, between the summit of the column and the base of the arms, is small, and composed of the pelvis, and the costal, and scapular plates, (see Pl. 51. Pl. 52. Fig. 1. 3. and Pl. 53. Fig. 2. 6. E. F. H.). The arms and fingers are long and spreading, and have numerous joints, or tentacula; each joint is armed at its margin with a small tubercle, or hook, (Pl. 53. Fig. 17.), the form of which varies in every joint, to act as an organ of prehension; these arms and fingers, when expanded, must have formed a net of greater capacity than that of the Encrinites.*

We have seen that Parkinson calculates the number of bones in the Lily Encrinite to exceed twenty-six thousand. The number of bones in the fingers and tentacula of the Briarean Pentacrinite amounts at least to a hundred thousand; if to these we add fifty thousand more for the ossicula of the side-arms, which is much too

* The place of the Pentacrinites in the family Echinoderms, would lead us to expect to find minute pores on the internal surface of the fingers, analogous to those of the more obvious ambulacra of Echini; they were probably observed by Guettard, who speaks of orifices at the terminating points of the fingers and tentacula.

Lamarck also, describing his generic character of Encrinus, says: "The branches of the Umbel are furnished with Polypes, or suckers, disposed in rows."

little, the total number of bones will exceed a hundred and fifty thousand. As each bone was furnished with at least two fasciculi of fibres, one for contraction, the other for expansion, we have a hundred and fifty thousand bones, and three hundred thousand fasciculi of fibres equivalent to muscles, in the body of a single Pentacrinite—an amount of muscular apparatus concerned in regulating the ossicula of the skeleton, infinitely exceeding any that has been yet observed throughout the entire animal creation.*

When we consider the profusion of care, and exquisite contrivance, that pervades the frame of every individual in this species of Pentacrinite, forming but one of many members, of the almost extinct family of Crinoïdeans—and when we add to this the amount of analogous mechanisms that characterize the other genera and species of this curious family,—we are almost lost in astonishment, at the microscopic attention that has been paid to the welfare of creatures, holding so low a place among the inhabitants of the ancient deep;† and we feel a no less irresistible conviction of the universal presence and eternal agency

* Tiedemann, in a monograph on Holothuria, Echini, and Asteriæ, states that the common Star-fish has more than three thousand little bones.

† A frequent repetition of the same parts is proof of the low place and comparative imperfection of the animal in which it occurs. The number of bones in the human body is but two hundred and forty-one, and that of the muscles two hundred and thirty-two pairs. South's Dissector's Manual.

of Creative care, in the lower regions of organic
life, than is forced upon us by the contemplation
of those highest combinations of animal mechanism, which occur in that paragon of animal
organization, the corporeal frame of Man.

SECTION II.

FOSSIL REMAINS OF POLYPES.

It was stated in our Chapter on Strata of the
Transition Series, that some of their most abundant animal remains are fossil Corals or Polyparies. These were derived from an order of
animals long considered to be allied to marine
plants, and designated by the name of Zoophytes;
they are usually fixed, like plants, to all parts
of the bottom of the sea in warm climates which
are not too deep to be below the influence of
solar heat and light, and in many species, send
forth branches, assuming in some degree the form
and aspect of vegetables. These coralline bodies
are the production of Polypes, nearly allied to
the common Actinia, or Sea Anemone of our own
shores. See Pl. 54. Fig. 4. Some of them, e. g.
the Caryophyllia, see Pl. 54. Figs. 9, 10. are soli-
tary, each forming its own independent stem and
support; others are gregarious, or confluent; living together on the same common base or Poly-

pary, which is covered by a thin gelatinous substance, on the surface of which are scattered tentacula, corresponding with the stars on the surface of the coral, (see Pl. 54. Fig. 5).

Le Sueur, who observed them in the West Indies, describes these Polypes, when expanded in calm weather at the bottom of the sea, as covering their stony receptacles with a continuous sheet of most brilliant colours.

The gelatinous bodies of these Polypes are furnished with the power of secreting carbonate of Lime, with which they form a basis of attachment, and cell of retreat. These calcareous cells not only endure beyond the life of the Polypes that secreted them, but approach so nearly to Limestone in their chemical composition, that at the death of the Polype they remain permanently attached to the bottom. Thus one generation establishes the basis whereon the next fixes its habitation, which is destined to form the foundation of a further and continual succession of similar constructions, until the mass, being at length raised to the surface of the sea, a limit is thereby put to its further accumulation.

The tendency of Polypes to multiply in the waters of warm climates is so great, that the bottom of our tropical seas swarms with countless myriads of these little creatures, ever actively engaged in constructing their small but enduring habitations. Almost every submarine rock, and submarine volcanic cone, and ridge, within these latitudes, has become the nucleus and foundation of a colony of Polypes, chiefly belonging to the genera Madrepora, Astrea, Caryophyllia, Meandrina, and Millepora. The calcareous secretions of these Polypes are accumulated into enormous banks or reefs of coral, sometimes extending to a length of many hundred miles; these continually rising to the surface in spots where they were unknown before, endanger the navigation of many parts of the tropical seas.*

If we look to the office these Polypes perform in the present economy of nature, we find them acting as scavengers of the lowest class, perpetually employed in cleansing the waters of the sea from the impurities which escape even the smaller Crustacea ; in the same manner as the Insect Tribes, in their various stages, are destined to find their food by devouring impurities caused by dead animal and vegetable matter upon the land.†

---

* Interesting accounts of the extent and mode of formation of these Coral Reefs may be found in the voyages of Peron, Flinders, Kotzebue, and Beechy ; and an admirable application of the facts connected with modern Corals to the illustration of geological phenomena has been made by Dr. Kidd in his Geological Essay, and by Mr. Lyell, in his Principles of Geology, 3rd edit. vol. iii.

† Mr. De la Beche observed that the Polypes of the Caryophyllia Smithii (Pl. 54, Figs. 9, 10, 11,) devoured portions of the flesh of fishes, and also small Crustacea, with which he fed several individuals at Torquay, seizing them with their tentacula, and digesting them within the central sac which forms their stomach.

The same system appears to have prevailed from the first commencement of life in the most ancient seas, throughout that long series of ages whose duration is attested by the varied succession of animal and vegetable exuviæ, which are buried in the strata of the earth. In all these strata the calcareous habitations of such minute and apparently unimportant creatures as Polypes, have formed large and permanent additions to the solid materials of the globe, and afford a striking example of the influence of animal life upon the mineral condition of the earth.*

If there be one thing more surprising than another in the investigation of natural phenomena, it is perhaps the infinite extent and vast importance of things apparently little and insignificant.

* Among the Corals of the Transition Series are many existing genera, and Mr. de la Beche has justly remarked (Manual of Geology, p. 454) that wherever there is an accumulation of Polypifers such as would justify the appellation of coral banks or reefs, the genera Astrea and Caryophyllia are present; genera which are among architects of coral reefs in the present seas.

A large part of the Limestone called *Coral Rag*, which forms the elevated plains of Bullington and Cunmer, and the hills of Wytham, on three sides of the valley of Oxford, is filled with continuous beds and ledges of petrified corals of many species, still retaining the position in which they grew at the bottom of an ancient sea; as coral banks, are now forming in the intertropical regions of the present ocean.

The same fossil coralline strata extend through the calcareous hills of the N.W. of Berkshire, and N. of Wilts; and again recur in equal or still greater force in Yorkshire, in the lofty summits on the W. and S. W. of Scarborough.

When we descry an insect, smaller than a mite, moving with agility across the paper on which we write, we feel as incapable of forming any distinct conception of the minutiæ of the muscular fibres, which effect these movements, and of the still smaller vessels by which they are nourished, as we are of fully apprehending the magnitude of the universe. We are more perplexed in attempting to comprehend the organization of the minutest Infusoria,* than that of a whale; and one

* Ehrenberg has ascertained that the Infusoria, which have heretofore been considered as scarcely organized, have an internal structure resembling that of the higher animals. He has discovered in them muscles, intestines, teeth, different kinds of glands, eyes, nerves, and male and female organs of reproduction. He finds that some are born alive, others produced by eggs, and some multiplied by spontaneous divisions of their bodies into two or more distinct animals. Their powers of reproduction are so great, that from one individual (Hydatina senta) a million were produced in ten days; on the eleventh day four millions, and on the twelfth sixteen millions. The most astonishing result of his observations is, that the size of the smallest coloured spots on the body of Monas Termo, (the diameter of which is only $\frac{1}{2000}$ of a line) is $\frac{1}{48000}$ of a line, and that the thickness of the skin of the stomach may be calculated at from $\frac{1}{4500000}$ to $\frac{1}{6400000}$ of a line. This skin must also have vessels of a still smaller size, the dimensions of which are too minute to be ascertained. *Abhandlungen der Academie der Wissenschaften zu Berlin*, 1831.

Ehrenberg has described and figured more than 500 species of these Animalcules; many of them are limited to a certain number of vegetable infusions; a few are found in almost every infusion. Many vegetables produce several species, some of which are propagated more readily than others in each particular infusion. The familiar case of the rapid appearance and propagation of animalcules in pepper water will suffice to illustrate the rest.

of the last conclusions at which we arrive, is a conviction that the greatest and most important operations of nature are conducted by the agency of atoms too minute to be either perceptible by the human eye, or comprehensible by the human understanding.

We cannot better conclude this brief outline of the history of fossil Polyparies, extending as they do, from the most early transition rocks to the present seas, than in the words with which

These most curious observations throw important light on the obscure and long-disputed question of equivocal generation; the well-known fact that animalcules of definite characters appear in infusions of vegetable and animal matter, even when prepared with distilled water, receives a probable explanation, and the case of Infusoria no longer appears to differ from that of other animals as to the principle on which their propagation is conducted. The chief peculiarity seems to consist in this, that their increase takes place both by the oviparous and viviparous manner of descent from parent animals, and also by division of the bodies of individuals. The great difficulty is, to explain the manner in which the eggs or bodies of preceding individuals can find access to each particular infusion. This explanation is facilitated by the analogous cases of various fungi which start into life, without any *apparent* cause, wherever decaying vegetable matter is exposed to certain conditions of temperature, humidity, and medium. Fries explains the sudden production of these plants, by supposing the light and almost invisible sporules of preceding plants, of which he has counted above 10,000,000 in a single individual, to be continually floating in the air, and falling every where. The greater part of these never germinate, from not falling on a proper matrix; those which find such matrix start rapidly into life, and begin to propagate.

A similar explanation seems applicable to the case of Infusoria; the extreme minuteness of the eggs and bodies of these animal-

Mr. Ellis expresses the feelings excited in his own mind by his elaborate and beautiful investigations of the history of living Corallines.

"And now, should it be asked, granting all this to be true, to what end has so much labour been bestowed in the demonstration? I can only answer, that as to me these disquisitions have opened new scenes of wonder and astonishment, in contemplating how variously, how extensively, life is distributed through the universe of things,

cules probably allows them to float in the air, like the invisible sporules of fungi; they may be raised from the surface of fluids by various causes of attraction, perhaps even by evaporation. From every pond or ditch that dries up in summer, these dessiccated eggs and bodies may be raised by every gust of wind, and dissipated through the atmosphere like smoke, ready to start into life whenever they fall into any medium admitting of their suscitation; Ehrenberg has found them in fog, in rain, and snow.

If the great aerial ocean which surrounds the earth be thus charged with the rudiments of life, floating continually amidst the atoms of dust we see twinkling in a sun-beam, and ever ready to return to life as soon as they find a matrix adapted to their development, we have in these conditions of the very air we breathe a system of provisions for the almost infinite dissemination of life throughout the fluids of the present Earth; and these provisions are in harmony with the crowded condition of the waters of the ancient world, which is manifested by the multitudes of fossil microscopic remains, to which we have before alluded. (See Sect. viii. page 384.)

Mr. Lonsdale has recently discovered that the Chalk at Brighton, Gravesend, and near Cambridge, is crowded with microscopic shells; thousands of these may be extracted from a small lump, by scrubbing it with a nail brush in water; among these he has recognized vast numbers of the Valves of a marine Cypris (Cytherina) and sixteen species of Foraminifers.

so it is possible, that the facts here related, and these instances of nature animated in a part hitherto unsuspected, may excite the like pleasing ideas in others ; and, in minds more capacious and penetrating, lead to farther discoveries, farther proofs, (should such yet be wanting,) that One infinitely wise, good, all-powerful Being has made, and still upholds, the Whole of what is good and perfect ; and hence we may learn, that, if creatures of so low an order in the great scale of Nature, are endued with faculties that enable them to fill up their sphere of action with such Propriety, we likewise, who are advanced so many gradations above them, owe to ourselves, and to Him who made us and all things, a constant application to acquire that degree of Rectitude and Perfection, to which we also are endued with faculties of attaining."—*Ellis on Corallines*, p. 103.

CHAPTER XVIII.

Proofs of Design in the Structure of Fossil Vegetables.

SECTION I.

GENERAL HISTORY OF FOSSIL VEGETABLES.

THE history of Fossil Vegetables has a twofold claim upon our consideration, in relation to the object of our present enquiry. The first regards the influence exerted on the actual condition of Mankind, by the fossil carbonaceous remains of Plants, which clothed the former surface of the Earth, and has been briefly considered in a former chapter; (Chap. VII. P. 63.) the second directs our attention to the history and structure of the ancient members of the vegetable kingdom.

It appears that nearly at the same points in the progress of stratification, where the most striking changes take place in the remains of Animal life, there are found also concurrent changes in the character of fossil Vegetables.

A large and new field of investigation is thus laid open to our enquiry, wherein we may compare the laws which regulated the varying sys-

tems of vegetation, on the earlier surfaces of our
earth, with those which actually prevail. Should
it result from this enquiry, that the families which
make up our fossil Flora were formed on principles, either identical with those that regulate the
development of existing plants, or so closely
allied to them, as to form connected parts of
one and the same great system of laws, for the
universal regulation of organic life, we shall
add another link to the chain of arguments
which we extract from the interior of the Earth,
in proof of the Unity of the Intelligence and of
the Power, which have presided over the entire
construction of the material world.

We have seen that the first remains of Animal
life yet noticed are marine, and as the existence
of any kind of animals implies the prior, or at least
the contemporaneous existence of Vegetables, to
afford them sustenance, the presence of *sea weeds*
in strata coeval with these most ancient animals,
and their continuance onwards throughout all
formations of marine origin, is a matter of *a priori*
probability, which has been confirmed by the
results of actual observation. M. Adolphe Brongniart, in his admirable History of Fossil Vegetables,* has shewn, that the existing *submarine
vegetation* seems to admit of three great divisions
which characterize, to a certain degree, the Plants
of the frigid, temperate, and torrid zones; and

* Histoire des Végétaux Fossiles, 4to. Paris, 1828.

that an analogous distribution of the fossil submerged Algæ appears to have placed in the
lowest and most ancient formations, genera allied
to those which now grow in regions of the greatest
heat, whilst the forms of marine vegetation that
succeed each other in the Secondary and Tertiary periods, seem to approximate nearer to those
of our present climate, as they are respectively
enclosed in strata of more recent formation.*

If we take a general review of the remains of
terrestrial Vegetables, that are distributed through
the three great periods of geological history, we
find a similar division of them into groups, each
respectively indicating the same successive dimi-

---

* See Ad. Brongniart's Hist. de Vég. Foss. 1 Liv. p. 47.—
Dr. Harlan in the Journal of the Academy of Nat. Sc. of Phila-
delphia, 1831, and Mr. R. C. Taylor in Loudon's Mag. Nat.
Hist. Jan. 1834, have published accounts of numerous deposits
of *fucoids*, as occurring in repeated thin layers among the Transition strata of N. America, and extending over a long tract on the
E. flank of the Alleghany chain.   The most abundant of these is
the Fucoides Alleghaniensis of Dr. Harlan.   Mr. R. C. Taylor has
found extensive deposits of fossil Fuci in the Grauwacke of central
Pennsylvania; in one place seven courses of Plants are laid bare
in the thickness of four feet, in another, one hundred courses
within a thickness of twenty feet.   (*Jameson's Journal, July,*
1835, p. 185.)   I have also seen Fucoids in great abundance in
the Grauwacke-slate of the Maritime Alps, in many parts of the
new road between Nice and Genoa.   I once found small Fucoids
dispersed abundantly through shale of the Lias formation, from
a well at Cheltenham.   The Fucoides granulatus occurs in Lias
at Lyme Regis, and at Boll in Wurtemberg; and F. Targionii in
the Upper Green-sand near Bignor in Sussex.

nutions of Temperature upon the Land, which
have been inferred from the remains of the vegetation of the Sea. Thus, in strata of the Transition
series, we have an association of a few *existing*
families of *Endogenous* Plants,* chiefly Ferns
and Equisetaceæ, with *extinct* families both *Endogenous* and *Exogenous*, which some modern
botanists have considered to indicate a Climate
hotter than that of the Tropics of the present day.

In the Secondary formations, the species of
these most early families become much less numerous, and many of their genera, and even of
the families themselves entirely cease ; and a
large increase takes place in two families, that
comprehend many existing forms of vegetables,
and are rare in the Coal formation, viz. *Cycadeæ*
and *Coniferæ*. The united characters of the groups
associated in this series, indicate a Climate,
whose temperature was nearly similar to that
which prevails within the present Tropics.

In the Tertiary deposits, the greater number of
the families of the first series, and many of those
of the second, disappear ; and a more complicated *dicotyledonous* † Vegetation takes place of

* Endogenous Plants are those, the growth of whose stems takes
place by addition from within. Exogenous are those in which
the growth takes place by addition from without.

† Monocotyledonous Plants are those, the embryo of whose
seed is made up of one cotyledon or lobe, like the seed of a Lily
or an Onion. Dicotyledonous Plants are those, the embryo of

the simpler forms which predominated through the two preceding periods. Smaller Equisetaceæ also succeed to the gigantic Calamites ; Ferns are reduced in size and number to the scanty proportions they bear on the southern verge of our temperate climates; the presence of Palms attests the absence of any severe degree of cold, and the general character marks a Climate nearly approaching to that of the Mediterranean.

We owe to the labours of Schlotheim, Sternberg and Ad. Brongniart the foundation of such a systematic arrangement of fossil plants, as enables us to enter, by means of the analogies of recent plants, into the difficult question of the Ancient Vegetation of the Earth, during those periods when the strata were under the process of formation.

Few persons are aware of the nature of the evidence, upon which we have at length arrived at a certain and satisfactory conclusion, respecting the long disputed question as to the vegetable origin of Coal. It is not unfrequent to find among

whose seed is made up of two lobes, as in the Bean and Coffee seed. The stems of Monocotyledonous Plants are all Endogenous, i. e. increase from within by the addition of bundles of vessels set in cellular substance, and enlarge their bulk by addition from the centre outwards, e. g. Palms, Canes, and Liliaceous plants. The stems of Dicotyledonous Plants are all Exogenous, i. e. increase externally by the addition of concentric layers from without; these form the rings, which mark the amount of annual growth in the Oak and other forest trees in our climate.

the cinders beneath our grates, traces of fossil plants, whose cavities, having been filled with silt, at the time of their deposition in the vegetable mass, that gave origin to the Coal, have left the impression of their forms upon clay and sand enclosed within them, sharp as those received by a cast from the interior of a mould.

A still more decisive proof of the vegetable origin, even of the most perfect bituminous Coal has recently been discovered by Mr. Hutton; he has ascertained that if any of the three varieties of Coal found near Newcastle be cut into very thin slices and submitted to the microscope, more or less of vegetable structure can be recognized.*

* " In these varieties of coal," says Mr. Hutton, " even in samples taken indiscriminately, more or less of Vegetable Texture could always be discovered, thus affording the fullest evidence, if any such proof were wanting, of the Vegetable Origin of Coal.

" Each of these three kinds of coal, besides the fine distinct reticulation of the original vegetable texture, exhibits other cells, which are filled with a light wine-yellow-coloured matter, apparently of a bituminous nature, and which is so volatile as to be entirely expelled by heat, before any change is effected in the other constituents of the coal. The number and appearance of these cells vary with each variety of coal. In caking coal, the cells are comparatively few, and are highly elongated.—In the finest portions of this coal, where the crystalline structure, as indicated by the rhomboidal form of its fragments, is most developed, the cells are completely obliterated.

" The slate-coal, contains two kinds of cells, both of which are filled with yellow bituminous matter. One kind is that already noticed in caking coal; while the other kind of cells constitutes groups of smaller cells, of an elongated circular figure.

We shall further illustrate this point, by a brief description of the manner in which the remains of vegetables are disposed in the Carboniferous strata of two important Coal fields, namely, those of Newcastle in the north of England, and of Swina in Bohemia, on the N. W. of Prague.

The Newcastle Coal field is at the present time supplying rich materials to the Fossil Flora of Great Britain, now under publication by Professor Lindley and Mr. Hutton. The plants of the Bohemian Coal field laid the foundation of Count Sternberg's *Flore du monde primitif*, the publication of which commenced at Leipsic and Prague in 1820.

" In those varieties which go under the name of Cannel, Parrot, and Splent Coal, the crystalline structure, so conspicuous in fine caking coal, is wholly wanting; the first kind of cells are rarely seen, and the whole surface displays an almost uniform series of the second class of cells, filled with bituminous matter, and separated from each other by thin fibrous divisions. Mr. Hutton considers it highly probable that these cells are derived from the reticular texture of the parent plant, rounded and confused by the enormous pressure, to which the vegetable matter has been subject."

The author next states that though the crystalline and uncrystalline, or, in other terms, perfectly and imperfectly developed varieties of coal generally occur in distinct strata, yet it is easy to find specimens which in the compass of a single square inch, contain both varieties. From this fact, as also from the exact similarity of position which they occupy in the mine, the differences in different varieties of coal are ascribed to original difference in the plants from which they were derived. *Proceedings of Geological Society. Lond. and Edin. Phil. Mag. 3rd Series, Vol.* 2. p. 302. *April*, 1833.

Lindley and Hutton state (Fossil Flora, Vol.
I. page 16) that " It is the beds of shale, or
argillaceous schistus, which afford the most abundant supply of these curious relics of a former
World ; the fine particles of which they are composed having sealed up and retained in wonderful
perfection, and beauty, the most delicate forms of
the vegetable organic structure. Where shale
forms the roof of the workable seams of coal, as
it generally does, we have the most abundant
display of fossils, and this, not perhaps arising
so much from any peculiarity in these beds, as
from their being more extensively known and
examined than any others. The principal deposit is not in immediate contact with the coal,
but about from twelve to twenty inches above it ;
and such is the immense profusion in this situation, that they are not unfrequently the cause of
very serious accidents, by breaking the adhesion
of the shale bed, and causing it to separate and
fall, when by the operation of the miner the coal
which supported it is removed. After an extensive fall of this kind has taken place, it is a curious sight to see the roof of the mine covered
with these vegetable forms, some of them of great
beauty and delicacy ; and the observer cannot
fail to be struck with the extraordinary confusion,
and the numerous marks of strong mechanical
action exhibited by their broken and disjointed
remains."

A similar abundance of distinctly preserved
vegetable remains, occurs throughout the other
Coal fields of Great Britain. But the finest
example I have ever witnessed, is that of the
coal mines of Bohemia just mentioned. The most
elaborate imitations of living foliage upon the
painted ceilings of Italian palaces, bear no comparison with the beauteous profusion of extinct
vegetable forms, with which the galleries of these
instructive coal mines are overhung. The roof
is covered as with a canopy of gorgeous tapestry,
enriched with festoons of most graceful foliage,
flung in wild, irregular profusion over every portion of its surface. The effect is heightened by
the contrast of the coal-black colour of these
vegetables, with the light ground work of the
rock to which they are attached. The spectator
feels himself transported, as if by enchantment,
into the forests of another world ; he beholds
Trees, of forms and characters now unknown
upon the surface of the earth, presented to his
senses almost in the beauty and vigour of their
primeval life ; their scaly stems, and bending
branches, with their delicate apparatus of foliage,
are all spread forth before him ; little impaired
by the lapse of countless Ages, and bearing
faithful records of extinct systems of vegetation,
which began and terminated in times of which
these relics are the infallible Historians.

Such are the grand natural Herbaria wherein

these most ancient remains of the vegetable kingdom are preserved, in a state of integrity, little
short of their living perfection, under conditions
of our Planet which exist no more.

SECTION II.

VEGETABLES IN STRATA OF THE TRANSITION SERIES.*

THE remains of plants of the Transition period
are most abundant in that newest portion of the
deposits of this era, which constitutes the Coal
Formation, and afford decisive evidence as to
the condition of the vegetable kingdom at this
early epoch in the history of Organic Life.

The Nature of our Evidence will be best illustrated, by selecting a few examples of the many
genera of fossil plants that are preserved in the
Strata of the Carboniferous Order, beginning
with those which are common both to the ancient
and existing states of Vegetable Life.

Equisetaceæ.†

Among existing vegetables, the Equisetaceæ
are well known in this climate in the common
Horse-tail of our swamps and ditches.   The extent of this family reaches from Lapland to the

---

* See Pl. 1. Figs. 1, to 13.
† See Pl. 1. Fig. 2.

Torrid Zone, its species are most abundant in the temperate zone, decrease in size and number as we approach the regions of cold, and arrive at their greatest magnitude in the warm and humid regions of the Tropics, where their numbers are few.

M. Ad. Brongniart* has divided fossil Equisetaceæ into two Genera; the one exhibits the characters of living *Equiseta*, and is of rare occurrence in a fossil state; the other is very abundant, and presents forms that differ materially from them, and often attain a size unknown among living Equisetaceæ; these have been arranged under the distinct genus *Calamites*,† they abound universally in the most ancient Coal formation, occur but sparingly in the lower strata of the Secondary series, and are entirely wanting in the Tertiary formations, and also on the actual surface of the earth.

The same increased development of size, which in recent Equisetaceæ accompanies their geogra-

---

* Histoire des Végétaux Fossiles, 2nd Livraison.

† Calamites are characterized by large and simple cylindrical stems, articulated at intervals, but either *without sheaths*, or presenting them under forms unknown among existing Equiseta; they have sometimes marks of verticillated Branches around their articulations, the leaves also are without joints. But the most obvious feature wherein they differ from Equiseta, is their bulk and height, sometimes exceeding six or seven inches in diameter, whilst the diameter of a living Equisetum rarely exceeds half an inch. A Calamite fourteen inches in diameter has lately been placed in the Museum at Leeds.

phical approximation to the Equator, is found in the fossil species of this order to accompany the higher degrees of Antiquity of the strata in which they occur; and this without respect to the latitude, in which these formations may be placed. M. Ad. Brongniart (Prodrome, p. 167) enumerates twelve species of *Calamites* and two of *Equiseta* in his list of plants found in strata of the Carboniferous order.

*Ferns.**

The family of Ferns, both in the living and fossil Flora, is the most numerous of vascular Cryptogamous plants.† Our knowledge of the geographical distribution of existing Ferns, as connected with Temperature, enables us in some degree to appreciate the information to be derived from the character of fossil Ferns, in regard to the early conditions and Climate of our globe.

* See Pl. I. No. 6. 7. 8. 37. 38. 39.

† Ferns are distinguished from all other vegetables by the peculiar division and distribution of the veins of the leaves; and in arborescent species, by their cylindrical stems without branches, and by the regular disposition and shape of the scars left upon the stem, at the point from which the Petioles, or leaf stalks, have fallen off. Upon the former of these characters M. Ad. Brongniart has chiefly founded his classification of fossil Ferns, it being impossible to apply to them the system adopted in the arrangement of living Genera, founded on the varied disposition of the fructification, which is rarely preserved in a fossil state.

The total known number of existing species of Ferns is about 1500. These admit of a threefold geographical distribution :

1. Those of the temperate and frigid zone of the northern hemisphere, containing 144 species.

2. Those of the southern temperate zone, including the Cape of Good Hope, parts of South America, and the extra-tropical part of New Holland, and New Zealand, 140 species.

3. Those which grow within 30 or 35 degrees on each side of the Equator, 1200 species.

If we compare the amount of Ferns with the united numbers of other tribes of plants, we may form some idea of the relative importance of this family in the vegetation of the district, or period to which we apply such comparison. Thus, in the entire number of known species of plants now existing on the globe, we have 1500 Ferns and 45,000 Phanerogamiæ, being in the proportion of 1 to 30. In Europe this proportion varies from 1:35 to 1:80, and may average 1:60. Between the Tropics, Humboldt estimates the number in Equinoxial America at 1:36, and Mr. Brown gives 1:20 as the proportion in those parts of intertropical Continents which are most favourable* to Ferns.

Mr. Brown (Appendix to Tuckey's Congo Expedition) states that the circumstances most

---

* Botany of Congo, p. 42.

favourable to the growth of Ferns are humidity, shade, and heat. These circumstances are most frequently combined in the highest degree in small and lofty tropical islands, where the air is charged with humidity, which it is continually depositing on the mountains, and thereby imparting freshness to the soil. Thus in Jamaica Ferns are to the Phanerogamiæ nearly in the proportion of 1 to 10 ; in New Zealand as 1 to 6 ; in Taiti as 1 to 4 ; in Norfolk Island as 1 to 3 ; in St. Helena as 1 to 2 ; in Tristan d'Acunha (extratropical) as 2 to 3. Ferns are also the most abundant Plants in the Islands of the Indian Archipelago.

It appears still further, that not only are certain Genera and Tribes of Ferns peculiar to certain climates, but that the enlarged size of the arborescent species depends in a great degree on Temperature, since Arborescent Ferns are now found chiefly within, or near the limit of the Tropics.*

From the above considerations as to the characters and distribution of living Ferns, M. Ad. Brongniart has applied himself with much ingenuity, to illustrate the varying condition and climate of our Globe, during the successive periods of geological formations. Finding that the fossil

---

* The few exceptions to this rule appear to be 'confined to the southern hemisphere, and one species is found in New Zealand as far south as lat. 46°. *See Brown in Appendix to Flinders's Voyage.*

remains of Ferns decrease continually in number, as we ascend from the most ancient to the
most recent strata, he founds upon this fact an
important conjecture, with respect to the successive diminutions of temperature, and changes of
climate, which the earth has undergone.   Thus,
in the great Coal formation there are about 120
known species of Ferns, forming almost one half
of the entire known Flora of this formation ;
these species represent but a small number of
the forms which occur among living Ferns, and
nearly all belong to the Tribe of Polypodiaceæ,
in which Tribe we find the greater number of
existing arborescent species.*   Fragments of the
stems of arborescent Ferns occur occasionally in
the same formation.   M. Brongniart considers
these circumstances as indicating a vegetation,
analogous to that of the Islands in the equinoctial
regions of the present Earth ; and infers that
the same conditions of Heat and Humidity which
favour the existing vegetation of these islands,
prevailed in still greater degree during the for-

---

* In plate 1, figs. 7, and 37, represent two of the graceful
forms of arborescent Ferns which adorn our modern tropics, where
they attain the height of forty and fifty feet.

An arborescent Fern forty-five feet high (Alsophila bruno-
niana), from Silhet in Bengal, may be seen in the staircase of the
British Museum.   The stems of these Ferns are distinguished
from those of all arborescent Monocotyledonous plants, by the
peculiar form and disposition of the scars, from which the Petioles

mation of the Carboniferous strata of the Transition Series.

In strata of the Secondary Series, the absolute and relative numbers of species of Ferns considerably diminishes, forming scarcely one third of the known Flora of these midway periods of geological history. (See Pl. 1. Figs. 37. 38. 39.)

In the Tertiary Strata, Ferns appear to bear to other vegetables nearly the same proportion as in the temperate regions of the present Earth.

or leaf stalks have fallen off. In Palms and other arborescent Monocotyledons, the leaves, or Petioles, embrace the stem and leave broad transverse scars, or rings, whose longer diameter is *horizontal*. In the case of Ferns alone, with the single exception of Angiopteris, the scars are either elliptic or rhomboidal, and have their longer diameter *vertical*.

M. Ad. Brongniart (Hist. des Veg. Foss. p. 261, Pl. 79. 80.) has described and figured the leaf and stem of an arborescent fern (Anomopteris, Mougeottii) from the variegated sand-stone of Heilegenberg in the Vosges. Beautiful leaves of this species, with their capsules of fructification sometimes adhering to the pinnules, abound in the New red sand-stone formation of this district.

M. Cotta has published an interesting Work on fossil Remains of arborescent ferns, which occur abundantly in the New red sand-stone of Saxony near Chemnitz. (*Dendrolithen. Dresden and Leipsig*, 1832.) These consist chiefly of sections of the Trunks of many extinct species, sufficiently allied in structure to that of existing arborescent Ferns, to leave little doubt that they are the remains of extinct species of arborescent Plants of this family, that grew in Europe at this Period of the Secondary formation.

*Lepidodendron.**

The genus Lepidodendron comprehends many species of fossil plants, which are of large size, and of very frequent occurrence in the Coal formation. In some points of their structure they have been compared to Coniferæ, but in other respects and in their general appearance, with the exception of their great size, they very much resemble the *Lycopodiaceæ*, or *Club Moss* Tribe. (See Pl. 1. Figs. 9. 10.). This tribe at the present day, contains no species more than three feet high, but the greater part of them are weak, or creeping plants, while their earliest fossil representatives appear to have attained the dimensions of Forest Trees.†

Existing Lycopodiaceæ follow nearly the same law as ferns and Equisetaceæ, in respect of geographical distribution; being largest and most abundant in hot and humid situations within the Tropics, especially in small islands. The belief that Lepidodendra were allied to the Lycopo-

---

* Pl. 1. Figs. 11. 12. and Pl. 55, Figs. 1. 2. 3.

† Prof. Lindley states that the affinities of existing Lycopodiaceæ are intermediate between Ferns and Coniferæ on the one hand, and Ferns and Mosses on the other; They are related to Ferns in the want of sexual apparatus, and in the abundance of annular ducts contained in their axis; to Coniferæ, in the aspect of the stems of some of the larger kinds; and to Mosses in their whole appearance.

diaceæ, and their size, and abundant occurrence among the fossils of the Coal Formation have led writers on fossil plants to infer that great heat, and moisture, and an insular Position were the conditions, under which the first forms of this family attained that gigantic stature, which they exhibit in deposits of the Transition period; thus corroborating the conclusion they had derived from the Calamites associated with them, as already mentioned.*

Lindley and Hutton state, that Lepidodendra are, after Calamites, the most abundant class of fossils in the Coal formation of the North of England; they are sometimes of enormous size, fragments of stems occurring from twenty to forty-five feet long; in the Jarrow colliery a compressed tree of this class measured four feet two inches in breadth. Thirty-four species of Lepi-

---

* The leaves of existing Lycopodiaceæ are simple, and arranged in spiral lines around the stem, and impress on its surface scars of rhomboidal, or lanceolate form, marked with prints of the insertions of vessels. In the fossil Lepidodendra, we find a large and beautiful variety of similar scars, arranged like scales in spiral order, over the entire surface of the stems. A large division of these are arborescent and dichotomous, and have their branches covered with simple lanceolate leaves. Our figure of Lepidodendron Sternbergii (Pl. 55. Figs. 1. 2. 3.) represents all these characters in a single Tree from the Coal mines of Swina in Bohemia.

The form of the scales varies at different parts of the same stem, those nearest the base are elongated in the vertical direction.

dodendron are enumerated in M. Ad. Brongniart's Catalogue of fossil plants of the coal formation.

The internal structure of the Lepidodendron has been shewn to be intermediate between Lycopodiaceæ and Coniferæ,[*] and the conclusions which Prof. Lindley draws from the intermediate condition of this curious extinct genus of fossil plants, are in perfect accordance with the inferences which we have had occasion to derive from analogous conditions in extinct genera of fossil animals. " To Botanists, this discovery is of very high interest, as it proves that those systematists are right, who contend for the possibility of certain chasms now existing between the gradations of organization, being caused by the extinction of genera, or even of whole orders ; the existence of which was necessary to complete the harmony which it is believed originally existed in the structure of all parts of the Vegetable kingdom. By means of Lepidodendron, a better passage is established from Flowering to Flowerless Plants, than by either Equisetum or Cycas, or any other known genus." *Lindley and Hutton's Fossil Flora, vol. ii. page* 53.

[*] See annual Report of the Yorkshire Phil. Society for 1832. Witham's Fossil Vegetables, 1833, Pl. 12. 13. and Lindley and Hutton's Fossil Flora. Pl. 98 and 99.

Sigillaria.[*]

Besides the above plants of the Coal formation which are connected with existing Families or Genera, there occur many others which can be referred to no known type in the vegetable kingdom. We have seen that the Calamites take their place in the existing family of Equiseteceæ; that many fossil Ferns are referrible to living genera of this extensive family; and that Lepidodendra approximate to living Lycopodiaceæ and Coniferæ. Together with these, there occur other groups of Plants unknown in modern vegetation, and of which the duration seems to have been limited to the Epochs of the Transition Period. Among the largest and tallest of these unknown forms of Plants, we find colossal Trunks of many species, which M. Ad. Brongniart has designated by the name of *Sigillaria*. These are dispersed throughout the sand-stones and shales that accompany the Coal, and can occasionally be detected in the Coal itself, to the substance of which they have largely contributed by their remains. They are sometimes seen in an erect position, where views of the strata are afforded by cliffs on the

* Pl. 56, Figs. 1. 2.

sea shore, or by inland sections of quarries, banks of rivers, &c.*

The vertical position of these trunks, however, is only occasional and accidental; they lie inclined at all degrees throughout all the strata of the carboniferous series; but are most frequently prostrate, and parallel to the lines of stratification, and, in this position are usually compressed. When erect, or highly inclined, they retain their natural shape, and their interior is filled with sand or clay, often different from that of the stratum in which their lower parts are fixed, and mixed with small fragments of various other plants. As this foreign matter has thus entirely

* On the coast of Northumberland, at Greswell hall, and Newbiggin, near Morpeth, many stems of Sigillaria may be seen, standing erect at right angles to the planes of alternating strata of shale and sand-stone; they vary from ten to twenty feet in height, and from one to three feet in diameter, and are usually truncated at their upper end; many terminate downwards in a bulb-shaped enlargement, near the commencement of the roots, but no roots remain attached to any of them. Mr. W. C. Trevelyan counted twenty portions of such Trees, within the length of half a mile; all but four or five of these were upright; the bark, which was seen when they were first uncovered, but soon fell off, was about half an inch in thickness, and entirely converted into coal. Mr. Trevelyan observed four varieties of these stems, and engraved a sketch of one of them in 1816, which is copied in our Pl. 56. fig. 1.

In September, 1834, I saw in one of the Coal Mines of Earl Fitzwilliam, at Elsecar, near Rotherham, many large Trunks of Sigillaria, in the sides of a gallery by which you walk into the mine, from the outcrop of a bed of Coal about six feet thick. These stems were inclined in all directions, and some of them

filled the interior of these trunks, it follows that they must have been without any transverse dissepiments, and hollow throughout, at the time when the sand, and mud, and fragments of other plants, found admission to their interior. The bark, which alone remains, and has been converted into coal, probably surrounded an axis composed of soft and perishable pulpy matter, like the fleshy interior of the stems of living Cacteæ; and the decay of this soft internal trunk, whilst the stems were floating in the water, probably made room for the introduction of the sand and clay.

These trunks usually vary from half a foot to three feet in diameter. When perfect, the height nearly vertical. The interior of those whose inclination exceeded 45° was filled with an indurated mixture of clay and sand; the lower extremity of several rested on the upper surface of the bed of Coal. None had any traces of Roots, nor could any one of them have grown in its present place.

M. Alex. Brongniart has engraved a section at St. Etienne, in which many similar stems are seen in an erect position, in sandstone of the Coal formation, and infers from this fact that they grew on the spot where they are now found. M. Constant Prevost justly objects to this inference, that, had they grown on the spot, they would all have been rooted in the same stratum, and not have had their bases in different strata. When I visited these quarries in 1826, there were other trunks, more numerous than the upright ones, inclined in various directions.

I have seen but one example, viz. that of Balgray quarry, three miles N. of Glasgow, of erect stumps of large trees fixed by their roots in sand-stone of the coal formation, in which, when soft, they appear to have grown, close to one another. See Lond. and Edin. Phil. Mag. Dec. 1835, p. 487.

of many of them must have been fifty or sixty feet, at least.*

Count Sternberg has applied the name *Syringodendron* to many species of Sigillaria, from the parallel pipe-shaped flutings that extend from the top to the bottom of their trunks. These trunks are without joints, and many of them attain the size of forest trees. The flutings on their surface bear dot-like, or linear impressions, of various figures, marking the points at which the leaves were inserted into the stem. This fluted portion of the Sigillariæ, formed their external covering, separable like true bark from the soft internal axis, or pulpy trunk ; it varied in thickness from an inch to one-eighth of an inch, and is usually converted into pure coal. (See Pl. 56, Fig. 2. *a, b, c.*)

A fleshy trunk surrounded and strengthened only by such thin bark, must have been incapable of supporting large and heavy branches at its summit. It therefore probably terminated abruptly at the top, like many of the larger species of living Cactus, and the abundant disposition of small leaves around the entire extent of the trunk seems to favour this hypothesis.

* M. Ad. Brongniart found in a coal mine in Westphalia near Essen, the compressed stem of a Sigillaria laid horizontally, to the length of forty feet ; it was about twelve inches in diameter at its lower, and six inches at its upper extremity, where it divided into two parts, each four inches in diameter. The lower end was broken off abruptly. Lindley and Hutton's Foss. Flora, vol. i. p. 153.

The impressions, or scars, which formed the articulations of leaves on the longitudinal flutings of the trunks of Sigillariæ, are disposed in vertical rows on the centre of each fluting from the top to the bottom of the trunk. Each of these scars marks the place from which a leaf has fallen off, and exhibits usually two apertures, by which bundles of vessels passed through the bark to connect the leaves with the axis of the tree. No leaf has yet been found attached to any of these trunks; we are therefore left entirely to conjecture as to what their nature may have been. This non-occurrence of a single leaf upon any one of the many thousand trunks that have come under observation, leads us to infer that every leaf was separated from its articulation, and that many of them perhaps, like the fleshy interior of the stems, had undergone decomposition, during the interval in which they were floating between their place of growth, and that of their final submersion.

M. Ad. Brongniart enumerates forty-two species of Sigillaria, and considers them to have been nearly allied to arborescent Ferns, with leaves very small in proportion to the size of the stems, and differently disposed from those of any living Ferns. He would refer to these stems many of the numerous fern leaves of unknown species, which resemble those of existing arborescent genera of this family. Lindley and Hutton shew strong reasons for considering that Sigillariæ were Dicotyledonous plants, entirely distinct from Ferns, and different from any thing that occurs in the existing system of vegetation.*

Favularia.   Megaphyton.   Bothrodendron. Ulodendron.†

The same group of fossil plants to which Lindley and Hutton have referred the genus Sigillaria, contains four other extinct genera, all of which exhibit a similar disposition of *scars arranged in vertical rows*, and indicating the places at which leaves, or cones, were attached to the trunk. The names of these are Favularia, Megaphyton, Bothrodendron, Ulodendron.‡ Our figures Pl. 56, Figs. 3, 4, 5, 6, represent portions of

---

* "There can be no doubt," say they, (Foss. Flora, vol. i. p. 155) that as far as external characters go, Sigillaria approached Euphorbiæ and Cacteæ more nearly than any other plants now known, particularly in its soft texture, in its deeply channelled stems, and what is of more consequence in its scars, placed in perpendicular rows between the furrows. It is also well known that both these modern tribes, particularly the latter, arrive even now at great stature; further, it is extremely probable, indeed almost certain, that Sigillaria was a dicotyledonous plant, for no others at the present day have a true separable bark. Nevertheless, in the total absence of all knowledge of the leaves and flowers of these ancient trees, we think it better to place the genus among other species, the affinity of which is at present doubtful."

† Pl. 56, Figs. 3. 4. 5. 6. 7.

‡ The genera composing this group are thus described, Foss. Flora, vol. ii. p. 96.

1. Sigillaria.   Stem furrowed.   Scars of leaves small, round, much narrower than the ridges of the stem.

the trunk and scars of some of these extraordinary Coniferæ.

Among existing vegetables, there are only a few succulent plants which present a similar disposition of leaves, one exactly above another in parallel rows; but in the fossil Flora of the Coal

2. Favularia. Stem furrowed. Scars of leaves small, square, as broad as the ridges of the stem.
3. Megaphyton. Stem not furrowed, dotted. Scars of leaves very large, of a horse shoe figure, much narrower than the ridges.
4. Bothrodendron. Stem not furrowed, covered with dots. Scars of cones, obliquely oval.
5. Ulodendron. Stem not furrowed, covered with rhomboidal marks. Scars of cones circular.

In the three first genera of this group, the scars appear to have given origin to leaves; in the two latter they indicate the insertion of large cones.

In the genus Favularia (Pl. 56, Fig. 7) the trunk was entirely covered with a mass of densely imbricated foliage, the bases of the leaves are nearly square, and the rows of leaves separated by intermediate grooves; whilst in Sigillaria the leaves were placed more loosely, and at various intervals in various species. (Foss. Flora, Pl. 73. 74. 75).

In the genus Megaphyton the stem is not furrowed, and the leaf scars are very large, and resemble the form of horse shoes disposed in two vertical rows, one on each side of the trunk. The minor impressions resembling horse shoes, in the middle of these scars, appear to indicate the figure of the woody system of the leaf stalk. (Foss. Flora, Pl. 116, 117.)

In the genus Bothrodendron (Foss. Flora, Pl. 80, 81) and the genus Ulodendron, (Foss. Flora, Pl. 5. 6.) the stems are marked with deep oval or circular concavities, which appear to have been made by the bases of large cones. These cavities are ranged in two vertical rows, on opposite sides of the trunk, and in some species are nearly five inches in diameter. (Pl. 56. figs. 3. 4. 5. 6.)

formation, nearly one half, out of eighty known species of *Arborescent* plants, have their leaves growing in parallel series. The remaining half are *Lepidodendra*, or extinct Coniferæ. (See Lindley and Hutton, Foss. Flora, vol. ii. p. 93.)

*Stigmaria.**

The recent discoveries of Lindley and Hutton have thrown much light upon this very extraordinary family of extinct fossil plants. Our figure, Pl. 56, Fig. 8, copied from their engraving of Stigmaria ficoides, (Foss. Flora, Pl. 31, Fig. 1) represents one of the best known examples of the genus.†

The centre of the plant presents a dome-shaped trunk or stem, three or four feet in diameter, the substance of which was probably yielding and fleshy; both its surfaces were slightly corrugated, and covered with indistinct circular spots. (Pl. 56, Fig. 8. 9.)

From the margin of this dome there proceed many horizontal branches, varying in number in different individuals from nine to fifteen; some of these branches become forked at unequal distances from the dome; they are all broken off

---

* Pl. 56, Fig. 8. 9. 10. 11.

† Seventeen specimens of this kind have been found within the space of 600 square yards, in the shale covering the Bensham seam of coal at Jarrow Colliery near Newcastle, at the depth of 1200 feet.

short, the longest yet found attached to the stem, was four feet and a half in length. The extent of these branches, when outstretched and perfect, was probably from twenty to thirty feet.* The surface of each branch is covered with spirally disposed tubercles, resembling the papillæ at the base of the spines of Echini. From each tubercle there proceeded a cylindrical and probably succulent leaf; these extended to the length of several feet from all sides of the branches. (Pl. 56, Figs. 10. 11.) The leaves, usually in a compressed state, are found penetrating in all directions into the sand-stone or shale which forms the surrounding matrix; they have been traced to the length of three feet, and have been said to be much longer.†

---

* It appears from sections of a branch of Stigmaria, engraved by Lindley and Hutton, (Foss. Flora, Pl. 166), that its interior was a hollow cylinder composed exclusively of spiral vessels, and containing a thick pith, and that the transverse section exhibits a structure something like that of Coniferæ, but without concentric circles, and with open spaces instead of the muriform tissue of medullary rays. No such structure is known among living plants.

These cylindrical branches are usually depressed on one side, probably the inferior side (Pl. 56, Figs. 8. *a b.* and 10. *b.*); adjacent to this depression there is found a loose internal eccentric axis, or woody core, (Pl. 56. Fig. 10. *a.*) surrounded with vascular fasciculi that communicated with the external tubercles, and resembled the internal axis within the stems of certain species of Cactus.

† All these are conditions, which a Plant habitually floating with the leaves distended in every direction, would not cease to maintain, when drifted to the bottom of an Estuary, and there gradually surrounded by sediments of mud and silt.

In many of the strata that accompany the coal, fragments of these plants occur in vast abundance; they have been long noticed in the sandstone called *Gannister* and *Crowstone*, in the Yorkshire and Derbyshire coal fields, and have been incorrectly considered to be fragments of the stems of Cacti.

The discovery of the dome-shaped centres above described, and the length and forms of the leaves and branches render it highly probable that the Stigmariæ were aquatic plants, trailing in swamps, or floating in still and shallow lakes, like the modern Stratiotes and Isoetes. From such situations they may have been drifted by the same inundations, that transported the Ferns and other land vegetables, with which they are associated in the coal formation. The form of the trunk and branches shews that they could not have risen upwards into the air; they must therefore either have trailed on the ground, or have floated in water.* The Stigmaria was probably dicotyledonous, and its internal structure seems to have borne some analogies to that of the Euphorbiaceæ.

* The place and form of the leaves, supposing them to have grown on all sides of branches suspended horizontally in water, would have been but little changed by being drifted into, and sinking to the bottom of, an estuary or sea, and there becoming surrounded by sediments of mud or sand. This hypothesis seems supported by the observations made at Jarrow, that the extremities of the branches *descend* from the dome towards the adjacent bed of coal.

Conclusion.

Besides these Genera which have been enumerated, there are many others whose nature is
still more obscure, and of which no traces have
been found among existing vegetables, nor in
any strata more recent than the Carboniferous
series.* Many years must elapse before the
character of these various remains of the primeval vegetation of the Globe can be fully understood. The plants which have contributed
most largely to the highly-interesting and important formation of Coal, are referrible principally
to the Genera whose history we have attempted
briefly to elucidate : viz. Calamites, Ferns, Lycopodiaceæ, Sigillariæ, and Stigmariæ. These
materials have been collected chiefly from the
carboniferous strata of Europe. The same kind
of fossil plants are found in the coal mines of
N. America, and we have reason to believe
that similar remains occur in Coal formations of
the same Epoch, under very different Latitudes,
and in very distant quarters of the Globe, e. g.
in India, and New Holland, in Melville Island,
and Baffin's Bay.

The most striking conclusions to which the
present state of our knowledge has led, respect-

* Some of the most abundant of these have been classed under
the names of Asterophyllites, (see Pl. 1, Figs. 4. 5.) from the
stellated disposition of the leaves around the branches.

ing the vegetables which gave origin to coal are, 1st, that a large proportion of these plants were vascular Cryptogamiæ, and especially Ferns ; 2dly, that among these Cryptogamic plants, the Equisetaceæ attained a gigantic size; 3rdly, that Dicotyledonous plants, which compose nearly two-thirds of living Vegetables, formed but a small proportion of the Flora of these early periods.* 4thly, that although many extinct genera, and certain families have no living representatives, and even ceased to exist after the deposition of the Coal formation, yet are they connected with modern vegetables by common principles

---

* The value to be attached to *numerical* proportions of fossil Plants, in estimating the entire condition of the Flora of these early periods, has been diminished by the result of a recent interesting experiment made by Prof. Lindley, on the *durability* of Plants immersed in water. (See Fossil Flora, No. xvii. vol. iii. p. 4.) Having immersed in a tank of fresh water, during more than two years, 177 species of plants, including representatives of all those which are either constantly present in the coal measures or universally absent, he found :

1. That the leaves and bark of most dicotyledonous Plants are wholly decomposed in 2 years, and that of those which do resist it, the greater part are *Coniferæ* and *Cycadeæ*.

2. That Monocotyledons are more capable of resisting the action of water, particularly Palms and Scitamineous Plants ; but that Grasses and Sedges perish.

3. That Fungi, Mosses, and all the lowest forms of Vegetation disappear.

4. That Ferns have a great Power of resisting water if gathered in a green state, not one of those submitted to the experiment having disappeared, but that their *fructification* perished.

Although the Results of this experiment in some degree in-

of structure, and by details of organization, which shew them all to be parts of One grand, and consistent, and harmonious Design.

We may end our account of the Plants to which we have traced the origin of Coal, with a summary view of the various Natural changes, and processes in Art and Industry, through which we can follow the progress of this curious and most important vegetable production.

Few persons are aware of the remote and wonderful Events in the economy of our Planet, and of the complicated applications of human Industry and Science, which are involved in the production of the Coal that supplies with fuel

validate the certainty of our knowledge of the *entire* Flora of each of the consecutive Periods of Geological History, it does not affect our information as to the number of the *enduring* Plants which have contributed to make up the Coal formation; nor as to the varying proportions, and changes in the species of Ferns and other plants, in the successive systems of vegetation that have clothed our globe.

It may be further noticed, that as both trunks and leaves of *Angiospermous dycotyledonous* Plants have been preserved abundantly in the Tertiary formations, there appears to be no reason why, if Plants of this Tribe had existed during the Secondary and Transition Periods, they should not also occasionally have escaped destruction in the sedimentary deposits of these earlier epochs.

In Loudon's Mag. Nat. Hist. Jan. 1834, p. 34, is an account of some interesting experiments by Mr. Lukis, on successive changes in the form of the cortical and internal parts of the stems of succulent plants, (e. g. Sempervivum arboreum) during various stages of decay, which may illustrate analogous appearances in many fossil plants of the coal formation.

the Metropolis of England. The most early stage to which we can carry back its origin, was among the swamps and forests of the primeval earth, where it flourished in the form of gigantic Calamites, and stately Lepidodendra, and Sigillariæ. From their native bed, these plants were torn away, by the storms and inundations of a hot and humid climate, and transported into some adjacent Lake, or Estuary, or Sea. Here they floated on the waters, until they sank saturated to the bottom, and being buried in the detritus of adjacent lands, became transferred to a new estate among the members of the mineral kingdom. A long interment followed, during which a course of Chemical changes, and new combinations of their vegetable elements, have converted them to the mineral condition of Coal. By the elevating force of subterranean Fires, these beds of Coal have been uplifted from beneath the waters, to a new position in the hills and mountains, where they are accessible to the industry of man. From this fourth stage in its adventures, our Coal has again been moved by the labours of the miner, assisted by the Arts and Sciences, that have co-operated to produce the Steam Engine and the Safety Lamp. Returned once more to the light of day, and a second time committed to the waters, it has, by the aid of navigation, been conveyed to the scene of its next and most considerable change by

fire; a change during which it becomes subservient to the most important wants and con-
veniences of Man. In this seventh stage of its
long eventful history, it seems to the vulgar eye
to undergo annihilation; its Elements are indeed released from the mineral combinations
they have maintained for ages, but their apparent destruction is only the commencement of
new successions of change and of activity. Set
free from their long imprisonment, they return
to their native Atmosphere, from which they
were absorbed to take part in the primeval vegetation of the Earth. To-morrow, they may contribute to the substance of timber, in the Trees
of our existing forests; and having for a while
resumed their place in the living vegetable kingdom, may, ere long be applied a second time
to the use and benefit of man. And when
decay or fire shall once more consign them to the
earth, or to the atmosphere, the same Elements
will enter on some further department, of their
perpetual ministration, in the economy of the
material world.

Fossil Coniferæ.*

The Coniferæ form a large and very important
tribe among living plants, which are charac-

* See Pl. 1. Figs. 1. 31. 32. 69

terized, not only by peculiarities in their fructification (as *Gymnospermous phanerogamiæ*),* but
also by certain remarkable arrangements in the
structure of their wood, whereby the smallest
fragment may be identified.

Recent microscopic examinations of fossil
woods have led to the recognition of an internal
structure, resembling that of existing Coniferæ,
in the trunks of large trees, both in the Carboniferous series,† and throughout the Secondary
formations ;‡ and M. Ad. Brongniart has enu-

* We owe to Mr. Brown, the important discovery, that Coniferæ and Cycadeæ are the only two families of plants that have
their seeds originally naked, and not enclosed within an Ovary.
(see Appendix to Captain King's Voyage to Australia). They
have for this reason been arranged in a distinct order, as *Gymnospermous Phanerogamiæ*. This peculiarity in the Ovulum is
accompanied throughout both these families, by peculiarities in
the internal structure of their stems, in which they differ from
almost all dicotyledonous plants, and in some respects also from
each other.

The recognition of these peculiar characters in the structure of
the stem, is especially important to the Geological Botanist, because the stems of plants are often the only parts which are found
preserved in a fossil state.

† The occurrence of large coniferous trees in strata of the
great Coal formation, was first announced in Mr. Witham's Fossil
Vegetables, 1831. It was here stated that the higher and more
complex organizations of Coniferæ exist in the Coal fields of
Edinburgh and Newcastle, in strata which till lately have been
supposed to contain only the simpler forms of vegetable structure.

‡ In the lower region of the Secondary strata, M. Ad. Brongniart has enumerated, among the fossil plants of the New red

merated twenty species of fossil Coniferæ in strata of the Tertiary series. Many of these last approach more closely to existing Genera than those in the Secondary strata, and some are referrible to them.

It has been further shewn by Mr. Nicol, (Edin. New. Phil. Journal, January, 1834) that some of the most ancient fossil Coniferæ may be referred to the existing genus Pinus, and others to that of Araucaria; the latter of these comprehends some of the tallest among living trees, (see Pl. 1, Fig. 1) and is best known in the Araucaria excelsa, or Norfolk Island Pine.

These discoveries are highly important, as they afford examples among the earliest remains of vegetable life, of identity in minute details of internal organization, between the most ancient

---

sandstone of the Vosges, four species of *Voltzia*, a new genus of Coniferæ, having near affinities to the Araucaria and Cunninghamia. Branches, leaves, and cones of this genus are most abundant at Sultz les Bains, near Strasburgh.

Mr. Witham reckons eight species of Coniferæ among the fossil woods of the Lias; and five species, of which four are allied to the existing genus Thuia, occur in the Oolite formation of Stonesfield. (See Ad. Brongniart's Prod. page 200). For figures of Cones from the Lias and Green-sand near Lyme Regis, and the Inferior oolite of Northamptonshire, see Lindley and Hutton's Fossil Flora, Plates 89, 135, 137.

Dr. Fitton has described and figured two very beautiful and perfect cones, one from Purbeck? and one from the Hastings sand. Geol. Trans. 2nd Series, Vol. iv. Pl. 22, Figs. 9, 10. p. 181 and 230.

trees of the primeval forests of our globe, and some of the largest living Coniferæ.*

The structure of Araucarias alone has been as yet identified in trees from the Carboniferous series of Britain.†   That of ordinary Pines oc-

* The transverse section of any coniferous wood in addition to the radiating and concentric lines represented Pl. 56ᵃ, Fig. 7, exhibits under the microscope a system of reticulations by which Coniferæ are distinguishable from other plants. The form of these reticulations magnified 400 times is given in Pl. 56ᵃ, Figs. 2, 4, 6. These apertures are transverse sections of the same vessels, which are seen in a longitudinal section at Pl. 56ᵃ, Fig. 8, cut from the centre towards the bark, and parallel to the medullary rays. These vessels exhibit a characteristic and beautiful structure, whereby a distinction is marked between true Pines and Araucarias. In such a section the small and uniform longitudinal vessels, (Pl. 56ᵃ, Fig. 8) which constitute the woody fibre, present at intervals a remarkable appearance of small, nearly circular figures disposed in vertical rows (See Pl. 56ᵃ, Figs. 1, 3, 5). These objects under the name of glands or discs, are differently arranged in different species; they are generally circular, but sometimes elliptical, and when near each other, become angular. Each of these discs has near its centre a smaller circular areola. Pl. 56ᵃ, Fig. 1, represents their appearance in the Pinus strobus of North America.

In some Coniferæ, the discs are in single rows; in others, in double as well as single rows, e. g. in Pinus strobus, Pl. 56ᵃ, Fig. 1.

Throughout the entire genus of living Pines, when double rows of discs occur in one vessel, the discs of both rows are placed side by side, and never alternate, and the number of the rows of discs is never more than two.

In the Araucarias the groups of discs are arranged in single, double, triple and sometimes quadruple rows, see Pl. 56 , Fig. 3. 5. They are much smaller than those in the true Pines, scarcely half their size, and in the double rows they always *alternate* with

curs in wood from the Coal formation of Nova Scotia and New Holland.

The same ordinary structure of Pines predominates in the fossil wood of the Lias at Whitby; trunks of Araucarias also are found there in the same Lias; and branches, with the leaves still adhering to them, in the Lias at Lyme Regis.‡

Professor Lindley justly remarks that it is an important fact, that at the period of the deposit of the Lias, the vegetation was similar to that of the Southern Hemisphere, not alone in the single fact of the presence of Cycadeæ, but that the Pines were also of the nature of species now found only to the south of the Equator.   Of

each other, and are sometimes circular, but mostly polygonal. Mr. Nicol has counted a row of not less than fifty discs in a length of the twentieth part of an inch, the diameter of each disc not exceeding the thousandth part of an inch; but even the smallest of these are of enormous size, when compared with the fibres of the partitions bounding the vessels in which they occur.

† A trunk of Araucaria forty-seven feet long was found in Cragleith Quarry near Edinburgh, 1830. (See Witham's Fossil Vegetables, 1833, Pl. 5). Another, three feet in diameter, and more than twenty-four feet long, was discovered in the same quarries in 1833. (See Nicol on Fossil Coniferæ, Edin. New Phil. Journal, Jan. 1834). The longitudinal sections of this Tree exhibit, like the recent Araucaria excelsa, small polygonal discs, arranged in double, and triple and quadruple rows within the longitudinal vessels; so also does a similar section from the Coal field of New Holland.

‡ See Lindley and Hutton's Fossil Flora, Pl. 88.   A fossil cone referrible to Coniferæ, and possibly to the genus Araucaria, from the Lias of Lyme Regis, is represented at Plate 89 of the same work.

the four recent species of Araucaria at present known, one is found on the east coast of New Holland, another in Norfolk Island, a third in Brazil, and the fourth in Chili. (Foss. Flora, vol. ii. p. 21.)

Whatever result may follow from future investigations, our present information shows that the largest and most perfect fossil Coniferæ, which have been as yet sufficiently examined from the Coal formation and the Lias, are referrible either to the genus Pinus, or Araucaria,* and that both these modifications of the existing Family of Coniferæ date their commencement from that very ancient period, when the Carboniferous strata of the Transition formation were deposited.

* Mr. Nicol states that in fossil woods from the Whitby Lias, when concentric layers are distinctly marked on their transverse section, (Pl. 56ᵃ, Fig. 2, a, a.) the longitudinal sections have also the structure of Pinus (Pl. 56ᵃ, Fig. 1.); but when the transverse section exhibits no distinct annual layers, (Pl. 56ᵃ, Fig. 4.) or has them but slightly indicated, (Pl. 56ᵃ, Fig. 6. a) the longitudinal section has the characters of Araucaria. (Pl. 56ᵃ, Fig. 3, 5.) So also those Coniferæ of the great Coal formation of Edinburgh and Newcastle, which exhibit the structure of Araucaria in their longitudinal section, have no distinct concentric layers; whilst in the fossil Coniferæ from the New Holland and Nova Scotia Coal field, both longitudinal and transverse sections agree with those of the recent tribe of Pinus.

Mr. Witham also observes that the Coniferæ of the Coal formation, and mountain limestone group, have few and slight appearances of the concentric lines, by which the annual layers of the wood are separated, which is also frequently the case with the Trees of our present tropical regions, and from this circumstance conjectures that, at the epochs of these formations, the changes of season, as to temperature at least were not abrupt.

Fragments of trunks of Coniferous wood, and occasionally leaves and cones occur through all stages of the Oolite formation, from the Lias to the Portland stone. On the upper surface of the Portland stone, we find the remains of an ancient forest, in which are preserved large prostrate silicified trunks, and silicified stumps of Coniferæ, having their roots still fixed in the black vegetable mould in which they grew. Fragments of coniferous wood are also frequent throughout the Wealden and Green-sand formations, and occur occasionally in Chalk.*

It appears that the Coniferæ are common to fossiliferous strata of all periods; they are least abundant in the Transition series, more numerous in the Secondary, and most frequent in the Tertiary series. Hence we learn that there has been no time since the commencement of terrestrial vegetation on the surface of our Globe, in which large Coniferous trees did not exist; but our present evidence is insufficient, to ascertain with accuracy the proportions they bore to the relative numbers of other families of plants, in each of the successive geological epochs, which are thus connected with our own, by a new and beautiful series of links, derived from one of the most important tribes of the vegetable kingdom.

* There is in the Oxford Museum a fragment of silicified coniferous wood, perforated by Teredines, found by Rev. Dr. Faussett, in a chalk flint at Lower Hardres, near Canterbury.

SECTION III.

VEGETABLES IN STRATA OF THE SECONDARY SERIES.[*]

Fossil Cycadeæ.

THE Flora of the Secondary Series[†] presents characters of an intermediate kind between the Insular vegetation of the Transition series, and the Continental Flora of the Tertiary formations. Its predominating feature consists in the abundant presence of Cycadeæ, (see Pl. 1, Figs. 33, 34, 35,) together with Coniferæ,[‡] and Ferns.[§] (See Pl. 1, Figs. 37, 38, 39.)

[*] See Pl. 1, Figs. 31 to 39.

[†] M. Ad. Brongniart, in his arrangement of fossil plants, has formed a distinct group out of the few species which have been found in the Red-sandstone formation (Gres bigarré) immediately above the Coal. In our division of the strata, this Red-sandstone is included, as an inferior member, in the Secondary series. Five Algæ, three Calamites, five Ferns, and five Coniferæ, two Liliaceæ, and three uncertain Monocotyledonous plants form the entire amount of species which he enumerates in this small Flora.

See also Jæger uber die Pflanzenversteinerungen in dem Bausandstein von Stuttgart, 1827.

[‡] We again refer to Witham's Account of Coniferæ from the Lias, in his observations on Fossil Vegetables, 1833.

[§] A very interesting account, accompanied by figures, shewing the internal structure of the stems of fossil arborescent Ferns of the Secondary period, is given in Cotta's Dendrolithen, Dresden, 1832; these appear to be chiefly from the New red sandstone of Chemnitz near Dresden.

M. Ad. Brongniart enumerates about seventy species of land plants in the Secondary formations, (from the Keuper to the Chalk inclusive;) one half of these are Coniferæ and Cycadeæ, and of this half, twenty-nine are Cycadeæ; the remaining half are chiefly vascular Cryptogamiæ, viz. Ferns, Equisetaceæ, and Lycopodiaceæ. In our actual vegetation, Coniferæ and Cycadeæ scarcely compose a three hundredth part.*

The family of Cycadeæ comprehends only two living Genera; viz. Cycas, (Pl. 58.) and Zamia. (Pl. 59.) There are five known living Species of Cycas and about seventeen of Zamia. Not a single species of the Cycadeæ grows at the present time in Europe; their principal localities are parts of equinoctial America, the West Indies, the Cape of Good Hope, Madagascar, India, the Molucca Islands, Japan, China, and New Holland.

Four or five genera, and twenty-nine species of Cycadeæ, occur in the fossil Flora of the Secondary period, but remains of this family are very

---

* The fossil vegetables in the Secondary series, although they present many kinds of Lignite, very rarely form beds of valuable Coal. The imperfect coal of the Cleveland Moorlands near Whitby, and of Brora in Sutherland, belong to the inferior region of the Oolite formation. So also does the bituminous coal of Bückeberg near Minden, in Westphalia.

The coal of Hoer in Scania is either in the Wealden formation, or in the Green-sand (*Ann. des Sciences Nat.* tom. iv. p. 200).

rare in strata of the Transition, and Tertiary series.†

The Cycadeæ form a beautiful family of plants whose external habit resembles that of Palms, whilst their internal structure approximates in several essential characters to that of Coniferæ. In a third respect, (viz. the *Gyrate Vernation*, or mode in which the leaves are curled up at their

† I learn by letter from Count Sternberg, (Aug. 1835.) that he has found Cycadeæ and Zamites in the Coal formation of Bohemia, of which he will publish figures in the 7th and 8th Cahier of his Flore du Monde primitif. This is, I believe, the first example of the recognition of plants of this family in strata of the Carboniferous series.

During a recent visit to the extensive and admirably arranged geological collection in the Museum at Strasbourg, I was informed by M. Voltz that the stem of a *Cycadites* in that Museum, described by M. Ad. Brongniart as a *Mantellia*, from the Muschelkalk of Luneville, is derived from the Lias near that Town. M. Voltz knows no example of any Cycadites from the Muschelkalk. Stems and leaves of Cycadeæ occur also in the Lias at Lyme Regis. (Lind. Foss. Fl. Pl. 143.)

The most abundant deposit of fossil leaves of Cycadeæ in England, is in the Oolitic formation on the coast of Yorkshire, between Whitby and Scarborough, (See Phillips' Illustrations of the Geology of Yorkshire.) Leaves of this family occur also in the Oolitic slate of Stonesfield. Lindley and Hutton, Foss. Flora, Pl. 172, 175.

In Lindley and Hutton's Fossil Flora, Pl. 136, Figures are given of Cones which he refers to the genus Zamia, from the sandstone of the Wealden formation at Yaverland on the South coast of the I. of Wight.

M. Ad. Brongniart has established a new fossil genus *Nilsonia*, in the family of Cycadeæ, which occurs at Hoer in Scania, in strata, either of the Wealden or Green-sand formation; and another genus, *Pterophyllum*, which is found from the New red sand-stone upwards to the Wealden formation.

points within the buds), they resemble Ferns. (See Pl. 1. F. 33, 34, 35, and Pl. 58, 59.)

I shall select the family of Cycadeæ from the Fossil Flora of the Secondary period, and shall enter into some details respecting its organization, with a view of showing an example of the method of analysis, by which Geologists are enabled to arrive at information as to the structure and economy of extinct species of fossil vegetables, and of the importance of the conclusions they are enabled to establish. Those who have attended to the recent progress of vegetable Physiology will duly appreciate the value of microscopic investigations, which enable us to identify the structure of vegetables of such remote antiquity, with that which prevails in the organization of living species.

The physiological discoveries that have lately been made with respect to living species of Cycadeæ, have shown them to occupy an intermediate place between Palms, Ferns, and Coniferæ, to each of which they bear certain points of resemblance; and hence a peculiar interest attends the recognition of similar structures in fossil plants, referrible to a family whose characters are so remarkable.

The figure of a Cycas revoluta (Pl. 58,)* represents the form and habit of plants belonging to this beautiful genus. In the magnificent crown

* Drawn from a Plant in Lord Grenville's Conservatory at Dropmore in 1832.

of graceful foliage surrounding the summit of a simple cylindrical trunk, it resembles a Palm. The trunk in the genus Cycas, is usually long. That of C. circinalis rises to 30 feet.* In the genus Zamia it is commonly short.

Our figure of a Zamia pungens,† (Pl. 59,) shews the mode of inflorescence in this Genus, by a single cone, rising like a Pine Apple, deprived of its foliaceous top, from within the crown of leaves at the summit of the stem.

The trunk of the Cycadeæ has no true bark, but is surrounded by a dense case, composed of persistent scales which have formed the bases of fallen leaves; these, together with other abortive scales, constitute a compact covering that supplies the place of bark. (See Pl. 58 and 59.)

In the Geol. Trans. of London (vol. iv. part 1. New Series) I have published, in conjunction with Mr. De la Beche, an account of the circumstances under which silicified fossil trunks of Cycadeæ are found in the Isle of Portland, immediately above the surface of the Portland stone, and below the Purbeck stone. They are lodged in the same beds of black mould in which they grew, and are accompanied by prostrate

---

* In Curtis's Botanical Magazine, 1828, Pl. 2826, Dr. Hooker has published an Engraving of a Cycas circinalis which in 1827 flowered in the Botanic Garden at Edinburgh. See Pl. 1. Fig. 33.

† Copied from an engraving published by Mr. Lambert, of a plant that bore fruit at Walton on Thames in the conservatory of Lady Tankerville, 1832.

trunks of large coniferous trees, converted to
flint, and by stumps of these trees standing erect
with their roots still fixed in their native soil.
(See Pl. 57, Fig. 1.)*

Pl. 57, Fig. 3, exhibits similar stumps of trees
rooted in their native mould, in the Cliff immediately east of Lulworth Cove.   Here the strata
have been elevated nearly to an angle of 45°,
and the stumps still retain the unnatural inclination into which they have been thrown by this
elevation.

The facts represented in these three last figures
are fully described and explained in the paper
above referred to; they prove that plants belonging to a family that is now confined to the
warmer regions of the earth, were at a former
period, natives of the southern coast of England.†

* The sketch, Pl. 57, Fig. 2, represents a triple series of circular undulations, marked in the stone, which surrounds a single
stump, rooted in the dirt-bed in the Isle of Portland.   This very
curious disposition has apparently resulted from undulations,
produced by winds, blowing at different times in different directions on the surface of the shallow fresh water, from the sediments of which the matter of this stratum was supplied, while the
top of this stem stood above the surface of the water.   See Geol.
Trans. Lond. N. S. vol. iv. p. 17.

† The structure of this district affords also a good example
of the proofs which Geology discloses, of alternate elevations and
submersions of the strata, sometimes gradually, and sometimes
violently, during the formation of the crust of our planet.

First. We have evidence of the rise of the Portland stone, till
it reached the surface of the sea wherein it was formed.

Secondly. This surface became for a time, dry land, covered by

As no leaves have yet been found with the fossil Cycadeæ under consideration, we are limited to the structure of their trunk and scales, in our search for their distinguishing characters.

I have elsewhere (Geol. Trans. London, N. S. vol. ii. part iii. 1828) instituted a comparison between the internal structure of two species of these fossil trunks, and that of the trunks of a recent Zamia and recent Cycas.*

a temporary forest, during an interval which is indicated by the thickness of a bed of black mould, called the Dirt bed, and by the rings of annual growth in large petrified trunks of prostrate trees, whose roots had grown in this mould.

Thirdly. We find this forest to have been gradually submerged, first beneath the waters of a freshwater lake, next of an estuary, and afterwards beneath those of a deep sea, in which Cretaceous and Tertiary strata were deposited, more than 2000 feet in thickness.

Fourthly. The whole of these strata have been elevated by subterranean violence, into their actual position in the hills of Dorsetshire.

We arrive at similar conclusions, as to the alternate elevation and depressions of the surface of the earth, from the erect position of the stems of Calamites, in sandstone of the lower Oolite formation on the eastern coast of Yorkshire. (See Murchison. Proceedings of Geol. Society of London, page 391.)

* M. Ad. Brongniart has referred these two fossil species to a new genus, by the name of Mantellia nidiformis and Mantellia cylindrica; in my paper, just quoted, I applied to them the provisional name of Cycadeoidea megalophylla and Cycadeoidea microphylla; but Mr. Brown is of opinion, that until sufficient reasons are assigned for separating them from the genus Cycas or Zamia, the provisional name of Cycadites is more appropriate, as expressing the present state of our knowledge upon this subject. The name Mantellia is already applied by Parkinson (Introduction to Fossil Org. Rem. p. 53) to a genus of Zoophytes, which is figured in Goldfuss, T. vi. p. 14.

I must refer to the memoir, in which these sections are described, for specific details as to the varied proportions and numerical distribution of these concentric circles of laminated wood and cellular tissue, in the trunks of living and fossil species of Cycadeæ.*

A strict correspondence is also exhibited in

* Plates 60, Fig. 1, and 61, Fig. 1, represent very perfect specimens of fossil Cycadites from Portland, now in the Oxford Museum; both having the important character of Buds protruding from the Axillæ of the leaf stalks.

The section given in Pl. 59, Fig. 2, of the trunk of a recent Zamia horrida, from the Cape of Good Hope, displays a structure similar to that in the section of the fossil Cycadites megalophyllus from the Isle of Portland; (Pl. 60, Fig. 2) each presents a single circle of radiating laminæ of woody fibre, B, placed between a central mass of cellular tissue, A, and an exterior circle of the same tissue, C. Around the trunk, thus constituted of three parts, is placed a case or false bark, D, composed of the persistent bases of fallen leaves, and of abortive scales. The continuation of the same structure is seen at the summit of the stem, Pl. 60, Fig. 1, A. B. C. D.

The Cycadites microphyllus, Pl. 61, Fig. 1, affords a similar approach to the internal structure of the stem in the recent Cycas. The summit of this fossil exhibits a central mass of cellular tissue (A), surrounded by two circles of radiating woody plates, B. b., between these laminated circles, is a narrow circle of cellular tissue, whilst a broader circle of similar cellular tissue (C) is placed between the exterior laminated circle, (b) and the leaf scales (D). This alternation of radiating circles of wood with circles of cellular tissue, is similar to the two laminated circles near the base of a young stem of Cycas revoluta, (Pl. 59, Fig. 3.) This section was communicated to me by Mr. Brown early in 1828, to confirm the analogy which had been suggested from the external surface, between these fossils, and the recent Cycadeæ; and is figured in Geol. Trans. N. S. vol. ii. Pl. 46.

G.   K K

the internal structure of the scales, or bases
of leaf stalks surrounding the trunks of our fossil
Cycadites, with that of the corresponding scales
in the recent species.*

* In Pl. 61, Figs. 2, 3, represent two vertical·sections of a
Cycadites microphyllus from Portland, converted to Chalcedony.
These slices are parallel to the axis of the trunk, and intersect
transversely the persistent bases of the Petioles or Leaf-stalks.
In each rhomboidal Petiole, we see the remains of three systems
of vegetable structure, of which magnified representations are
given Pl. 62, Fig. 1, 2, 3. We have, first, the principal mass of
cellular tissue (f); secondly, sections of gum vessels (h) irregularly dispersed through this cellular tissue ; thirdly, bundles
of vessels, (c), placed in a somewhat rhomboidal form, parallel
to, and a little within, the integument of each petiole. These
bundles of vessels are composed of vascular woody fibres proceeding from the trunk of the plant towards the leaf. See magnified
section of one bundle at Pl. 62, Fig. 3, c'.

A similar arrangement of nearly all these parts exists in the
transverse section of the leaf stalks of recent Cycadeæ. In Cycas
circinalis, and C. revoluta, and Zamia furfuracea, the bundles
of vessels are placed as in our fossil, nearly parallel to the integument. In Zamia spiralis, and Z. horrida, their disposition
within the Petiole, is less regular, but the internal structure of
each bundle is nearly the same. In Pl. 62, Fig. A shews the
place of these bundles of vessels in a transverse section of the
leaf stalk of Zamia spiralis ; Fig. A. c'. is the magnified appearance of one of the bundles in this section ; Fig. B. c″ is the magnified transverse section of a similar bundle of vessels in the
petiole of Zamia horrida. In this species the vascular fibres
are smaller and more numerous than in Z. spiralis, and the opake
lines less distinct. Both in recent and fossil Cycadeæ the component vascular fibres of these bundles are in rows approximated
so closely to each other, that their compressed edges give an
appearance of opake lines between the rows of vascular fibres,
(see Pl. 62, Fig. 1, c'. Fig. B, c″. and Fig. 3, c'.) These bundles

Mode of increase by Buds the same in recent and fossil Cycadeæ.

The Cycas revoluta figured in Pl. 58* possesses a peculiar interest in relation to both our

of vessels seem to partake of the laminated disposition of the woody circle within the trunk.

An agreement is found also in the longitudinal sections of the Petioles of recent and fossil Cycadeæ. Pl. 62, Fig. 1, is the longitudinal section of part of the base of a Petiole of Zamia spiralis, magnified to twice the natural size. It is made up of cellular tissue, (f), interspersed with gum vessels, and with long bundles of vascular fibres, (c) proceeding from the trunk towards the leaf. On the lower integument, (b') is a dense coating of minute curling filaments of down or cotton, (a) which being repeated on each scale, renders the congeries of scales surrounding the trunk, impervious to air and moisture.

A similar disposition is seen in the longitudinal section of the fossil Petiole of Cycadites microphyllus represented at Pl. 62, Fig. 2, and magnified four times. At f, we have cellular tissue interspersed with gum vessels, h. Beneath c, are longitudinal bundles of vessels; at b, is the integument; at a, a most beautiful petrifaction of the curling filaments of down or cotton, proceeding from the surface of this integument.

In the vascular bundles within the fossil Petioles, (c) Mr. Brown has recognized the presence of spiral, or scalariform vessels (Vasa scalariformia) such as are found in the Petioles of recent Cycadeæ; he has also detected similar vessels, in the laminated circle within the trunk of the fossil Buds next to be described. The existence of vessels with discs peculiar to recent Cycadeæ and Coniferæ, such as have been described in speaking of fossil Coniferæ, has not yet been ascertained.

* This plant had been living many years, in Lord Grenville's conservatory at Dropmore. In the autumn of 1827, the external part of the scales was cut away to get rid of insects : in the fol-

fossil species, in consequence of its protruding a series of buds from the axillæ of many of the scales around its trunk. These buds explain analogous appearances at the axillæ of many fossil scales on Cycadites megalophyllus, and Cycadites microphyllus, (see Pl. 60, Fig. 1, and Pl. 61, Fig. 1,) and form an important point of agreement in the Physiology of the living and fossil Cycadeæ.*

lowing spring the buds began to protrude. Similar buds appeared also in the same conservatory on a plant of the Zamia spiralis from New Holland. In vol. vi. p. 501, Horticult. Trans. leaves are stated to have protruded from the scales of a decayed trunk of Zamia horrida in a conservatory at Petersburgh.

I learn from Professor Henslow, that the trunk of a Cycas revoluta, which in 1830 produced a Cone loaded with ripe drupæ, in Earl Fitzwilliam's hothouse at Wentworth, threw out a number of buds, from the axillæ of the leaf-scales soon after the Cone was cut off from its summit. In Linn. Trans. vol. vi. tab. 29, is a figure of a similar cone which bore fruit at Farnham Castle, 1799.

It is stated in Miller's Gardener's Dictionary, that the Cycas revoluta was introduced into England about 1758, by Captain Hutchinson; his ship was attacked, and the head of the plant shot off, but the stem being preserved, threw out several new heads, which were taken off, and produced as many plants.

* In the fossil trunk of Cycadites microphyllus, Pl. 61, Fig. 1, we see fourteen Buds protruding from the axillæ of the leaf stalks, and in Pl. 60, Fig. 1, we have three Buds in a similar position in Cycadites megalophyllus.

In Pl. 61, Figs. 2, 3, exhibit transverse sections of three Buds of Cycadites microphyllus. The section of the uppermost bud, Fig. 3, g, passes only through the leaf stalks near its crown. The section of the bud, Fig. 3, 'd, being lower down in the embryo trunk, exhibits a double woody circle, arranged in radiating

Thus, we see that our fossil Cycadites are closely allied by many remarkable characters of structure, to existing Cycadeæ.

1. By the internal structure of the trunk, containing a radiating circle, or circles, of woody fibre, embedded in cellular tissue.   2. By the structure of their outer case, composed of persistent bases of petioles, in place of a bark ; and by all the minute details in the internal organization of each Petiole.   3. By their mode of increase by Buds protruded from germs in the Axillæ of the Petioles.

However remote may have been the time when

plates, resembling the double woody circle in the mature trunk, Pl. 61, 1, B, b.   But in Pl. 61, Fig. 2, the laminated circle within the embryo trunk near d, is less distinctly double, as might be expected in so young a state.

At Pl. 62, Fig. 3, d, and d', we see magnified representations of a portion of the embryo circle within the Bud, Pl. 61. Fig. 3, 'd. These woody circles within the buds, are placed between an exterior circle of cellular tissue, interspersed with gum vessels, and a central mass of the same tissue, as in the mature stems.

On the right of the lower bud, Pl. 61, Fig. 3, above b, and in the magnified representation of the same at Pl. 62, Fig. 3, e, we have portions of a small, imperfect laminated circle.   Similar imperfect circles occur also near the margin of the sections, Pl. 61, Figs. 2, 3, at e, e', e''; these may be imperfectly developed Buds, crowded like the small Buds near the base of the living Cycas, Pl. 58 : or they may have resulted from the confluence of the bundles of vessels, in the Bases of leaves, forced together by pressure, connected with a diminution or decay of their cellular substance.   The normal position of these bundles of vessels is seen magnified in Pl. 62. Fig. 3. c. and in nearly all the Sections of Bases of petioles in Pl. 61. Fig. 2.

these Prototypes of the family of Cycadeæ
ceased to exist, the fact of their containing so
many combinations of peculiarities identical with
those of existing Cycadeæ, connects these ancient arrangements in the Physiology of fossil
Botany, with those which now characterize
one of the most remarkable families among
existing plants.   In virtue of these peculiar
structures, the living Cycadeæ form an important link, which no other Tribe of plants
supplies, connecting the great family of Coniferæ,
with the families of Palms and Ferns, and thus
fill up a blank, which would otherwise have
separated these three great natural divisions of
dicotyledonous, monocotyledonous, and acotyle-
donous plants.

The full development of this link in the Secondary periods of Geological history, affords an
important evidence of the Uniformity of Design
which now pervades, and ever has pervaded, all
the laws of vegetable life.

Facts like these are inestimably precious to
the Natural Theologian ; for they identify, as it
were, the Artificer, by details of manipulation
throughout his work.   They appeal to the Physiologist, in language more commanding than
human Eloquence ; the voice of very stocks and
stones, that have been buried for countless ages
in the deep recesses of the earth, proclaiming
the universal agency of One all-directing, all-

sustaining Creator, in whose Will and Power,
these harmonious systems originated, and by
whose Universal Providence, they are, and have
at all times been, maintained.

Fossil Pandaneæ.

The Pandaneæ, or Screw-Pines, form a monocotyledonous family which now grows only in
the warmer zones, and chiefly within the influence of the sea; they abound in the Indian
Archipelago, and the islands of the Pacific Ocean.
Their aspect is that of gigantic Pine apple
plants having arborescent stems. (See Pl. 63,
Fig. 1.)

This family of Plants seems destined, like the
Cocoa nut Palm, to be among the first vegetable
Colonists of new lands just emerging from the
ocean; they are found together almost universally
by navigators on the rising Coral islands of tropical seas. We have just been considering the
history of the fossil stems of Cycadeæ in the Isle
of Portland, from which we learn that Plants of
that now Extra-european family were natives of
Britain, during the period of the Oolite formation.
The unique and beautiful fossil fruit represented
in our figures (Plate 63, Figs. 2, 3, 4,) affords
probable evidence of the existence of another
tropical family nearly allied to the Pandaneæ at

the commencement of the great Oolitic series in the Secondary formations.*

In structure this fossil Fruit approaches nearer to *Pandanus* than to any other living plant, and viewing the peculiarities of the fruit of Pandaneæ,† in connection with the office assigned

---

* This fossil was found by the late Mr. Page, of Bishport near Bristol, in the lower region of the Inferior Oolite formation on the E. of Charmouth, Dorset, and is now in the Oxford Museum. The size of this Fruit is that of a large orange, its surface is occupied by a stellated covering or Epicarpium, composed of hexagonal Tubercles, forming the summits of cells, which occupy the entire circumference of the fruit. (Figs. 2, *a*. 3, *a*. 4, *a*. 8, *a*.) Within each cell is contained a single seed, resembling a small grain of Rice more or less compressed, and usually hexagonal, Figs. 5, 6, 7, 8, 10. Where the Epicarpium is removed, the points of the seeds are seen, thickly studded over the surface of the fruit, (Fig. 2, 3, e.) The Bases of the cells (Fig. 3 and 10 c.) are separated from the receptacle, by a congeries of foot-stalks (d) formed of a dense mass of fibres, resembling the fibres beneath the base of the seeds of the modern Pandanus (Fig. 13, 14, 15, d.) As this position of the seeds upon foot-stalks composed of long rigid fibres, at a distance from the receptacle, is a character that exists in no other family than the Pandaneæ, we are hereby enabled to connect our fossil fruit with this remarkable tribe of plants, as a new genus, *Podocarya*. I owe the suggestion of this name, and much of my information on this subject, to the kindness of my friend, Mr. Robert Brown.

† The large spherical fruit of Pandanus, hanging on its parent tree is represented at Pl. 63, Fig. 1. Fig. 11 is the summit of one of the many Drupes into which this fruit is usually divided. Each cell when not barren contains a single oblong slender seed ; the cells in each drupe vary from two to fourteen in number, and many of them are abortive, (Fig. 13.) The seeds within each

in the Economy of nature, to this family of sea-side plants, viz. to take the first possession of new-formed land, just emerging from the water, we see in the disposition of light buoyant fibres within the interior of these fruits, an arrangement peculiarly adapted to the office of vegetable colonization.* The sea-side locality of the Pandaneæ, causes many of their fruits to fall into the water, wherein they are drifted by the winds and waves, until they find a resting place upon some distant shore. A single drupe of Pandanus, thus charged with seeds, transports the elements of vegetation to the rising

drupe of Pandanus are enclosed in a hard nut, of which sections are given at Figs. 14, 15. These nuts are wanting in the Podocarya, whose seeds are smaller than those of Pandaneæ, and not collected into drupes, but dispersed uniformly in single cells over the entire circumference of the fruit. (See Pl. 63, Figs. 3, 8, 10.) The collection of the seeds into drupes surrounded by a hard nut, in the fruit of Pandanus, forms the essential difference between this genus, and our new genus, Podocarya.

In the fruit of Pandanus, Pl. 63, Figs. 11, 16, 17, the summit of each cell is covered with a hard cap or tubercle, irregularly hexagonal, and crowned at its apex with the remains of a withered stigma. We have a similar covering of hexagonal tubercles over the cells of Podocarya (Pl. 63, Figs. 2, *a*. 8, *a*. 10, *a*.) The remains of a stigma appear also in the centre of these hexagons above the apex of each seed. (Figs. 8, *a*. 10, *a*.)

* There is a similar provision for transporting to distant regions of the ocean, the seeds of the other family of sea-side plants which accompanies the Pandanus, in the buoyant mass of fibrous covering that surrounds the fruit of the Cocoa-nut.

volcanic and coral islands of the modern Pacific. The seed thus stranded upon new formed land, produces a plant which has peculiar provision for its support on a surface destitute of soil, by long and large aerial roots protruded above the ground around the lower part of its trunk. (See Pl. 63. Fig. 1.) These roots on reaching the ground are calculated to prop up the plant as buttresses surrounding the basis of the stem, so that it can maintain its erect position, and flourish in barren sand on newly elevated reefs, where little soil has yet accumulated.

We have as yet discovered no remains of the leaves, or trunk of Pandaneæ in a fossil state, but the presence of our unique fruit in the Inferior Oolite formation near Charmouth, carries us back to a point of time, when we know from other evidence that England was in the state of new-born land, emerging from the seas of a tepid climate; and shews that combinations of vegetable structure such as exist in the modern Pandaneæ, adapted in a peculiar manner to the office of vegetable colonization, prevailed also at the time when the Oolite rocks were in process of formation.

This fruit also adds a new link to the chain of evidence, which makes known to us the Flora of the Secondary periods of geology, and therein discloses fresh proofs of Order, and Harmony, and of Adaptation of peculiar means to peculiar ends; extending backwards from the actual condition of our Planet through the manifold stages of change, which its ancient surface has undergone.*

SECTION IV.

VEGETABLES IN STRATA OF THE TERTIARY SERIES.†

IT has been stated that the vegetation of the Tertiary period presents the general character of that of our existing Continents within the Temperate Zone. In Strata of this Series, Dicotyledonous Plants assume nearly the same proportions as at present, and are four or five times more numerous than the Monocotyledonous; and the greater number of fossil Plants, although of extinct species, have much resemblance to living Genera.

This third great change in the vegetable kingdom is considered to supply another argument in favour of the opinion, that the temperature of the Atmosphere, has gone on continually diminishing from the first commencement of life upon our globe.

* Fruits of another genus of Pandaneæ, to which Mr. Ad. Brongniart has given the name of Pandanocarpum, (*Prodrome*, p. 138,) occur together with fruits of Cocoa nut, at an early period of the Tertiary formations, among the numerous fossil fruits that are found in the London-clay of the Isle of Sheppey.

† See Pl. 1, Figs. 66 to 72.

The number of species of plants in the various divisions of the Tertiary strata, is as yet imperfectly known. In 1828, M. Ad. Brongniart considered the number then discovered, but not all described, to be 166. Many of these belonging to Genera at that time not determined. The most striking difference between the vegetables of this and of the preceding periods is the abundance in the Tertiary series, of existing forms of Dicotyledonous Plants and large trees, *e. g.* Poplars, Willows, Elms, Chestnuts, Sycamores, and many other Genera whose living species are familiar to us.

Some of the most remarkable accumulations of this vegetation are those, which form extensive beds of *Lignite* and *Brown-coal*.* In some parts of Germany this Brown-coal occurs in strata of more than thirty feet in thickness, chiefly composed of trees which have been drifted, apparently by fresh water, from their place of growth, and spread forth in beds, usually alternating with sand and clay, at the bottom of then existing lakes or estuaries.

The Lignite, or beds of imperfect and stinking Coal near Poole in Dorset, Bovey in Devon, and Soissons in France, have been referred to the first, or *Eocene* period of the Tertiary formations.

---

* See an admirable article on Lignites by Alexandre Brongniart in the 26th vol. of the Dictionnaire des Sciences Naturelles.

To the same period probably belongs the Surturbrand of Iceland, (see Henderson's Iceland, vol.
ii. p. 114.) and the well-known examples of Brown-
coal on the Rhine near Cologne and Bonn, and
of the Meisner mountain, and Habichtswald near
Cassel. These formations occasionally contain
the remains of Palms, and Professor Lindley has
lately recognized, among some specimens found
by Mr. Horner in the Brown-coal near Bonn
(See Ann. Phil. Lond. Sept. 1833, V. 3, p. 222),
leaves closely allied to the Cinnamomum of our
modern tropics, and to the Podocarpus of the
southern hemisphere.*

---

* At Pützberg near Bonn, six or seven beds of Brown-coal
alternate with beds of sandy clay and plastic clay. The trees in
the Brown-coal are not all parallel to the planes of the strata,
but cross one another in all directions, like the drifted trees now
accumulated in the alluvial plains, and Delta of the Mississippi;
(see Lyell's Geology, 3rd edit. vol i. p. 272.) some of them
are occasionally forced even into a vertical position. In one
vertical tree at Pützberg, which was three yards in diameter,
M. Nöggerath counted 792 concentric rings. In these rings we
have a chronometer, which registers the lapse of nearly eight
centuries, in that early portion of the Tertiary period which gave
birth to the forests, that supplied materials for the formation of
the Brown-coal.

The fact mentioned by Faujas that neither roots, branches, or
leaves are found attached to the trunks of trees in the Lignite at
Bruhl and Liblar near Cologne, seems to show that these trees
did not grow on the spot, and that their more perishable parts
have been lost during their transport from a distance.

In the Brown-coal Formation near Bonn, and also with the
Surturbrand of Iceland, are found Beds that divide into Laminæ

In the Molasse of Switzerland, there are many similar deposits affording sometimes Coal of considerable purity, formed during the second, or *Miocene* period of this series, and usually containing fresh water shells. Such are the Lignites of Vernier near Geneva, of Paudex and Moudon near Lausanne, of St. Saphorin near Vevay, of Kæpfnach near Horgen on the lake of Zurich, and of Œningen near Constance.

The Brown-coal at Œningen forms thin beds of little importance for fuel, but very perfect remains of vegetables are dispersed in great abundance through the marly slates and limestone quarries which are worked there, and afford the most perfect history of the vegetation of the *Miocene* Period, which has yet come within our reach.*

as thin as paper (*Papier Kohle*) and are composed entirely of a congeries of many kinds of leaves. Henderson mentions the leaves of two species of Poplar, resembling the P. tremula and P. balsamifera, and a Pine, resembling the Pinus abies as occurring in the Surturbrand of Iceland.

Although we have followed Mr. Brongniart in referring the deposits here enumerated to the first, or *Eocene* period of the Tertiary series, it is not improbable that some of them may be the products of a later era, in the *Miocene* or *Pliocene* periods. Future observations on the Species of their animal and vegetable remains will decide the exact place of each, in the grand Series of the Tertiary formations.

* I have recently been favoured by Professor Braun of Carlsruhe, with the following important and hitherto unpublished catalogue, and observations on the fossil plants found in the

No distinct catalogues of plants found in the *Pliocene*, or most recent periods of the Tertiary series, have yet been published.

Freshwater formation of Œningen, which has been already spoken of in our account of fossil fishes. The plants enumerated in this catalogue, were collected during a long series of years by the inmates of a monastery near Œningen, on the dissolution of which they were removed to their present place in the Museum of Carlsruhe. It appears by this catalogue that the plants of Œningen afford examples of thirty-six species belonging to twenty-five genera of the following families.

" Families.	Genera.	Species.		Genera.	Species.
Polypodiaceæ	2	2			
Equisitaceæ	1	1	Cryptogamiæ, total 4		4
Lycopodiaceæ	1	1			
Coniferæ	2	2	Gymnospermiæ	2	2
Gramineæ	1	1	Moncotyledons	3	3
Najadeæ	2	2			
Amentaceæ	5	10			
Juglandeæ	1	2			
Ebenaceæ	1	1			
Tiliaceæ	1	1			
Acerineæ	1	5	Dicotyledons	16	27
Rhamneæ	1	2			
Leguminosæ	2	2			
Dicotyledons of doubtful families	4	4			

This table shews the great preponderance of Dicotyledonous plants in the Flora of Œningen, and affords a standard of comparison with those of the Brown-coal of other localities in the Tertiary series. The greater number of the species found here correspond with those in the Brown-coal of the Wetteraw and vicinity of Bonn.

Amid this predominance of dicotyledonous vegetables, not a single herbaceous plant has yet been found excepting some frag-

Fossil Palms.

The discovery of the remains of Palm Trees in the Brown-coal of Germany has been already

ments of Ferns and Grasses, and many remains of aquatic plants : all the rest belong to Dicotyledonous, and Gympospermous ligneous plants.

Among these remains are many single leaves, apparently dropped in the natural course of vegetation; there are also branches with leaves on them, such as may have been torn from trees by stormy weather; ripe seed vessels; and the persistent calix of many blossoms.

The greater part of the fossil plants at Œningen (about two thirds) belong to Genera which still grow in that neighbourhood; but their species differ, and correspond more nearly with those now living in North America, than with any European species, the fossil Poplars afford an example of this kind.

On the other hand, there are some Genera, which do not exist in the present Flora of Germany, e. g. the Genus Diospyros ; and others not in that of Europe, e. g. Taxodium, Liquidambar, Juglans, Gleditschia.

Judging from the proportions in which their remains occur, Poplars, Willows, and Maples were the predominating foliaceous trees in the former Flora of Œningen. Of two very abundant fossil species, one, (Populus latior,) resembles the modern Canada Poplar; the other, (Populus ovalis) resembles the Balsam Poplar of North America.

The determination of the species of fossil Willows is more difficult. One of these (Salix angustifolia) may have resembled our present Salix viminalis.

Of the genus Acer, one species may be compared with Acer campestre, another with Acer pseudoplatanus; but the most frequent species, (Acer protensum,) appears to correspond most nearly with the Acer dasycarpon of North America ; to another

noticed; and the more frequent occurrence of similar remains of this interesting family, in the Tertiary formations of France, Switzerland, and

species, related to Acer negundo, Mr. Braun gives the name of Acer trifoliatum. A fossil species of Liquidambar (L. europeum, Braun.) differs from the living Liquidambar styracifluum of America, in having the narrower lobes of its leaf terminated by longer points, and was the former representative of this genus in Europe. The fruit of this Liquidambar is preserved, and also that of two species of Acer and one Salix.

The fossil Linden Tree of Œningen resembled our modern large leaved Linden tree (Tilia grandiflora.)

The fossil Elm resembled a small leaved form of Ulmus campestris.

Of two species of Juglans, one (J. falcifolia) may be compared with the American J. nigra; the other, with J. Alba, and, like it, probably belonged to the division of nuts with bursting external shells, (Carya Nuttal.)

Among the scarcer plants at Œningen, is a species of Diospyros (D. brachysepala.) A remarkable calyx of this plant is preserved and shews in its centre, the place where the fruit separated itself: it is distinguished from the living Diospyros lotus of the South of Europe by blunter and shorter sections.

Among the fossil shrubs are two species of Rhamnus; one of these (R. multinervis, Braun) resembles the R. alpinus, in the costation of its leaf. The second and most frequent species, (R. terminalis, Braun) may with regard to the position and costation of its leaves, be compared in some degree with R. catharticus, but differed from all living species in having its flowers placed at the tips of the plant.

Among the fossil Leguminous plants is a leaf more like that of a fruticose Cytisus than of any herbaceous Trefoil.

Of a Gleditschia, (G. podocarpa, Braun) there are fossil pinnated leaves and many pods; the latter seem, like the G. monasperma of North America, to have been single seeded, and are small and short, with a long stalk contracting the base of the pod.

With these numerous species of foliaceous woods, are found also

G.   L L

England, whilst they are comparatively rare in strata of the Secondary and Transition series,

a few species of Coniferæ. One species of Abies is still undetermined; branches and small cones of another tree of this family (Taxodium europeum Ad. Brong.) resemble the Cypress of Japan (Taxodium Japonicum.)

Among the remains of aquatic plants are a narrow-leaved Potamogeton; and an Isoetes, similar to the I. lacustris now found in small lakes of the Black Forest, but not in the Lake of Constance.

The existence of Grasses at the period when this formation was deposited, is shewn by a well preserved impression of a leaf, similar to that of a Triticum, turning to the right, and on which the costation is plainly expressed.

Fragments of fossil Ferns occur here, having a resemblance to Pteris aquilina and Aspidium Filix mas.

The remains of Equisetum indicate a species resembling E. palustre.

Among the few undetermined remains are the five-cleft and beautiful veined impressions of the Calyx of a blossom, which are by no means rare at Œningen.

No remains of any Rosaceæ have yet been discovered at this place." *Letter from Prof. Braun to Dr. Buckland*, Nov. 25, 1835.

In addition to these fossil Plants, the strata at Œningen contain many species of freshwater Shells, and a remarkable collection of fossil Fishes, which we have before mentioned, P. 285. In the family of Reptiles they present a very curious Tortoise, and a gigantic aquatic Salamander, more than three feet long, the *Homo Diluvii testis* of Scheuchzer. A Lagomys and fossil Fox have also been found here. (See Geol. Trans. Lond. N. S. vol. iii. p. 287.

In Oct. 1835, I saw in the Museum at Leyden, a living Salamander three feet long, the first ever brought alive to Europe, of a species nearly allied to the fossil Salamander of Œningen. This animal was brought by Dr. Siebold from a lake within the crater of an extinct volcano, on a high mountain in Japan. It fed greedily on small fishes, and frequently cast its epidermis.

suggests the propriety of consigning to this part of our subject the few observations we have to make on their history.

The existing family of Palms* is supposed to consist of nearly a thousand species, of which the greater number are limited to peculiar regions of the torrid Zone. If we look to the geological history of this large and beautiful family, we shall find that although it was called into existence, together with the most early vegetable forms of the Transition period, it presents very few species in the Coal formation, (See Lindley's Foss. Flora, No. XV, Pl. 142, P. 163,) and occurs sparingly in the Secondary series ;† but in the Tertiary formation we have abundant stems and leaves, and fruits, derived from Palms.‡

Fossil Trunks of Palm Trees.

The fossil stems of Palms are referrible to many species; they occur beautifully silicified in the Tertiary deposits of Hungary, and in the Calcaire Grossier of Paris.§   Trunks of Palms are found

---

* See Pl. 1, Figs. 66, 67, 68.

† See Sprengel's Account of Endogenites Palmacites in New red sandstone, near Chemnitz, (Halle, 1828.) and Cotta's Dendrolithen, (Dresden and Leipsig, 1832, Pl. ix, x.)

‡ Eight species in the family of Palms are given in Ad. Brongniart's list of the fossils of the Tertiary Series.

§ Our figure, Pl. 64, Fig. 2, represents the summit of a beau-

also in the Fresh water formation of Mont
Martre.‖—It is stated, that at Liblar, near Cologne, they have been seen in a vertical position.¶ Beautifully silicified stems of Palm Trees
abound in Antigua, and in India, and on the
banks of the Irawadi, in the kingdom of Ava.

It is not surprising to find the remains of Palms
in warm latitudes where plants of this family are
now indigenous, as in Antigua or India; but
their occurrence in the Tertiary formations of
Europe, associated with the remains of Crocodiles and Tortoises, and with marine shells, nearly
allied to forms which are at present found in seas

tiful fossil Trunk in the Museum at Paris, allied to the family of
Palms, and nearly four feet in circumference, from the lower
region of the Calcaire Grossier at Vaillet near Soissons. M.
Brongniart has applied to this fossil the name of *Endogenites
echinatus.* The projecting bodies that surround it, like the
foliage of a Corinthian Capital, are the persistent portions of
fallen Petioles which remain adhering to the stem after the leaves
themselves have fallen off. They have a dilated base embracing
one-fourth or one-third of the stem; the form of these bases, and
the disposition of their woody tissue in fasciculi of fibres, refer
this fossil to some arborescent Monocotyledonous Tree allied to
Palms.

‖ Prostrate trunks of Palm trees of considerable size are found
in the argillaceous marl beds above the Gypsum strata of the
Paris Basin, together with shells of Lymnea and Planorbis; as
these Trunks occur here in fresh water deposits they can not
have been drifted by marine current from distant regions, but
were probably natives of Europe, and of France.

¶ It is not shewn whether these Palm trees were drifted into
this position, or are still standing in the spot whereon they grew
like the Cycadites and Coniferæ in the Isle of Portland.

of a warm temperature, seems to indicate that the climate of Europe during the Tertiary period, was warmer than it is at present.

Fossil Palm leaves.

We have seven known localities of fossil Palm leaves, in the Tertiary strata of France, Switzerland, and the Tyrol; and among them at least three species, of flabelliform leaves, all differing not only from that of the *Chamærops humilis*, the only native palm of the South of Europe, but also from every known living species.* These leaves are too well preserved to have endured transport by water from a distant region, and must apparently be referred to extinct species, which, in the Tertiary period, were indigenous in Europe.

No pinnated Palm leaf has yet been found in the Tertiary Strata, although the number of these forms among existing palms, is more than double that of the flabelliform leaves.†

---

* The leaf represented in Pl. 64. fig. 1. is that of a *flabelliform* Palm (Palmacites Lamanonis), from the Gypsum of Aix in Provence; similar leaves have been found in three other parts of France, near Amiens, Mans, and Angers, all in strata of the Tertiary epoch. Another species (Palmacites Parisiensis) has been found in the Calcaire Grossier, near Versailles (*Cuvier and Brongniart, Geognosie des Environs de Paris*, Pl. 8, fig. 1. E.) A third species of Palm leaf (Palmacites flabellatus) occurs in the Molasse of Switzerland, near Lausanne, and in the Lignite of Hœring, in Tyrol. See Pl. 1. figs. 13. 66.

† The Date, Cocoa-nut Palm, and Areca are familiar examples of Palms having pinnated leaves. See Pl. 1. figs. 67. 68.

Fossil Fruits of Palms.

Many fossil fruits of the Tertiary period belong to the family of Palms, all of which, according to M. Ad. Brongniart, seem derived from
Genera that have pinnated leaves. Several such
fruits occur in the Tertiary clay of the Island of
Sheppey; among which are the Date,* now pe-
culiar to Africa and India; the Cocoa-nut,†
which grows universally within the tropics; the
Bactris, which is limited to America; and the
Areca, which is found only in Asia. Not one of
these can be referred to any flabelliform palm.
Fossil Cocoa-nuts occur also at Brussels, and at
Liblar near Cologne, together with fruits of the
Areca.

Although all these fruits belong to Genera
whose leaves are pinnated, no fossil pinnated
Palm leaves (as we have just stated), have yet
been found in Europe. It seems therefore most
likely, from the mode in which so large a number
of miscellaneous fruits are crowded together in
the Isle of Sheppey, mixed with marine shells
and fragments of timber, almost always perforated by Teredines, that the fruits in question
were drifted by marine currents from a warmer

---

* See Parkinson's Org. Rem. Vol. i. Pl. VI. fig. 4, 9.

† See Parkinson's Org. Rem. Vol. i. Pl. VII. fig. 1—5.
M. Brongniart says, these fruits are undoubtedly of the Genus
Cocos, near to Cocos lapidea, of Gærtner.

climate than that which Europe presented after the commencement of the Tertiary Epoch; in the same manner as tropical seeds and logs of mahogany are now drifted from the Gulf of Mexico to the Coasts of Norway and Ireland.

Besides the fruits of Palms, the Isle of Sheppey presents an assemblage of many hundred species of other fruits,* most of them apparently tropical; these could scarcely have been accumulated, as they are, without a single leaf of the tree on which they grew, and have been associated with drifted timber bored by Teredines, by any other means than a sea current.

We have no decisive information as to the number of species of these fossil fruits; they have been estimated at from six to seven hundred.† In the same clay with them are found

---

* According to M. Ad. Brongniart, many of these have near relation to the aromatic fruits of the Amomum *(cardomom)*, they are triangular, much compressed, and umbilicated at the summit, which presents a small circular areola, apparently the cicatrix of an adherent calyx; within are three valves. A slight furrow passes along the middle of each plain surface, similar to that on the fruit of many *scitamineous* plants. These Sheppey fruits, however, cannot be identified with any known Genus of that Family, but approach so nearly to it, that Ad. Brongniart gives them the name of Amomocarpum.

† See Parkinson's Organic Remains, Vol. i. Pl. 6, 7. Jacob's Flora Favershamensis. And Dr. Parsons, in Phil. Trans. Lond. 1757, Vol. 50, page 396, Pl. XV. XVI. An immense collection of these fruits is preserved in the British Museum, another in the Museum at Canterbury, and a third in that of Mr. Bowerbank, in London.

great numbers of fossil Crustaceans, and also the remains of many fishes, and of Crocodiles, and aquatic Tortoises.

As the drifted seeds that occur in Sheppey seem to have been collected by the action of marine currents, the history of European vegetation during the Tertiary period, must be sought for in those other remains of plants, whose state and circumstances show that they have grown at no great distance from the spot in which they are now found.*

Conclusion.

The following is a summary of what is yet known, respecting the varying conditions of the Flora of the three great periods of Geological history we have been considering.

The most characteristic distinctions between the vegetable remains of these periods are as follows. In the first period, the predominance of vascular Cryptogamic, and comparative rarity of Dicotyledonous plants. In the second, the approximation to equality of vascular Cryptogamic, and Dicotyledonous plants.† In the third, the predominance of Dicotyledonous, and rarity of

---

* The beautiful Amber, which is found on the eastern shores of England, and on the Coasts of Prussia and Sicily, and which is supposed to be fossil resin, is derived from beds of Lignite in Tertiary strata. Fragments of fossil gum were found near London in digging the tunnel through the London clay at Highgate.

† The dicotyledonous plants of the Transition and Secondary

vascular Cryptogamic plants. Among existing vegetables almost two thirds are Dicotyledonous.

The Remains of Monocotyledonous Plants occur, though sparingly, in each Period of Geological formations.

The number of fossil plants as yet described is about five hundred; nearly three hundred of these are from strata of the Transition series; and almost entirely from the Coal formation. About one hundred are from strata of the Secondary series, and more than a hundred from formations of the Tertiary series. Many additional species have been collected from each of these series, but are not yet named.

As the known species of living vegetables are more than fifty thousand, and the study of fossil botany is as yet but in its infancy, it is probable that a large amount of fossil species lies hid in the bowels of the earth, which the discoveries of each passing year will be continually bringing to light.

The plants of the First period are in a great measure composed of Ferns, and gigantic Equisetaceæ; and of families, of intermediate character between existing forms of Lycopodiaceæ and Coniferæ, e. g. Lepidodendriæ, Sigillariæ, and Stigmariæ; with a few Coniferæ.

Of plants of the Second period, about one third

---

formations present only that peculiar tribe of this class, which is made up of Cycadeæ and Coniferæ, viz. Gymnospermous Phanerogamiæ.

are Ferns; and the greatest part of the remainder are, Cycadeæ and Coniferæ, with a few Liliaceæ. More species of Cycadeæ occur among the fossils of this period, than are found living on the present surface of the earth. They form more than one third of the total known fossil Flora of the Secondary formations; whilst of our actual vegetation, Cycadeæ are not one two-thousandth part.

The vegetation of the Third period approximated closely to that of the existing surface of the globe.

Among living families of plants, Sea weeds, Ferns, Lycopodiaceæ, Equisetaceæ, Cycadeæ and Coniferæ, bear the nearest relations to the earliest forms of vegetation that have existed upon our planet.

The family which has most universally pervaded every stage of vegetation is that of Coniferæ; increasing in the number and variety of its genera and species, at each successive change in the climate and condition of the surface of the earth. This family forms about one three-hundredth part of the total number of existing vegetables.

Another family which has pervaded all the Series of formations, though in small proportions, is that of Palms.

The view we have taken, of the connexions between the extinct and living systems of the vegetable kingdom, supplies an extensive fund of arguments, and lays open a new and large field of enquiry, both to the Physiologist, and to the student in Physico-Theology.

In the fossil Flora, we have not only the existing fundamental distinctions between Endogenous and Exogenous plants, but we have also agreement in the details of structure, throughout numerous families, which indicates the influence of the same Laws, that regulate the development of the living members of the vegetable kingdom.

The remains of Fructification, also ; found occasionally with the plants of all formations, shew still further, that the principles of vegetable Reproduction have at all times been the same.

The exquisite organizations which are disclosed by the microscope, in that which to the naked eye is but a log of Lignite, or a lump of Coal, not only demonstrate the adaptation of means to ends, but the application also of similar means, to effect corresponding ends, throughout the several Creations which have modified the changing forms of vegetable life.

Such combinations of contrivances, varying with the varied conditions of the earth, not only prove the existence of a Designer from the existence of method, and design ; but from the Connexion of parts, and Unity of purpose, which pervade the entirety of one vast, and complex, but harmonious Whole, shew that One, and the same Mind gave origin and efficacy to them all.

CHAPTER XIX.

Proofs of Design in the Dispositions of Strata of the Carboniferous Order.

In reviewing the History and geological position of vegetables which have passed into the state of mineral coal, we have seen that our grand supplies of fossil fuel are derived almost exclusively from strata of the Transition series. Examples of Coal in any of the Secondary strata are few and insignificant; whilst the Lignites of the Tertiary formations, although they occasionally present small deposits of compact and useful fuel, exert no important influence on the economical condition of mankind.*

* Before we had acquired by experiment some extensive knowledge of the contents of each series of formations which the Geologist can readily identify, there was no *a priori* reason to expect the presence of coal in any one Series of strata rather than another. Indiscriminate experiments in search of coal, in strata of every formation, were therefore desirable and proper, in an age when even the name of Geology was unknown; but the continuance of such Experiments in districts which are now ascertained to be composed of the non-carboniferous strata of the Secondary and Tertiary Series, can no longer be justified, since the accumulated experience of many years has proved, that it is only in those strata of the Transition Series which have been designated as the *Carboniferous Order*, that productive Coal mines on a large scale have ever been discovered.

It remains to consider some of the physical operations on the surface of the Globe, to which we owe the disposition of these precious Relics of a former world, in a state that affords us access to inestimable treasure of mineral Coal.

We have examined the nature of the ancient vegetables from which Coal derives its origin, and some of the processes through which they passed in their progress towards their mineral state. Let us now review some further important geological phenomena of the carboniferous strata, and see how far the utility arising from the actual condition of this portion of the crust of the globe, may afford probable evidence that it is the result of Foresight and Design.

It was not enough that these vegetable remains should have been transported from their native forests, and buried at the bottom of ancient lakes and estuaries and seas, and there converted into coal; it was further necessary that great and extensive changes of level should elevate, and convert into dry and habitable land, strata loaded with riches, that would for ever have remained useless, had they continued entirely submerged beneath the inaccessible depths, wherein they were formed; and it required the exercise of some of the most powerful machinery in the Dynamics of the terrestrial globe, to effect the changes that were requisite to render these Elements of Art and Industry accessible to the

labour and ingenuity of man. Let us briefly examine the results that have been accomplished.

The place of the great Coal formation, in relation to the other series of strata, is shewn in our
first section (Pl. 1. Fig. 14.) This ideal section
represents an Example of dispositions which are
repeated over various areas upon the crust of
the Globe.*

The surface of the Earth is found to be covered
with a series of irregular depressions or Basins,
divided from one another, and sometimes wholly
surrounded by projecting portions of subjacent
strata, or by unstratified crystalline rocks, which
have been raised into hills and mountains, of
various degrees of height, direction and continuity. On either side of these more elevated
regions, the strata dip with more or less inclination, towards the lower spaces between one
mountain range and another. (See Pl. 1.)

This disposition in the form of Troughs or
Basins, which is common to all formations, has
been more particularly demonstrated in the Carboniferous Series, (See Pl. 65. Fig. 1, 2, 3.) because the valuable nature of beds of Coal often
causes them to be wrought throughout their whole
extent.

---

* The Coal Formation is here represented as having partaken
of the same elevatory movements, which have raised the strata of
all formations towards the mountain Ridges, that separate one
basin from another basin.

One highly beneficial result of the basin-shaped disposition of the Carboniferous strata has been, to bring them all to the surface around the circumference of each Basin, and to render them accessible, by sinking mines in almost every part of their respective areas ; (See Pl. 65. Figs. 1, 2, 3.). An uninterrupted inclination in one direction only, would have soon plunged the lower strata to a depth inaccessible to man.

The Basin of London, (Pl. 67.) affords an example of a similar disposition of the Tertiary strata reposing on the Chalk. The Basins of Paris, Vienna, and of Bohemia, afford other examples of the same kind. (See Pl. 1. Figs. 24—28.)

The Secondary and Transition strata of the central and North Western districts of England, are marginal portions of the great geological Basin of Northern Europe ; and their continuations are found in the plains, and on the flanks of mountain regions on the Continent.*

These general dispositions of all strata in the

---

* The section (Pl. 66. Fig. 1.) shews the manner in which the Strata of the Transition Series are continued downwards between the Coal formation and the older members of the Grauwacke formation, through a series of deposits, to which, Mr. Murchison has recently assigned the name of the " *Silurian system.*" This Silurian System is represented by No. 11, in our Section, Fig. 1. The recent labours of Mr. Murchison in the border counties of England and Wales have ably filled up what has hitherto been a blank page, in the history of this portion of the vast and important Systems of Rocks, included under the Transition series ; and

form of Troughs or Basins have resulted from two distinct systems of operations, in the economy of the terraqueous globe ; the first producing sedimentary deposits, (derived from the materials of older rocks, and from chemical precipitates,) on those lower spaces into which the detritus of ancient elevated regions was transported by the force of water ; the second raising these strata from the sub-aqueous regions in which they were deposited, by forces analogous to those whose effect we occasionally witness, in the tremendous movements of land, that form one of the phenomena of modern Earthquakes.

have shewn us the links which connect the Carboniferous system with the older Slaty rocks. The large group of deposits to which he has given the appropriate name of *Silurian system*, (as they occupy much of the Territory of the ancient Silures,) admits of a four-fold division, which is expressed in the section Pl. 66. Fig. 1.   This section represents the exact order of succession of these Strata in a district, which must henceforth be classic in the Annals of Geology.

In September, 1835, I found the three uppermost divisions of this system, largely developed in the same relative order of succession on the south frontier of the Ardennes, between the great Coal formation and the Grauwacke.   See Proceedings of the Meeting of the Geological Society of France at Mézières and Namur, Sep. 1835, (*Bulletin de la Société Géologique de France, Tom. VII.*) The same subdivisions of the Silurian system, maintain their relative place and importance over, a large extent of the mountainous district of the Eifel, between the Ardennes and the Valley of the Rhine ; and are continued East of the Rhine through great part of the duchy of Nassau. (See Stiffts Gebirgs-Karte, von dem Herzothum-Nassau. Wiesbaden, 1831.)

I am relieved from the necessity of entering into details respecting the history of the Coal Fields of our own country, by the excellent summary of what is known upon this interesting subject, which has recently been given in a judicious and well selected anonymous publication, entitled *The History and Description of Fossil Fuel, the Collieries, and Coal Trade of Great Britain. London*, 1835.

The most remarkable accumulations of this important vegetable production in England are in the Wolverhampton and Dudley Coal Field, (Pl. 65, Fig. 1,) where there is a bed of coal, ten yards in thickness. The Scotch Coal field near Paisley presents ten beds, whose united thickness is one hundred feet. And the South Welsh Coal Basin (Pl. 65, Fig. 2,) contains, near Pontypool, twenty-three beds of coal, amounting together to ninety-three feet.

In many Coal fields, the occurrence of rich beds of iron ore in the strata of slaty clay, that alternate with the beds of coal, has rendered the adjacent districts remarkable as the site of most important Iron foundries; and these localities, as we have before stated, (p. 65,) usually present a further practical advantage, in having beneath the Coal and Iron ore, a substratum of Limestone, that supplies the third material required as a flux to reduce this ore to a metallic state.

Our section, Pl. 65, Fig. 1, illustrates the re-

G.   M M

sult of these geological conditions in enriching an important district in the centre of England, near Birmingham, with a continuous succession of Coal mines, and Iron foundries. A similar result has followed from the same causes, on the north-east frontier of the enormous Coal basin of South Wales, in the well-known Iron foundries, near Pontypool and Merthyr Tydfil,* (See Pl. 65, Fig. 2.) The beds of shale in the lower re-

---

* In the Transactions of the Natural History Society of Northumberland, Durham, and Newcastle, vol. i. p. 114, it is stated by Mr. Forster, that the quantity of iron annually manufactured in Wales is about 270,000 tons, of which about three-fourths are made into bars, and one-fourth sold as pigs and castings. The quantity of coal required for its manufacture will be about five tons and a half, for each ton of iron. The annual consumption of coals by the iron works will therefore be about 1,500,000 tons. The quantity used in the smelting of copper ore imported from Cornwall, in the manufacture of tin plate, forging of iron for various purposes, and for domestic uses, may be calculated at 350,000 tons, which makes altogether the annual consumption of coal in Wales 1,850,000 tons. The quantity of iron manufactured in Great Britain in the year 1827 was 690,000 tons. The production of this immense quantity was thus distributed,

	TONS.	FURNACES.
In Staffordshire	216,000	95
Shropshire	78,000	31
S. Wales	272,000	90
N. Wales	24,000	12
Yorkshire	43,000	24
Derbyshire	20,500	14
Scotland	36,500	18
	690,000	284

gion of this coal field are abundantly loaded with nodules of argillaceous iron ore, and below these is a bed of millstone grit capable of enduring the fire, and used in constructing the furnaces; still lower is the limestone necessary to produce the fusion of the ore. Pl. 65, Figs. 1, 2.

The great iron foundries of Derbyshire, Yorkshire, and the South of Scotland, afford other examples of the beneficial results of a similar juxtaposition, of rich argillaceous iron ore and coal.

" The occurrence of this most useful of metals," says Mr. Conybeare,* " in immediate connexion with the fuel requisite for its reduction, and the limestone which facilitates that reduction, is an instance of arrangement so happily suited to the purposes of human industry, that it can hardly be considered as recurring unnecessarily to final causes, if we conceive that this distribution of the rude materials of the earth was determined with a view to the convenience of its inhabitants."

Let us briefly consider what is the effect of mineral fuel, on the actual condition of mankind. The mechanical power of coals is illustrated in a striking manner, in the following statement in Sir J. F. W. Herschel's admirable Discourse on the study of Natural Philosophy, 1831, p. 59.

" It is well known to modern engineers that

* Geology of England and Wales, p. 333.

there is virtue in a bushel of coals, properly consumed, to raise seventy millions of pounds weight
a foot high.   This is actually the average effect of
an engine at this moment working in Cornwall.

The ascent of Mont Blanc from Chamouni
is considered, and with justice, as the most toil-
some feat that a strong man can execute in two
days.   The combustion of two pounds of coal
would place him on the summit."

The power which man derives from the use of
mineral coal, may be estimated by the duty *

* The number of pounds raised, multiplied by the number
of feet through which they are lifted, and divided by the number
of bushels of coal (each weighing eighty-four pounds) burnt in
raising them, gives what is termed the *duty* of a steam engine, and
is the criterion of its power.   (See an important paper on improvements of the steam engine, by Davies Gilbert, Esq. Phil.
Trans. 1830, p. 121.)

It is stated by Mr. J. Taylor, in his paper on the duty of steam
engines, published in his valuable *Records of Mining*, 1829, that
the power of the steam engine has within the last few years been
so advanced by a series of rapid improvements, that whereas, in
early times, the duty of an atmospheric engine was that of
5,000,000 pounds of water, lifted one foot high by a bushel of
coal, the duty of an engine lately erected at Wheal Towan in
Cornwall, has amounted to 87,000,000 pounds; or, in other
words, that a series of improvements has enabled us to extract
as much power from one bushel, as originally could be done
from seventeen bushels of coal.   Thus, through the instrumentality of coal as applied in the steam engine, the power of man
over matter has been increased seventeen fold since the first in-
vention of these engines; and increased nearly threefold within
twenty years.

There is now an engine at the mines called the Fowey Consols

done by a pound, or any other given weight of coal consumed in working a steam engine ; since the quantity of water that the engine will raise to a given height, or the number of quarters of corn that it will grind, or, in short, the amount of any other description of work that it will do, is proportionate to that duty. As the principal working of mineral veins can only be continued by descending deeper every year, the difficulty of extracting metals is continually on the increase, and can only be overcome by those en-

in Cornwall, of which Mr. Taylor considers the average duty, under ordinary circumstances, to be above 90,000,000; and which has been made to lift 97,000,000 lbs of water one foot high, with one bushel of coals.

The effect of these improvements on the operations of mines, in facilitating their drainage, has been of inestimable importance in extracting metals from depths which otherwise could never have been reached. Mines which had been stopped from want of power, have been reopened, others have been materially deepened, and a mass of mineral treasure has been rendered available, which without these engines must have been for ever inaccessible.

It results from these rapid advances in the application of coal to the production of power, and consequently of wealth, that mining operations of vast importance, have been conducted in Cornwall at depths till lately without example, e. g. in Wheal Abraham, at 242 fathoms, at Dolcoath at 235 fathoms, and in the Consolidated Mines in Gwennap at 290 fathoms, the latter mines giving daily employment to no less than 2,500 persons.

In the Consolidated Mines, the power of nine steam engines, four of which are the largest ever made, having cylinders ninety inches in diameter, lifts from thirty to fifty hogsheads of water per minute, (varying according to the season) from an average

larged powers of draining which Coal, and the steam engine, alone supply. It would be quite impossible to procure the fuel necessary for these engines, from any other source than mineral coal.

The importance of Coal should be estimated, not only by the pecuniary value of the metals thus produced, but by their further and more important value, when applied to the infinitely varied operations and productions of machinery and of the arts.

It has been calculated that in this country about 15,000 steam engines are daily at work ; one of those in Cornwall is said to have the power of a thousand horses,* the power of each horse, according to Mr. Watt, being equal to that of five and a half men ; supposing the average power of each steam engine to be that of twenty-five horses, we have a total amount of steam power equal to that of about two millions of men. When we con-

depth of 230 fathoms. The produce of these mines has lately amounted to more than 20,000 tons of ore per annum, yielding about 2,000 tons of fine copper, being more than one seventh of the whole quantity raised in Britain. The levels or galleries in these mines extend in horizontal distance a length of about 43 miles. (See J. Taylor's account of the depths of mines, third report of British Association, 1833, p. 428.)

Mr. J. Taylor further states, (Lond. Edin. Phil. Mag. Jan. 1836, p. 67) that the steam engines now at work in draining the mines in Cornwall, are equal in power to at least 44,000 horses, one sixteenth part of a bushel of coals performing the work of a horse.

* When Engineers speak of a 25 horse Engine, they mean one

sider, that a large proportion of this power is applied to move machinery, and that the amount of work now done by machinery in England, has been supposed to be equivalent to that of between three and four hundred millions of men by direct labour, we are almost astounded at the influence of Coal and Iron and Steam, upon the fate and fortunes of the human race. " It is on the rivers," (says Mr. Webster,) " and the boatman may repose on his oars ; it is in high ways, and begins to exert itself along the courses of land conveyances ; it is at the bottom of mines, a thousand (he might have said, 1800) feet below the earth's surface ; it is in the mill, and in the workshops of the trades. It rows, it pumps, it excavates, it carries, it draws, it lifts, it hammers, it spins, it weaves, it prints."*

We need no further evidence to shew that the presence of coal is, in an especial degree, the

which would do the work of that number of horses *constantly* acting, but supposing that the same horses could work only 8 hours in every 24, there must be 75 horses kept at least to produce the effect of such an Engine.

The largest Engine in Cornwall may, if worked to the full extent, be equal to from a 300 to 350 horse power, and would therefore require 1000 horses to be kept to produce the same constant effect.   In this way it has been said that an Engine was of 1000 horse power, but this is not according to the usual computation.
Letter from J. Taylor, Esq. to Dr. Buckland.

* As there is no reproduction of Coal in this country, since no natural causes are now in operation to form other beds of it ; whilst, owing to the regular increase of our population, and the

foundation of increasing population, riches, and power, and of improvement in almost every Art which administers to the necessities and comforts of Mankind. And, however remote may

new purposes to which the steam engine is continually applied, its consumption is advancing at a rapidly accelerating rate; it is of most portentous interest to a nation, that has so large a portion of its inhabitants dependent for existence on machinery, kept in action only by the use of coal, to economize this precious fuel. I cannot, therefore, conclude this interesting subject without making some remarks upon a practice which can only be viewed in the light of a national calamity, demanding the attention of the legislature.

We have, during many years witnessed the disgraceful and almost incredible fact, that more than a million chaldrons per annum, being nearly one third part of the best coals produced by the mines near Newcastle, have been condemned to wanton waste, on a fiery heap perpetually blazing near the mouth of almost every coal pit in that district.

This destruction originated mainly in certain legislative enactments, providing that Coal in London should be sold, and the duty upon it be rated, by *measure*, and not by *weight*. The smaller Coal is broken, the greater the space it fills ; it became, therefore, the interest of every dealer in Coal, to buy it of as large a size, and to sell it of as small a size as he was able. This compelled the Proprietors of the Coal-mines to send the large Coal only to market, and to consign the small Coal to destruction.

In the year 1830, the attention of Parliament was called to these evils ; and pursuant to the Report of a Committee, the duty on Coal was repealed, and Coal directed to be sold by *weight* instead of *measure*. The effect of this change has been, that a considerable quantity of Coal is now shipped for the London Market, in the state in which it comes from the pit; that after landing the cargo, the small coal is separated by skreening from the rest, and answers as fuel for various ordinary purposes, as well as much of the Coal which was sold in London before the alteration of the law.

have been the periods, at which these materials of future beneficial dispensations were laid up in store, we may fairly assume, that, besides the immediate purposes effected at, or before the time

The destruction of Coals on the fiery heaps near Newcastle, although diminished, still goes on, however, to a frightful extent, that ought not to be permitted; since the inevitable consequence of this practice, if allowed to continue, must be, in no long space of time, to consume all the beds nearest to the surface, and readiest of access to the coast; and thus enhance the price of Coal in those parts of England which depend upon the Coal-field of Newcastle for their supply; and finally to exhaust this Coal-field, at a period, nearer by at least one third, than that to which it would last, if wisely economized. (See Report of the Select Committee of the House of Commons, on the state of the Coal Trade, 1830, page 242. and Bakewell's Introduction to Geology, 1833, page 183 and 543.)

We are all fully aware of the impolicy of needless legislative interference; but a broad line has been drawn by nature between commodities annually or periodically reproduced by the Soil on its surface, and that subterranean treasure, and sustaining foundation of Industry, which is laid by Nature in strata of mineral Coal, whose amount is limited, and which, when once exhausted, is gone for ever. As the Law most justly interferes to prevent the wanton destruction of life and property, it should seem also to be its duty to prevent all needless waste of mineral fuel; since the exhaustion of this fuel would irrecoverably paralyze the industry of millions. The Tenant of the soil may neglect, or cultivate his lands, and dispose of his produce, as caprice or interest may dictate; the surface of his fields is not consumed, but remains susceptible of tillage by his successor; had he the physical power to annihilate the Land, and thereby inflict an irremediable injury upon posterity, the legislature would justly interfere to prevent such destruction of the future resources of the nation. This highly favoured Country, has been enriched with mineral treasures in her strata of Coal, incomparably more precious than

of their deposition in the strata of the Earth, an ulterior prospective view to the future uses of Man, formed part of the design, with which they were, ages ago, disposed in a manner so admirably adapted to the benefit of the Human Race.

mines of silver or of gold. From these sustaining sources of industry and wealth let us help ourselves abundantly, and liberally enjoy these precious gifts of the Creator; but let us not abuse them, or by wilful neglect and wanton waste, destroy the foundations of the Industry of future Generations.

Might not an easy remedy for this evil be found in a Legislative enactment, that all Coals from the Ports of Northumberland and Durham, should be shipped in the state in which they come from the Pit, and forbidding by high penalties the screening of any Sea-borne Coals before they leave the Port at which they are embarked. A Law of this kind would at once terminate that ruinous competition among the Coal owners, which has urged them to vie with each other in the wasteful destruction of small Coal, in order to increase the Profits of the Coal Merchants, and gratify the preference for large Coals on the part of rich consumers; and would also afford the Public a supply of Coals of every price and quality, which the use of the screen would enable him to accommodate to the demands of the various Classes of the Community.

A farther consideration of national Policy should prompt us to consider, how far the duty of supporting our commercial interests, and of husbanding the resources of posterity should permit us to allow any extensive exportation of Coal, from a densely peopled manufacturing country like our own; a large proportion of whose present wealth is founded on machinery, which can be kept in action only by the produce of our *native* Coal Mines, and whose prosperity can never survive the period of their exhaustion.

Chapter XX.

Proofs of Design in the Effects of Disturbing Forces on the Strata of the Earth.

In the proofs of the agency of a wise, and powerful, and benevolent Creator, which we have derived from the Animal and Vegetable kingdoms, the evidence has rested chiefly on the prevalence of Adaptations and Contrivances, and of Mechanisms adapted to the production of certain ends, throughout the organic remains of a former world.

An argument of another kind may be founded on the Order, Symmetry, and Constancy, of the Crystalline forms of the unorganized Mineral ingredients of the Earth. But in considering the great geological phenomena which appear in the disposition of the strata, and their various accidents, a third kind of evidence arises from conditions of the earth, which are the result of *disturbing forces*, that appear to a certain degree to have acted at random and fortuitously.

Elevations and subsidences, inclinations and contortions, fractures and dislocations, are phenomena, which, although at first sight they present only the appearance of disorder and con-

fusion, yet when fully understood, demonstrate
the existence of Order, and Method, and Design,
even in the operations of the most turbulent,
among the many mighty physical forces which
have affected the terraqueous globe.*

Some of the most important results of the action of these forces have been already noticed in

* " Notwithstanding the appearances of irregularity and confusion in the formation of the crust of our globe, which are
presented to the eye in the contemplation of its external features,
Geologists have been able in numerous instances to detect, in
the arrangement and position of its stratified masses, distinct
approximations to geometrical laws.   In the phenomena of anticlinal lines, faults, fissures, mineral veins, &c. such laws are
easily recognized." Hopkin's Researches in Physical Geology.
Trans. Cambridge Phil. Soc. v. 6. part 1. 1835.

" It scarcely admits of a doubt," says the author of an able
article in the Quarterly Review, (Sept. 1826, p. 537,) " that the
agents employed in effecting this most perfect and systematic
arrangement have been *earthquakes*, operating with different
degrees of violence, and at various intervals of time, during a
lapse of ages.   The order that now reigns has resulted therefore,
from causes which have generally been considered as capable
only of defacing and devastating the earth's surface, but which
we thus find strong grounds for suspecting were, in the primeval
state of the globe, and perhaps still are, instrumental in its perpetual renovation.   The effects of these subterranean forces
prove that they are governed by general laws, and that these
laws have been conceived by consummate wisdom and fore-
thought."

" Sources of apparent derangement in the system appear, when
their operation throughout a series of ages is brought into one
view, to have produced a great preponderance of good, and to be
governed by fixed general laws, conducive, perhaps essential, to
the habitable state of the globe."   Ibid. p. 539.

our fourth and fifth chapters; and our first Section, Pl. 1, illustrates their beneficial effect, in elevating and converting into habitable Lands, strata of various kinds that were formed at the bottom of the ancient Waters; and in diversifying the surface of these lands with Mountains, Plains, and Valleys, of various productive qualities, and variously adapted to the habitation of Man, and the inferior tribes of terrestrial animals.

In our last Chapter we considered the advantages of the disposition of the Carboniferous strata in the form of Basins. It remains to examine the further advantages that arise from other disturbances of these strata by *Faults* or *Fractures*, which are of great importance in facilitating the operations of Coal mines; and to extend our inquiry into the more general effect of similar Dislocations of other strata, in producing convenient receptacles for many valuable Metallic ores, and in regulating the supplies of Water from the interior of the earth, through the medium of Springs.

I have elsewhere observed* that the occurrence of *Faults*, and the *Inclined position* in which the strata composing the Coal measures are usually laid out, are facts of the highest importance, as connected with the accessibility of their mineral contents. From their *inclined*

* Inaugural Lecture, Oxford, 1819.

position, the thin strata of Coal are worked with greater facility than if they had been horizontal; but as this inclination has a tendency to plunge their lower extremities to a depth that would be inaccessible, a series of Faults, or Traps, is interposed, by which the component portions of the same formation are arranged in a series of successive tables, or stages, rising one behind another, and elevated continually upwards towards the surface, from their lowest points of depression. (See Pl. 65. Fig. 3. and Pl. 66. Fig. 2.) A similar effect is often produced by *Undulations* or contortions of the strata, which give the united advantage of inclined position and of keeping them near the surface. The *Basin-shaped* structure which so frequently occurs in coal fields, has a tendency to produce the same beneficial consequences. (See Pl. 65. Figs. 1. 2. 3.)

But a still more important benefit results from the occurrence of *Faults* or *Fractures*,* without which the contents of many deep and rich mines

---

* " Faults," says Mr. Conybeare, " consist of fissures traversing the strata, extending often for several miles, and penetrating to a depth, in very few instances ascertained; they are accompanied by a subsidence of the strata on one side of their line, or (which amounts to the same thing) an elevation of them on the other; so that it appears, that the same force which has rent the rocks thus asunder, has caused one side of the fractured mass to rise, or the other to sink.—The fissures are usually filled by clay." *Geology of England and Wales*, Part I. p. 348.

would have been inaccessible. (See Pl. 65. Fig.
3. and Pl. 66. Fig. 2.)   Had the strata of Shale
and Grit, that alternate with the Coal, been continuously united without fracture, the quantity of
water that would have penetrated from the surrounding country, into any considerable excavations that might be made in the porous grit beds,
would have overcome all power of machinery
that could profitably be applied to the drainage
of a mine; whereas by the simple arrangement
of a system of *Faults*, the water is admitted only
in such quantities as are within control. Thus
the component strata of a Coal field are divided
into insulated masses, or sheets of rock, of irregular form and area, not one of which is continuous in the same plane over any very large
district; but each is usually separated from its
next adjacent mass, by a dam of clay, impenetrable to water, and filling the fissure produced
by the fracture which caused the Fault. (See
Pl. 66. Fig. 2. and Pl. 1. Figs. *l,—l,* 7.)

If we suppose a thick sheet of Ice to be broken
into fragments of irregular area, and these fragments again united, after receiving a slight degree of irregular inclination to the plane of the
original sheet, the reunited fragments of ice will
represent the appearance of the component portions of the broken masses, or sheets of Coal
measures we are describing. The intervening
portions of more recent Ice, by which they are

held together, represent the clay and rubbish that fill the Faults, and form the partition walls that insulate these adjacent portions of strata, which were originally formed, like the sheet of Ice, in one continuous plane. Thus, each sheet or inclined table of Coal measures, is inclosed by a system of more or less vertical walls of broken clay, derived from its argillaceous shale beds, at the moment in which the Fracture and Dislocation took place; and hence have resulted those joints and separations, which, though they occasionally interrupt at inconvenient positions, and cut off suddenly the progress of the collier, and often shatter those portions of the strata that are in immediate contact with them, yet are in the main his greatest safeguard, and are indeed essential to his operations.*

These same Faults also, whilst they prevent the Water from flowing in excessive quantities in

---

* " If a field of coal (says Mr. Buddle) abounding in water, was not intersected with slip Dykes, the working of it might be impracticable, as the whole body of water which it might contain would flow uninterruptedly into any opening which might be made into it; these Faults operate as Coffer Dams, and separate the field of coal into districts."—*Letter from Mr. John Buddle, an eminent Engineer and experienced Coal Viewer at Newcastle, to Prof. Buckland, Nov.* 30, 1831.

In working a Coal Pit, the Miner studiously avoids coming near a Fault, knowing that if he should penetrate this natural barrier, the Water from the other side will often burst in, and inundate the works he is conducting on the dry side of it.

A shaft was begun about the year 1825 at Gosforth, near New-

situations where it would be detrimental, are at the same time of the greatest service, in converting it to purposes of utility, by creating on the surface a series of Springs along the line of Fault, which often give notice of the Fracture that has taken place beneath.   This important effect of Faults on the *hydraulic* machinery of the globe extends through stratified rocks of every formation.   (See Pl. 69. Fig. 2.)   It is also probable that most of the Springs, that issue from unstratified rocks, are kept in action through the instrumentality of the Faults by which they are intersected.

A similar interruption of continuity in the masses of Primary rocks, and in the rocks of intermediate age between these and the Coal formation, is found to occur extensively in the working of metallic veins. A vein is often cut off suddenly by a Fault, or fracture, crossing it trans-

castle, on the wet side of the 90 fathom Dyke, and was so inundated with water that it was soon found necessary to abandon it.   Another shaft was then begun on the dry side of the dyke, only a few yards from the former, and in this they descended *nearly* 200 fathoms without any impediment from water.

Artificial dams are sometimes made in coal mines to perform the office of the natural barriers which Dykes and Faults supply.   A dam of this kind was lately made near Manchester, by Mr. Hulton, to cut off water that descended from the upper region of porous strata, which dipped towards his excavations in a lower region of the same strata, the continuity of which was thus artificially interrupted.

G.   N N

versely, and its once continuous portions are thrown to a considerable distance from each other. This line of fracture is usually marked by a wall of clay, formed probably by the abrasion of the rocks whose adjacent portions have been thus dislocated. Such faults are known in the mines of Cornwall by the term *flucan*, and they often produce a similar advantage to those that traverse the Coal measures, in guarding the miner from inundation, by a series of natural dams traversing the rocks in various directions, and intercepting all communication between that mass in which he is conducting his operations, and the adjacent masses on the other side of the fluckan or dam.*

It may be added also, that the Faults in a Coal field, by interrupting the continuity of the beds of coal, and causing their truncated edges to abut against those of uninflammable strata of shale or

---

* " My object is rather to suggest whether the arrangement of veins, &c. does not argue design and a probable connection with other phenomena of our Globe.

" Metalliferous veins, and those of quartz, &c. appear to be channels for the circulation of the subterraneous water and vapour; and the innumerable clay veins, or " flucan courses" (as they are termed in Cornwall), which intersect them, and are often found contained in them, being generally impervious to water, prevent their draining the surface of the higher grounds as they otherwise would, and also facilitate the working of mines to a much greater depth than would be practicable without them." *R. W. Fox on the Mines of Cornwall, Phil. Trans.* 1830, *p.* 404.

grit, afford a preservative against the ravages of accidental Fire beyond the area of that sheet in which it may take its beginning; but for such a provision, entire Coal fields might be occasionally burnt out and destroyed.

It is impossible to contemplate a disposition of things, so well adapted to afford the materials essential to supply the first wants, and to keep alive the industry of the Inhabitants of our earth; and entirely to attribute such a disposition to the blind operation of *Fortuitous* causes. Although indeed it be dangerous hastily to have recourse to *Final* causes, yet since in many branches of physical knowledge, (more especially in those which relate to organized matter,) the end of many a contrivance is better understood, than the contrivance itself, it would surely be as unphilosophical to hesitate at the admission of final Causes, when the general tenor and evidence of the Phenomena naturally suggest them, as it would be to introduce them gratuitously unsupported by such evidence. We may surely therefore feel ourselves authorized to view, in the Geological arrangements above described, a system of wise and benevolent Contrivances, prospectively subsidiary to the wants and comforts of the future inhabitants of the globe; and extending onwards, from its first Formation, through the subsequent Revolutions and Convulsions that have affected the surface of our Planet.

CHAPTER XXI.

Advantageous Effect of Disturbing Forces in giving Origin to Mineral Veins.*

A further result attending the Disturbances of the surface of the Earth has been, to produce Rents or Fissures in the Rocks which have been subjected to these violent movements, and to convert them into receptacles of metallic ores, accessible by the labours of man. The greater part of metalliferous veins originated in enormous cracks and crevices, penetrating irregularly and obliquely downwards to an unknown depth, and resembling the rents and chasms which are produced by modern Earthquakes. The general disposition of mineral veins within these narrow fissures, will be best understood by reference to our first Section. (Pl. 1. Figs. k 1.—k 24.) The narrow lines which pass obliquely from the lower to the upper portion of this Section, represent the manner in which Rocks of various ages are intersected by fissures, which have become the Receptacles of rich Treasures of Metallic Ore. These fissures are more or less filled with various forms of metalliferous and earthy minerals,

* See Pl. 1. Figs. *k* 1.—*k* 24, and Pl. 67. Fig. 3.

deposited in successive, and often corresponding layers on each side of the vein.

Metallic Veins are of most frequent occurrence in rocks of the Primary and Transition series, particularly in those lower portions of stratified rocks which are nearest to unstratified crystalline rocks. They are of rare occurrence in Secondary formations, and still more so in Tertiary strata.*

* M. Dufrénoy has recently shewn that the mines of Hæmatite and Spathic iron in the Eastern Pyrenees, which occur in Limestones of three ages, referrible severally to the Transition Series, to the Lias, and to the Chalk, are all situated in parts, where these Limestones are in near contact with the Granite; and he considers that they have all most probably been filled by the sublimation of mineral matter into cavities of the limestones, at, or soon after the time of the Elevation of the Granite of this part of the Pyrenees. The period of this elevation was posterior to the deposit of the Chalk formation, and anterior to that of the Tertiary Strata. These Limestones have all become crystalline where they are in contact with the Granite; and the Iron is in some places mixed with Copper pyrites, and Argentiferous galena. (Mémoire sur la Position des Mines de Fer de la Partie orientale des Pyrénées, 1834.)

According to the recent observations of Mr. C. Darwin, the Granite of the Cordilleras of Chili (near the Uspellata Pass) which forms peaks of a height probably of 14,000 feet, has been fluid in the Tertiary period; and Tertiary strata which have been rendered crystalline by its heat, and are traversed by dykes from the granitic mass, are now inclined at high angles, and form regular, and complicated anticlinal lines. These same sedimentary strata, and also lavas are there traversed by very numerous true metallic veins of iron, copper, arsenic, silver, and gold, and these can be traced to the underlying granite. (Lond. and Edin. Phil. Mag. N. S. Vol. 8, p. 158.)

A few metals are occasionally, though rarely, found *disseminated* through the substance of Rocks. Thus Tin is sometimes found disseminated through Granite, and Copper through the cupriferous slate at the base of the Hartz, at Mansfeld, &c.

The most numerous and rich of the metallic veins in Cornwall, and in many other mining districts, are found near the junction of the Granite with the incumbent Slates. These vary in width from less than an inch to thirty feet and upwards ; but the prevailing width, both of Tin and Copper Veins in that county, is from one to three feet; and in these narrower veins, the Ore is less intermixt with other substances, and more advantageously wrought.*

Several hypotheses have been proposed to

* An excellent illustration of the manner in which metallic veins are disposed in the Rocks which form their matrix, may be found in Mr. R. Thomas's geological Report, accompanied by a Map and Sections of the mining district near Redruth. This map comprehends the most interesting spot of all the mining districts in Cornwall, and exhibits in a small compass the most important phenomena of metallic veins, slides, and cross courses, all of them penetrating to an unknown depth, and continuing uninterruptedly through Rocks of various ages. In Pl. 67, Fig. 3, I have selected from this work a section, which exhibits an unusually dense accumulation of veins producing Tin, Copper, and Lead.

Much highly valuable information on these subjects may shortly be expected from the Geological Survey of Cornwall, now in progress by Mr. De la Beche, under the appointment of the Board of Ordnance.

explain the manner in which these chasms in solid rocks have become filled with metallic ores, and with earthy minerals, often of a different nature from the rocks containing them. Werner supposed that veins were supplied by matter descending into them from above, in a state of aqueous solution; whilst Hutton, and his followers, imagined that their contents were injected from below, in a state of igneous fusion. A third hypothesis has been recently proposed, which refers the filling of veins to a process of *Sublimation* from subjacent masses of intensely heated mineral matter, into apertures and fissures of the superincumbent Rocks.* A fourth hypothesis considers veins to have been slowly filled by *Segregation*, or infiltration; sometimes into contemporaneous cracks and cavities, formed

---

* In the London and Edin. Phil. Mag. March, 1829, p. 172, Mr. Patterson has published the result of his experiments in making artificial Lead Ore (Galena) in an Earthen tube, highly heated in the middle. After causing the steam of water to pass over a quantity of Galena, placed in the hottest portion of this tube, the water was decomposed, and all the Galena had been sublimed from the heated part, and deposited again in colder parts of the tube, in cubes which exactly resembled the original Ore. No pure Lead was formed. From this deposition of Galena, in a highly crystalline form, from its vapour in contact with steam, he draws the important conclusion, that Galena might, in some instances, have been supplied to mineral veins by *sublimation* from below.

Dr. Daubeny has found by a recent experiment that if steam be passed through heated Boracic Acid, it takes up and carries along with it a portion of the Acid, which *per se* does not sublime. This experiment illustrates the sublimation of Boracic Acid in volcanic craters.

during the contraction and consolidation of the
originally soft substances of the rocks themselves;
and more frequently into fissures produced by
the fracture and dislocation of the solid strata.
Segregation of this kind may have taken place
from electro-chemical agency, continued during
long periods of time.*

The total quantity of all metals known to
exist near the surface of the Earth (excepting Iron,) being comparatively small, and their
value to mankind being of the highest order,
as the main instruments by the aid of which he
emerges from the savage state, it was of the
utmost importance, that they should be disposed
in a manner that would render them accessible
by his industry; and this object is admirably
attained through the machinery of metallic veins.

---

* The observations of Mr. Fox on the electro-magnetic properties of metalliferous veins in Cornwall, (Phil. Trans. 1830, &c.)
seem to throw new light upon this obscure and difficult subject.
And the experiments of M. Becquerel on the artificial production
of crystallized insoluble compounds of Copper, Lead, Lime, &c.
by the slow and long continued reaction and transportation of
the elements of soluble compounds, (see Becquerel, Traité de
l'Electricité, T. i. c. 7, page 547, 1834,) appear to explain many
chemical changes that may have taken place under the influence
of feeble electrical currents in the interior of the earth, and more
especially in Veins.

I have been favoured by Professor Wheatstone with the following brief explanation of the experiments here quoted.

"When two bodies, one of which is liquid, react very feebly
on each other, the presence of a third body, which is either a
conductor of electricity, or in which capillary action supplies the

Had large quantities of metals existed through-out Rocks of all formations, they might have been noxious to vegetation; had small quan-tities been disseminated through the Body of the Strata, they would never have repaid the cost of separation from the matrix. These inconveni-ences are obviated by the actual arrangement, under which these rare substances are occasion-ally collected together in the natural Magazines afforded by metallic veins.

In my Inaugural Lecture (page 12) I have spoken of the evidences of design and benevolent contrivance, which are apparent in the original formation and disposition of the repositories of minerals; in the relative quantities in which they are distributed; in the provisions that are made to render them accessible, at a certain expence

place of conductibility, opens a path to the electricity resulting from the chemical action, and a voltaic current is formed which serves to augment the energy of the chemical action of the two bodies. In ordinary chemical actions, combinations are effected by the direct reaction of bodies on each other, by which all their constituents simultaneously concur to the general effect; but in the mode considered by Becquerel the bodies in the nascent state, and excessively feeble forces, are employed, by which the molecules are produced, as it were, one by one, and are disposed to assume regular forms, even when they are insoluble, because the *number* of the molecules cannot occasion any disturbance in their arrangement. By the application of these principles, that is, by the long-continued action of very feeble electrical currents, this author has shewn that many crystallized bodies, hitherto found only in nature, may be artificially obtained."

of human skill and industry, and at the same time secure from wanton destruction, and from natural decay; in the more general dispersion of those metals which are most important, and the comparatively rare occurrence of others which are less so; and still further in affording the means whereby their compound ores may be reduced to a state of purity.*

The argument, however, which arises from the utility of these dispositions, does not depend on the establishment of any one or more of the explanations proposed to account for them. Whatever may have been the means whereby

---

* I owe to my friend Mr. John Taylor the suggestion of another argument, arising from the phenomena of mines, which derives much value from being a result of the long experience of a practical man of science.

"There is one argument," says Mr. Taylor, "which has always struck me with considerable force, as proving wise and beneficent design, to be drawn from the position of the metals. I should say that they are so placed as to be out of the reach of immediate and improvident exhaustion, exercising the utmost ingenuity of man, first to discover them, then to devise means of conquering the difficulties by which the pursuit of them is surrounded.

"Hence a continued supply through successive ages, and hence motives to industry and to the exercise of mental faculties, from which our greatest happiness is derived. The metals might have been so placed as to have been all easily taken away, causing a glut in some periods and a dearth in others, and they might have been accessible without thought, or ingenuity.

"As they are, there appears to me to be that accordance with the perfect arrangements of an allwise Creator, which it is so beautiful to observe and to contemplate."

mineral veins were charged with their precious contents; whether Segregation, or Sublimation, were the *exclusive* method by which the metals were accumulated; or, whether each of the supposed causes may have operated simultaneously or consecutively in their production; the existence of these veins remains a fact of the highest importance to the human race: and although the Disturbances, and other processes in which they originated, may have taken place at periods long antecedent to the creation of our species, we may reasonably infer, that a provision for the comfort and convenience of the last, and most perfect creatures He was about to place upon its surface, was in the providential contemplation of the Creator, in his primary disposal of the physical forces, which have caused some of the earliest, and most violent Perturbations of the globe.*

---

* That part of the History of Metals which relates to their various Properties and Uses, and their especial Adaptation to the Physical condition of Man, has been so ably and amply illustrated by two of my Associates in this Series of Treatises, that I have more satisfaction in referring my readers to the Chapters of Dr. Kidd and Dr. Prout upon these subjects than in attempting myself to follow the history of the productions of metallic veins, beyond the sources from which they are derived within the body of the Earth.

A summary of the all-important Uses of Metals to Mankind is thus briefly given, by one of our earliest and most original writers on Physico-theology.

" As for Metals, they are so many ways useful to mankind, and those Uses so well known to all, that it would be lost labour

Chapter XXII.

Adaptations of the Earth to afford supplies of water through the medium of Springs.

As the presence of water is essential both to animal and vegetable existence, the adjustment of the Earth's surface to supply this necessary fluid, in due proportion to the demand, affords one of the many proofs of Design, which arise out of the investigation of its actual condition, and of its relations to the organized beings which are placed upon it.

to say anything of them : without the use of these we could have nothing of culture or civility: no Tillage or Agriculture; no Reaping or Mowing; no Ploughing or Digging; no Pruning or Loping; no Grafting or Insition; no mechanical Arts or Trades; no Vessels or Utensils of Household-stuff; no convenient Houses or Edifices; no Shipping or Navigation. What a kind of barbarous and sordid life we must necessarily have lived, the Indians in the Northern part of America are a clear demonstration. Only it is remarkable that those which are of most frequent and necessary use, as Iron, Brass and Lead, are the most common and plentiful : others that are more rare, may better be spared, yet are they thereby qualified to be made the common measure and standard of the value of all other commodities, and so to serve for Coin or Money, to which use they have been employed by all civil Nations in all Ages." Ray's Wisdom of God in the Creation. Pt. i. 5th ed. 1709, p. 110.

Nearly three fourths of the Earth being covered with Sea, whilst the remaining dry land is in need of continual supplies of water, for the sustenance of the animal and vegetable kingdoms, the processes by which these supplies are rendered available for such important purposes, form no inconsiderable part of the beautiful and connected mechanisms of the terraqueous Globe.

The great Instrument of communication between the surface of the Sea, and that of the Land, is the Atmosphere, by means of which a perpetual supply of fresh water is derived from an Ocean of salt water, through the simple process of evaporation.

By this process, water is incessantly ascending in the state of Vapour, and again descending in the form of Dew and Rain.

Of the water thus supplied to the surface of the land, a small portion only returns to the Sea directly *in seasons of flood* through the channels of Rivers ;*

A second portion is re-absorbed into the Atmosphere by Evaporation ;

A third portion enters into the composition of Animal and Vegetable bodies ;

* It is stated by M. Arago, that one third only of the water which falls in rain, within the basin of the Seine, flows by that river into the sea: the remaining two thirds either return into the atmosphere by evaporation, or go to the support of vegetable and animal life, or find their way into the sea by subterraneous passages. Annuaire, pour l'An 1835.

A fourth portion descends into the strata, and is accumulated in their interstices into subterraneous sheets and reservoirs of water, from which it is discharged gradually at the surface in the form of perennial Springs, that form the *ordinary supply* of Rivers.

As soon as Springs issue from the Earth, their waters commence their return towards the Sea; rills unite into streamlets, which, by further accumulation form rivulets and rivers, and at length terminate in estuaries, where they mix again with their parent ocean. Here they remain, bearing part in all its various functions, until they are again evaporated into the Atmosphere, to pass and repass through the same Cycles of perpetual circulation.

The adaptations of the Atmosphere to this important service in the economy of the Globe belong not to the province of the geologist. Our task is limited to the consideration of the mechanical arrangements in the solid materials of the Earth, by means of which they co-operate with the Atmosphere, in administering to the circulation of the most important of all fluids.

There are two circumstances in the condition of the strata, which exert a material influence in collecting subterraneous stores of water, from which constant supplies are regularly giving forth in the form of springs; the first consists in the *Alternation* of porous beds of sand and stone, with strata of clay that are impermeable by water;* the second circumstance is the *Dislocation* of these strata, resulting from Fractures and Faults.

The simplest condition under which water is collected within the Earth, is in superficial beds of Gravel which rest on a sub-stratum of any kind of Clay. The Rain that falls upon a bed of gravel sinks down through the interstices of the gravel, and charges its lowest region with a subterraneous sheet of water, which is easily penetrated by wells, that seldom fail except in seasons of extreme drought. The accumulations of this water are relieved by Springs, overflowing from the lower margin of each bed of gravel.

A similar result takes place in almost all kinds of permeable strata, which have beneath them a bed of clay, or of any other impermeable material. The Rain water descends and accumulates in the lower region of each porous stratum next above the clay, and overflows in the same manner by perennial springs. Hence the numerous alternations of porous beds with beds impenetrable to water, that occur throughout the entire series of stratified rocks, produce effects of the highest consequence in the hydraulic condition of the Earth, and maintain an universal system of natural Reservoirs, from which water

* See pp. 70, 71.

overflows incessantly in the form of Springs, that carry with them fertility into the adjacent valleys. (See Pl. 67, fig. 1, S.)

The discharges of water from these reservoirs are much facilitated, and increased in number, by the occurrence of *Faults* or *Fractures* that intersect the strata.*

There are two systems of Springs which have their origin in Faults, the one supplied by water *descending* from the higher regions of strata adjacent to a fault, by which it is simply intercepted in its descent, and diverted to the surface in the form of perennial springs; (see Pl. 67, fig. 1, H.) the other maintained by water *ascending* from below by Hydrostatic pressure, (as in Artesian Wells,) and derived from strata, which at their contact with the fault, are often at a great depth;

---

* Mr. Townsend, in his Chapter on Springs, states, that there are six distinct systems of springs in the neighbourhood of Bath, which issue from as many regular strata of subterraneous water, formed by filtration through either sand or porous rocks, and placed each upon its subjacent bed of clay. From these, one system of springs is produced by overflowing in the direction towards which the strata are inclined, or have their *dip;* whilst another system results from the dislocation of the strata, and breaks out laterally through the fractures by which they are intersected.

It is stated by Mr. Hopkins, (Phil. Mag. Aug. 1834, p. 131), that all the great springs in the Lime-stone District of Derbyshire are found in conjunction with great Faults, "I do not recollect (says he) a single exception to this rule, for I believe in every instance where I observed a powerful spring, I had independent evidence of the existence of a great fault."

the water is conducted to this depth either by percolation through pores and crevices, or by small subterraneous channels in these strata, from more elevated distant regions, whence it *descends*, until its progress is arrested by the Fault. (See Pl. 67. Fig. 2, d, and Pl. 69. Fig. 2, H. L.)

Besides the advantages that arise to the whole of the Animal Creation, from these dispositions in the structure of the Earth, whereby *natural* supplies of water are multiplied almost to infinity over its surface, a further result, of vast and peculiar importance to Man, consists in the facilities which are afforded him of procuring *artificial* wells, throughout those parts of the world which are best adapted for human habitation.

The Causes of the rise of water in ordinary artificial wells, are the same that regulate its discharge from the natural apertures which give origin to springs ; and as both these effects will be most intelligibly exemplified, by a consideration of the causes of the remarkable ascent of water to the surface, and often above the surface, in those peculiar perforations which are called Artesian Wells, our attention may here be profitably directed to their history.

Artesian Wells.

The name of Artesian Wells is applied to perpetually flowing artificial fountains, obtained by boring a small hole, through strata that are des-

titute of water, into lower strata loaded with subterraneous sheets of this important fluid, which
ascends by hydrostatic pressure, through pipes
let down to conduct it to the surface. The name
is derived from Artois (the ancient Artesium,)
where the practice of making such wells has for
a long time extensively prevailed.*

* The manner of action of an Artesian Well is explained by
the Section Pl. 69. Fig. 3, copied from M. Héricart de Thury's
representation of a double Fountain at St. Ouen, which brings
up water, from two water-bearing strata at different levels below
the surface. In this double fountain, the ascending forces of
the water in the two strata A and B are different; the water
from the lowest stratum B rising to the highest level b''; that
from the upper stratum A rising only to a'. The water from
both strata is thus brought to the surface by one Bore Hole of
sufficient size to contain a double pipe, viz. a smaller pipe included within a larger one, with an interval between them for
the passage of water; thus, the smaller pipe b brings up the
water of the lower stratum B, to the highest level of the fountain
b'', whilst the larger pipe a brings up the water from stratum A
to the lower level a': both these streams are employed to supply
the Canal-basin at St. Ouen, above the level of the Seine.
Should the lower stratum B contain pure water, and that in the
upper stratum A be tainted, the pure water might by this apparatus be brought to the surface through the impure, without
contact or contamination.

In common cases of Artesian wells, where a single pipe alone
is used, if the Boring penetrates a bed containing impure water;
it is continued deeper until it arrives at another stratum containing pure water; the bottom of the pipe being plunged into this
pure water, it ascends within it, and is conducted to the surface
through whatever impurities may exist in the superior strata.
The impure water, through which the boring may pass in its descent, being excluded by the pipe from mixing with the pure
water ascending from below.

Artesian Wells are most available, and of the greatest use, in low and level districts where water cannot be obtained from superficial springs, or by ordinary wells of moderate depth. Fountains of this kind are known by the name of *Blow wells*, on the Eastern coast of Lincolnshire, in the low district covered by clay between the Wolds of Chalk near Louth, and the Sea shore. These districts were without any springs, until it was discovered that by boring through this clay to the subjacent Chalk, a fountain might be obtained, which would flow incessantly to the height of several feet above the surface.

In the King's well at Sheerness sunk in 1781 through the London clay, into sandy strata of the Plastic clay formation, to the depth of 330 feet, the water rushed up violently from the bottom, and rose within eight feet of the surface. (*See Phil. Trans.* 1784.) In the years 1828 and 1829 two more perfect Artesian wells were sunk nearly to the same depth in the Dock yards at Portsmouth and Gosport.

Wells of this kind have now become frequent in the neighbourhood of London, where perpetual Fountains are in some places obtained by deep perforations through the London clay, into porous beds of the Plastic clay formation, or into the Chalk.*

* One of the first Artesian wells near London was that of Norland House on the N. W. of Holland House, made in 1794,

Important treatises upon the subject of Artesian Wells have lately been published by M. Héricart de Thury and M. Arago in France, and by M. Von

and described in Phil. Trans. London. 1797. The water of this well was derived from sandy strata of the plastic clay formation, but so much obstruction by sand attends the admission of water to the pipes from this formation, that it is now generally found more convenient to pass lower through these sandy strata, and obtain water from the subjacent chalk. Examples of wells that rise to the surface of the lowest tract of land on the W. of London may be seen in the Artesian fountain in front of the Episcopal palace at Fulham, and in the garden of the Horticultural Society. Many such fountains have been made in the Town of Brentford, from which the water rises to the height of a few feet above the surface.

This height is found to diminish as the number of perpetually flowing fountains increases; and a general application of them would discharge the subjacent water so much more rapidly than it arrives through the interstices of the chalk, that fountains of this kind when numerous would cease to overflow, although the water within them would rise and maintain its level nearly at the surface of the land.

The Section, Pl. 68 is intended to explain the cause of the rise of water in Artesian Wells in the Basin of London, from permeable strata in the Plastic-clay formation, and subjacent Chalk. The water in all these strata is derived from the rain, which falls on those portions of their surface that are not covered by the London Clay, and is upheld by clay beds of the Gault, beneath the Chalk and Fire-stone. Thus admitted and sustained, it accumulates in the joints and crevices of these strata to the line A.B. at which it overflows by springs, in valleys, such as that represented in our section under C. Below this line, all the permeable strata must be permanently filled with a subterranean sheet of water, except where faults and other disturbing causes afford local sources of relief. Where these reliefs do not interfere, the horizontal line A, B, represents the level to which water would rise by hydrostatic pressure, in any perforations

Bruckmann in Germany.* It appears that there are extensive districts in various parts of Europe, where, under certain conditions of geological structure, and at certain levels, artificial fountains will rise to the surface of strata which throw out no natural springs,† and will afford abundant supplies of water for agricultural and domestic

through the London Clay, either into sandy beds of the Plastic Clay formation, or into the Chalk; such as those represented at D. E. F. G. H. I. If the Perforation be made at G. or H. where the surface of the country is below the line A. B. the water will rise in a perpetually flowing Artesian fountain, as it does in the valley of the Thames between Brentford and London.

* See Héricart de Thury's Considérations sur la cause du Jaillissement des Eaux des puits forés, 1829.

Notices scientifiques par M. Arago. Annuaire, pour l'An. 1835.

Von Bruckmann über Artesische Brunnen. Heilbronn am Neckar, 1833.

† The Diagrams in Pl. 69, Figs. 1 and 2. are constructed to illustrate the causes of the rise of water in natural, or artificial springs, within basin-shaped strata that are intersected by the sides of Valleys, or traversed by Faults.

Supposing a Basin (Pl. 69 Fig. 1.) composed of permeable strata, E. F. G. alternating with impermeable strata, H. I. K. L. to have the margin of all these strata continuous in all directions at one uniformly horizontal level A, B, the water which falls in rain upon the extremities of the strata E, F, G, would accumulate within them, and fill all their interstices with water up to the line A, B; and if a Pipe were passed down through the upper, into either of the lower strata, at any point within the circumference of this basin, the water would rise within it to the horizontal line A, B, which represents the general level of the margin of the Basin. A disposition so regular never exists in nature, the extremities or *outcrops* of each stratum are usually at different levels, (Fig. 1. a. c. e. g.) In such cases the line a. b.

purposes and sometimes even for moving machi-
nery. The quantity of water thus obtained in
Artois is often sufficient to turn the wheels of
Corn mills.

In the Tertiary basin of Perpignan and the
chalk of Tours, there are almost subterranean
rivers having enormous upward pressure. The
Water of an Artesian well in Roussillon rises
from 30 to 50 feet above the surface. At Perpignan and Tours, M. Arago states that the
water rushes up with so much force, that a

represents the water level within the stratum G; below this line,
water would be permanently present in G; it could never rise
above it, being relieved by springs that would overflow at a.
The line, c. d. represents the level above which the water could
never rise in the stratum F; and the line e, f, represents the
highest water level within the stratum E. The discharge of all
rain waters that percolated the strata E, F, G, thus being effected
by overflowing at e. c. a.

If common wells were perforated from the surface, i. k. l. into
the strata G. F. E, the water would rise within them only to the
horizontal lines a b, c d, e f.

The upper porous stratum C, also, would be permanently
loaded with water below the horizontal line, g, h, and permanently dry above it.

The theoretical section, Pl. 69. fig. 2. represents a portion of a
basin intersected by the fault H, L, filled with matter impermeable
to water. Supposing the lower extremities of the inclined and
permeable strata N, O, P, Q, R, to be intersected by the
fault or dyke H, L, the rain water which enters the uncovered
portions of these strata, between the impermeable clay beds,
A, B, C, D, E, would accumulate in the permeable strata up to
the horizontal lines, A A″, B B″, C C″, D D″, E E″. If an Artesian well was perforated into each of these strata to A′, B′, C′,

Cannon Ball placed in the Pipe of an Artesian Well is violently ejected by the ascending stream.

In some places application has been made to economical purposes, of the higher temperature of the water rising from great depths. In Wurtemberg Von Bruckmann has applied the warm water of Artesian wells to heat a paper manufactory at Heilbronn, and to prevent the freezing of common water around his mill wheels. The same practice is also adopted in Alsace, and at Canstadt near Stutgardt. It has even been

D′, E′, through the clay beds A, B, C, D, E, the water from these beds would rise within a pipe ascending from the perforation, to the levels A″, B″, C″, D″, E″.

These theoretical Results can never occur to the extent here represented, in consequence of the intersections of the strata by valleys of Denudation, the irregular interposition of Faults, and the varying conditions of the matter composing Dykes.

If a valley were excavated in the stratum M below A″, the water of this stratum would overflow into the bottom of this valley, and would never rise on the side of the fault so high as the level H.

Wherever the contact of the Dyke H, L, with the strata M, N, O, P, Q, R, that are intersected by it, is imperfect, an issue is formed, through which the water from these inclined strata will be discharged at the surface by a natural Artesian well; hence a series of Artesian springs will mark the line of contact of the Dyke with the fractured edges of the strata from which the water rises, and the level of the water within these strata will be always approximating to that of the springs at H; but as the permeability of Dykes varies in different parts of their course, their effect in sustaining water within the strata adjacent to them, must be irregular, and the water line within these strata will vary according to circumstances, between the highest possible levels, A, B, C, D, E, and the lowest possible level H.

proposed to apply the heat of ascending springs to the warming of green houses. Artesian wells have long been used in Italy, in the duchy of Modena; they have also been successfully applied in Holland, China,* and N. America. By means of similar wells, it is probable that water may be raised to the surface of many parts of the sandy deserts of Africa and Asia, and it has been in contemplation to construct a series of

---

* An economical and easy method of sinking Artesian Wells and boring for coal, &c, has recently been practised near Saarbrück, by M. Sellow. Instead of the tardy and costly process of boring with a number of Iron Rods screwed to each other, one heavy Bar of cast Iron about six feet long and four inches in diameter, armed at its lower end with a cutting Chisel, and surrounded by a hollow chamber, to receive through valves, and bring up the detritus of the perforated stratum, is suspended from the end of a strong rope, which passes over a wheel or pulley fixed above the spot in which the hole is made. As this rope is raised up and down over the wheel, its tortion gives to the Bar of Iron a circular motion, sufficient to vary the place of the cutting Chisel at each descent.

When the chamber is full, the whole apparatus is raised quickly to the surface to be unloaded, and is again let down by the action of the same wheel. This process has been long practised in China, from whence the report of its use has been brought to Europe. The Chinese are said to have bored in this manner to the depth of 1000 feet. M. Sellow has with this instrument lately made perforations 18 inches in diameter, and several hundred feet deep, for the purpose of ventilating coal mines at Saarbrück. The general substitution of this method for the costly process of boring with rods of iron, may be of much public importance, especially where water can only be obtained from great depths.

these wells along the main road which crosses the Isthmus of Suez.

I have felt it important thus to enter into the theory of Artesian Wells, because their more frequent adoption will add to the facilities of supplying fresh Water in many regions of the Earth, particularly in low and level districts, where this prime necessary of Life is inaccessible by any other means ; and because the theory of their mode of operation explains one of the most important and most common contrivances in the subterraneous economy of the Globe, for the production of natural springs.

By these compound results of the original disposition of the strata and their subsequent disturbances, the entire Crust of the Earth has become one grand and connected Apparatus of Hydraulic Machinery, cooperating incessantly with the Sea and with the Atmosphere, to dispense unfailing supplies of fresh Water over the habitable surface of the Land.*

Among the incidental advantages arising to Man from the introduction of Faults and Dislocations of the strata, into the system of curious arrangements that pervade the subterranean eco-

---

* The causes of intermitting Springs, and ebbing and flowing wells, and many minor irregularities in the Hydraulic Action of natural vents of water, depend on local Accidents, such as the interposition of Syphons, Cavities, &c., which are scarcely of sufficient importance to be noticed, in the general view we are here taking of the Causes of the Origin of Springs.

nomy of the Globe, we may further include the circumstance, that these fractures are the most frequent channels of issue to *mineral* and *thermal* waters, whose medicinal virtues alleviate many of the diseases of the Human Frame.*

" Thus in the whole machinery of Springs and Rivers, and the apparatus that is kept in action for their duration, through the instrumentality of a system of curiously constructed hills and valleys, receiving their supply *occasionally* from the rains of heaven, and treasuring it up in their everlasting storehouses to be dispensed *perpetually* by thousands of never-failing fountains, we see a provision not less striking, than it is important. So also in the adjustment of the relative quantities of Sea and Land, in such due proportions as to supply the earth by constant evaporation, without diminishing the waters of the ocean ; and in the appointment of the Atmosphere to be the vehicle of this wonderful and unceasing circulation ; in thus separating these waters from their native salt, (which though of the highest utility to preserve the purity of the

* Dr. Daubeny has shewn that a large proportion of the thermal springs with which we are acquainted, arise through fractures situated on the great lines of dislocation of the strata. *See Daubeny on Thermal Springs, Edin. Phil. Jour. April,* 1832, p. 49.

Professor Hoffman has given examples of these fractures in the axis of *valleys of elevation,* through which chalybeate waters rise at Pyrmont, and in other valleys of Westphalia. See Pl. 67, fig. 2.

sea, renders them unfit for the support of terrestrial animals or vegetables), and transmitting them in genial showers to scatter fertility over the earth, and maintain the never-failing reservoirs of those springs and rivers by which they are again returned to mix with their parent ocean; in all these circumstances we find such evidence of nicely balanced adaptation of means to ends, of wise foresight, and benevolent intention, and infinite power, that he must be blind indeed, who refuses to recognize in them proofs of the most exalted attributes of the Creator."*

Chapter XXIII.

Proofs of Design in the Structure and Composition of unorganized Mineral Bodies.

Much of the physical history of the compound forms of unorganized mineral bodies, has been anticipated in the considerations given in our early chapters to the unstratified and crystalline rocks. It remains only to say a few words respecting the *simple minerals* that form the ingredients of these rocks, and the *elementary bodies* of which they are composed.†

* Buckland, Inaug. Lecture, p. 13.

† The term *simple mineral* is applied not only to uncombined mineral substances, which are rare in Nature, such as pure native

" In crossing a heath," (says Paley,) "suppose I pitched my foot against a *stone*, and were asked how the stone came to be there ; I might possibly answer, that, for any thing I knew to the contrary, it had lain there for ever : nor would it perhaps be very easy to show the absurdity of this answer."*

Nay says the Geologist, for if the stone were a pebble, the adventures of this pebble may have been many and various, and fraught with records of physical events, that produced important changes upon the surface of our planet ; and its rolled condition implies that it has undergone considerable locomotion by the action of water.

gold or silver, but also to all kinds of compound mineral bodies that present a regular crystalline structure, accompanied by definite proportions of their chemical ingredients.   The difference between a simple mineral and a simple substance may be illustrated by the case of calcareous spar, or crystallized carbonate of lime.   The ultimate elements, viz. Calcium, Oxygen, and Carbon, are simple substances ; the crystalline compound resulting from the union of these elements, in certain definite proportions, forms a simple mineral, called Carbonate of lime.   The total number of simple minerals hitherto ascertained according to Berzelius is nearly six hundred, that of simple substances, or elementary principles, is fifty-four.

* I have quoted this passage, not in disparagement of the general argument of Paley, which is altogether independent of the incidental and needless comparison with which he has prefaced it, but to show the importance of the addition, that has been made by the discoveries of Geology and Mineralogy, to the evidence of the non-eternity of the earth, which so great a master pronounced to be imperfect, for lack of such information as these modern sciences have recently supplied.

Or, should the stone be Sandstone, or part of any Conglomerate, or fragmentary stratum, made up of the rounded detritus of other rocks, the ingredients of such a stone would bear similar evidence of movements by the force of water, which reduced them to the state of sand, or pebbles, and transported them to their present place, before the existence of the stratum of which they form a part; consequently no such stratum can have lain in its present place for ever.

Again, should the supposed stone contain within it the petrified remains of any fossil Animal or fossil Plant, these would not only show that animal and vegetable life had preceded the formation of the rock in which they are embedded; but their organic structure might afford examples of contrivance and design, as unequivocally attesting the exercise of Intelligence and Power, as the mechanism of a Watch or Steam engine, or any other instrument produced by human art, bears evidence of intention and skill in the workman who invented and constructed them.

Lastly, should it even be Granite, or any crystalline Primary Rock, containing neither organic remains, nor fragments of other rocks more ancient than itself, it can still be shown that there was a time when even stones of this class had not assumed their present state, and consequently that there is not one of them, which can have existed, where they now are, for ever. The

Mineralogist has ascertained that Granite is a compound substance, made up of three distinct and dissimilar simple mineral bodies, Quartz, Felspar, and Mica, each presenting certain regular combinations of external form and internal structure, with physical properties peculiar to itself. And Chemical Analysis has shewn that these several bodies are made up of other bodies, all of which had a prior existence in some more simple state, before they entered on their present union in the mineral constituents of what are supposed to be the most ancient rocks accessible to human observation. The Crystallographer also has further shewn that the several ingredients of Granite, and of all other kinds of Crystalline Rocks are composed of Molecules which are invisibly minute, and that each of these Molecules is made up of still smaller and more simple Molecules, every one of them combined in fixed and definite proportions, and affording at all the successive stages of their analysis, presumptive proof that they possess determinate geometrical figures. These combinations and figures are so far from indicating the fortuitous result of accident, that they are disposed according to laws the most severely rigid, and in proportions mathematically exact.*

* The above Paragraphs of this Chapter excepting the first, are taken almost verbatim from the Author's MS. Notes of his Lectures on Mineralogy, bearing the date of June 1822, and he has adhered more closely to the form under which they ap-

The Atheistical Theory assuming the gratuitous postulate of the eternity of matter and motion, would represent the question thus. All matter, it would contend, must of necessity have assumed some form or other, and therefore may *fortuitously* have settled into any of those under which it actually appears. Now, on this hypothesis, we ought to find all kinds of substances presented occasionally under an infinite number of external forms, and combined in endless varieties of indefinite proportions; but observation has shewn that crystalline mineral bodies occur under a fixed and limited number of external forms called *secondary*, and that these are constructed on a series of more simple *primary* forms, which are demonstrable by cleavage and mechanical division, without chemical analysis: the *integrant* molecules* of these primary forms of crystals are

pear, than he might otherwise have done, for the sake of showing that no part of them has been suggested by any recent publications; and that the views here taken have not originated in express considerations called forth by the occasion of the present Treatise, but are the natural result of ordinary serious attention to the phenomena of Geology and Mineralogy, viewed in their conjoint relations to one another, and of enquiry pursued a few steps further beyond the facts towards the causes in which they originated.

* Ce que j'ai dit de la forme deviendra encore plus évident, si, en pénétrant dans le mecanisme intime de la structure, on conçoit tous ces cristaux comme des assemblages de molécules intégrantes parfaitement semblables par leurs formes, et subordonnées, à un arrangement régulier. Ainsi, au lieu qu'une

usually compound bodies, made up of an ulterior series of *constituent* molecules, i. e. molecules of the first substances obtained by chemical analysis; and these in many cases are also compound bodies, made up of the *elementary* molecules, or final indivisible atoms,* of which the ultimate particles of matter are probably composed.†

étude superficielle des cristaux n'y laissait voir que des singularités de la nature, une étude approfondie nous conduit à cette conséquence que le même Dieu dont la puissance et la sagesse ont soumis la course des astres à des lois qui ne se démentent jamais, en a aussi établi auxquelles ont obéi avec la même fidélité les molécules qui se sont réunies pour donner naissance aux corps cachés dans les retraites du globe que nous habitons. *Haüy. Tableau comparatif des Résultats de la Cristallographie et de l'Analyse Chimique.* P. xvii.

* " We seem to be justified in concluding, that a limit is to be assigned to the divisibility of matter, and consequently that we must suppose the existence of certain ultimate particles, stamped, as Newton conjectured, in the beginning of time by the hands of the Almighty with permanent characters, and retaining the exact size and figure, no less than the other more subtle qualities and relations which were given to them at the first moment of their creation.

" The particles of the several substances existing in nature may thus deserve to be regarded as the alphabet, composing the great volume which records the wisdom and goodness of the Creator."
Daubeny's Atomic Theory, p. 107.

† We may once for all illustrate the combinations of exact and methodical arrangements under which the ordinary crystalline forms of minerals have been produced, by the phenomena of a single species ; viz. the well-known substance of Carbonate of Lime.

We have more than five hundred varieties of secondary forms presented by the crystals of this abundant earthy mineral. In each of these we trace a five-fold series of subordinate relations

When we have in this manner traced back all kinds of mineral bodies, to the first and most simple condition of their component Elements, we find these Elements to have been at all times regulated by the self-same system of fixed and universal laws, which still maintains the mechanism of the material world.   In the operation of these laws we recognize such direct and constant subserviency of means to ends, so much of harmony, and order, and methodical arrangement, in the physical properties and proportional quantities, and chemical functions of the inorganic

of one system of combinations to another system, under which every individual crystal has been adjusted by laws, acting correlatively to produce harmonious results.

Every crystal of Carbonate of Lime is made up of millions of particles of the same compound substance, having one invariable primary form, viz. that of a rhomboidal solid, which may be obtained to an indefinite extent by mechanical division.

The *integrant molecules* of these rhomboidal solids form the smallest particles to which the Limestone can be reduced without chemical decomposition.

The first result of chemical analysis divides these integrant molecules of Carbonate of Lime into two compound substances, namely, Quick Lime and Carbonic Acid, each of which is made up of an incalculable number of *constituent* molecules.

A further analysis of these constituent molecules shews that they also are compound bodies, each made up of two elementary substances, viz. the Lime made up of *elementary molecules* of the metal Calcium, and Oxygen ; and the Carbonic Acid, of elementary molecules of Carbon and Oxygen.

These ultimate molecules of Calcium Carbon, and Oxygen, form the final indivisible atoms into which every secondary crystal of Carbonate of Lime can be resolved.

G.   P P

Elements, and we further see such convincing evidence of intelligence and foresight in the adaptation of these primordial Elements to an infinity of complex uses, under many future systems of animal and vegetable organizations, that we can find no reasonable account of the existence of all this beautiful and exact machinery, if we accept not that which would refer its origin to the antecedent Will and Power of a Supreme Creator; a Being, whose nature is confessedly incomprehensible to our finite faculties, but whom the " things which do appear" proclaim to be supremely Wise, and Great, and Good.

To attribute all this harmony and order to any fortuitous causes that would exclude Design, would be to reject conclusions founded on that kind of evidence, on which the human mind reposes with undoubting confidence in all the ordinary business of life, as well as in physical and metaphysical investigations. " Si mundum efficere potest concursus atomorum, cur porticum, cur templum, cur domum, cur urbem non potest? quæ sunt minus operosa et multo quidem faciliora."*

Such was the interrogatory of the Roman Moralist, arising from his contemplation of the obvious phenomena of the natural world; and the conclusion of Bentley from a wider view of more recondite phenomena, in an age remarkable for the advancement of some of the highest

* Cicero de Natura Deorum, lib. ii. 37.

branches of Physical Science, has been most abundantly confirmed by the manifold discoveries of a succeeding century. We therefore of the present age have a thousand additional reasons to affirm with him, that "though universal matter should have endured from everlasting, divided into infinite particles in the Epicurean way, and though motion should have been coeval and coeternal with it; yet those particles or atoms could never of themselves, by omnifarious kinds of motion, whether fortuitous or mechanical, have fallen, or been disposed into this or a like visible system."*—*Bentley, Serm.* vi. *of Atheism*, p. 192.

* Dr. Prout has pursued this subject still further in the third Chapter of his Bridgewater Treatise, and shewn that the molecular constitution of matter with its admirable adaptations to the economy of the natural world, cannot have endured from eternity, and is by no means a necessary condition of its existence; but has resulted from the Will of some intelligent and voluntary Agent, possessing power commensurate with his Will.

In the first Section of his fourth Chapter the same author has also so clearly shown the great extent to which several of the most common mineral substances e. g. lime, magnesia, and iron, enter into the composition of animal and vegetable bodies, and has so fully set forth the evidences of design in the constitution and properties of the few simple substances, viz. fifty-four Elementary principles, into some one or more of which the component materials of all the three great kingdoms of Nature can be resolved, that I deem it superfluous to repeat in another form, the substance of arguments which have been so well and fully drawn by my learned Colleague, from those phenomena of the mineral Elements, which form no small part of the evidence afforded by the Chemistry of Mineralogy, in proof of the Wisdom, and Power, and Goodness of the Creator.

Chapter XXIV.

Conclusion.

IN our last Chapter we have considered the Nature of the Evidence afforded by unorganized mineral bodies, in proof of the existence of design in the original adaptation of the material Elements to their various functions, in the inorganic and organic departments of the Natural World, and have seen that the only sufficient Explanation we can discover, of the orderly and wonderful dispositions of the material Elements " in measure and number and weight," throughout the terraqueous globe, is that which refers the origin of every thing above us, and beneath us, and around us, to the will and workings of One Omnipotent Creator.   If the properties imparted to these Elements at the moment of their Creation, adapted them beforehand to the infinity of complicated useful purposes, which they have already answered, and may have further still to answer, under many successive Dispensations in the material World, such an aboriginal constitution so far from superseding an intelligent Agent, would only exalt our conceptions of the consummate skill and power, that could comprehend

such an infinity of future uses under future systems, in the original groundwork of his Creation.

In an early part of our Enquiry, we traced
back the history of the Primary rocks, which
composed the first solid materials of the Globe, to
a probable condition of universal Fusion, incompatible with the existence of any forms of organic life, and saw reason to conclude that as
the crust of the Globe became gradually reduced
in temperature, the unstratified crystalline rocks,
and stratified rocks produced by their destruction,
were disposed and modified, during long periods
of time, by physical forces, the same in kind with
those which actually subsist, but more intense in
their degree of operation ; and that the result has
been to adapt our planet to become the receptacle of divers races of vegetable and animal
beings, and finally to render it a fit and convenient habitation for Mankind.

We have seen still further that the surface of
the Land, and the Waters of the Sea have during
long periods, and at distant intervals of time,
preceding the Creation of our species, been
peopled with many different races of Vegetables and Animals, supplying the place of other
races that had gone before them ; and in all
these phenomena, considered singly, we have
found evidence of Method and Design. We
have moreover seen such a systematic recurrence
of analogous Designs, producing various ends by

various combinations of Mechanism, multiplied almost to infinity in their details of application, yet all constructed on the same few common fundamental principles which pervade the living forms of organized Beings, that we reasonably conclude all these past and present contrivances to be parts of a comprehensive and connected whole, originating in the Will and Power of one and the same Creator.

Had the number or nature of the material Elements appeared to have been different under former conditions of the Earth, or had the Laws which have regulated the phenomena of inorganic matter, been subjected to change at various Epochs, during the progress of the many formations of which Geology takes cognizance, there might indeed have been proofs of Wisdom and Power in such unconnected phenomena, but they would have been insufficient to demonstrate the Unity and Universal Agency of the *same* eternal and supreme First Cause of all things.

Again, had Geology gone no further than to prove the existence of multifarious examples of Design, its evidences would indeed have been decisive against the Atheist; but if such Design had been manifested only by distinct and dissimilar systems of Organization, and independent Mechanisms, connected together by no analogies, and bearing no relations to one another, or to any existing types in the Animal or Vegetable king-

doms, these demonstrations of Design, although
affording evidence of Intelligence and Power,
would not have proved a common origin in the
Will of *one* and *the same* Creator ; and the Polytheist might have appealed to such non-ac-
cordant and inharmonious systems, as affording
indications of the agency of *many* independent
Intelligences, and as corroborating his theory of
a *plurality* of Gods.

But the argument which would infer an Unity
of cause, from unity of effects, repeated through
various and complex systems of organization
widely remote from each other in time and place
and circumstances, applies with accumulative
force, when we not only can expand the details of
facts on which it is founded, over the entire surface of the present world, but are enabled to
comprehend in the same category all the various
extinct forms of many preceding systems of organization, which we find entombed within the
bowels of the Earth. It was well observed by
Paley, respecting the variations we find in living
species of Plants and Animals, in distant regions
and under various climates, that " We never get
amongst such original or totally different modes
of Existence, as to indicate that we are come into
the province of a different Creator, or under the
direction of a different Will."* And the very

* Paley Nat. Theol. p. 450. Chap. on the Unity of the Deity.

extensive subterranean researches that have more recently been made, have greatly enlarged the range of Facts in accordance with those on which Paley grounded this assertion.

In all the numerous examples of Design which we have selected from the various animal and vegetable remains, that occur in a fossil state, there is such a never failing Identity in the fundamental principles of their construction, and such uniform adoption of analogous means, to produce various ends, with so much only of departure from one common type of mechanism, as was requisite to adapt each instrument to its own especial function, and to fit each Species to its peculiar place and office in the scale of created Beings, that we can scarcely fail to acknowledge in all these facts, a Demonstration of the Unity of the Intelligence, in which such transcendant Harmony originated; and we may almost dare to assert that neither Atheism nor Polytheism would ever have found acceptance in the World, had the evidences of high Intelligence and of Unity of Design, which are disclosed by modern discoveries in physical science, been fully known to the Authors, or the Abettors of Systems to which they are so diametrically opposed. "It is the same hand writing that we read, the same system and contrivance that we trace, the same unity of object, and relation to final causes, which we see maintained throughout, and constantly proclaiming the Unity of the great divine Original."*

It has been stated in our Sixth Chapter, on primary stratified rocks, that Geology has rendered an important service to Natural Theology, in demonstrating by evidences peculiar to itself, that there was a time when none of the existing forms of organic beings had appeared upon our Planet, and that the doctrines of the derivation of living species either by *Development* and *Transmutation* † from other species, or by an *Eternal Succession* from preceding individuals of the same species, without any evidence of a Be-

---

* Buckland's Inaug. Lect. 1819, p. 13.

† As a misunderstanding may arise in the minds of persons not familiar with the language of physiology, respecting the import of the word *Development*, it may be proper here to state, that in its primary sense, it is applied to express the organic changes which take place in the bodies of every animal and vegetable Being, from their embryo state, until they arrive at full maturity.   In a more extended sense, the term is also applied to those progressive changes in fossil genera and species, which have followed one another during the deposition of the strata of the earth, in the course of the gradual advancement of the grand system of Creation.   The same term has been adopted by Lamarck, to express his hypothetical views of the derivation of existing species from preceding species, by successive *Transmutations* of one form of organization into another form, independent of the influence of any creative Agent.   It is important that these distinctions should be rightly understood, lest the frequent application of the word *Development*, which occurs in the writings of modern physiologists, should lead to a false inference, that the use of this term implies an admission of the theory of *Transmutation* with which Lamarck has associated it.

ginning or prospect of an End, has no where been met by so full an answer, as that afforded by the phenomena, of fossil Organic Remains.

In the course of our enquiry, we have found abundant proofs, both of the Beginning and the End of several successive systems of animal and vegetable life; each compelling us to refer its origin to the direct agency of Creative Interference ; " We conceive it undeniable, that we see, in the transition from an Earth peopled by one set of animals to the same Earth swarming with entirely new forms of organic life, a distinct manifestation of creative power transcending the operation of known laws of nature : and, it appears to us, that Geology has thus lighted a new lamp along the path of Natural Theology."[*]

Whatever alarm therefore may have been excited in the earlier stages of their development, the time is now arrived when Geological discoveries appear to be so far from disclosing any phenomena, that are not in harmony with the arguments supplied by other branches of physical Science, in proof of the existence and agency of One and the same all-wise and all-powerful Creator, that they add to the evidences of Natural Religion links of high importance that have confessedly been wanting, and are now filled up by facts which the investigation of the structure of the Earth has brought to light.

* British Critic, No. XVII. Jan. 1831, p. 194.

" If I understand Geology aright, (says Professor Hitchcock,) so far from teaching the eternity of the world, it proves more directly than any other science can, that its revolutions and races of inhabitants had a commencement, and that it contains within itself the chemical energies, which need only to be set at liberty, by the will of their Creator, to accomplish its destruction. Because this science teaches that the revolutions of nature have occupied immense periods of time, it does not therefore teach that they form an eternal series. It only enlarges our conceptions of the Deity; and when men shall cease to regard Geology with jealousy and narrow minded prejudices, they will find that it opens fields of research and contemplation as wide and as grand as astronomy itself."* †

" There is in truth, (says Bishop Blomfield) no opposition nor inconsistency between Religion and Science, commonly so called, except

* Hitchcock's Geology of Massachusetts, P. 395.

† " Why should we hesitate to admit the existence of our Globe through periods as long as geological researches require; since the sacred word does not declare the time of its original creation; and since such a view of its antiquity enlarges our ideas of the operations of the Deity in respect to duration, as much as astronomy does in regard to space? Instead of bringing us into collision with Moses, it seems to me that Geology furnishes us with some of the grandest conceptions of the Divine Attributes and Plans to be found in the whole circle of human knowledge." Hitchcock's Geology of Massachusetts, 1835, p. 225.

that which has been conjured up by injudicious zeal or false philosophy, mistaking the ends of a divine revelation." And again in another passage of the same powerful discourse, after defining the proper objects for the exercise of the human understanding, his Lordship most justly observes, " Under these limitations and corrections we may join in the praises which are lavished upon philosophy and science, and fearlessly go forth with their votaries into all the various paths of research, by which the mind of man pierces into the hidden treasures of nature; harmonizes its more conspicuous features, and removes the veil which to the ignorant or careless observer, obscures the traces of God's glory in the works of his hands."*

The disappointment which many minds experience, at finding in the phenomena of the natural world no indications of the will of God, respecting the moral conduct or future prospects of the human race, arises principally from an indistinct and mistaken view of the respective provinces of Reason and Revelation.

By the exercise of our Reason, we discover abundant evidences of the Existence, and of some of the Attributes of a supreme Creator, and apprehend the operations of many of the second causes or instrumental agents, by which

* Sermon at the opening of King's College, London, 1831, pp. 19. 14.

He upholds the mechanism of the material World; but here its province ends: respecting the subjects on which, above all others, it concerns mankind to be well informed, namely, the will of God in his moral government, and the future prospects of the human race, Reason only assures us of the absolute need in which we stand of a Revelation. Many of the greatest proficients in philosophy have felt and expressed these distinctions. "The consideration of God's Providence (says Boyle) in the conduct of things corporeal may prove to a well-disposed Contemplator, a Bridge, whereon he may pass from Natural to Revealed Religion."†‡

" Next (says Locke) to the knowledge of one God, Maker of all things, a clear knowledge of their duty was wanting to mankind."

And He, whose name, by the consent of nations, is above all praise, the inventor and founder

† Christian Virtuoso, 1690. P. 42.

‡ "Natural Religion, as it is the first that is embraced by the mind, so it is the foundation upon which revealed religion ought to be superstructed, and is as it were, the stock upon which Christianity must be engrafted. For though I readily acknowledge natural religion to be *insufficient*, yet I think it very *necessary*. It will be to little purpose to press an infidel with arguments drawn from the worthiness, that appears in the Christian doctrine to have been revealed by God, and from the miracles its first preachers wrought to confirm it; if the unbeliever be not already persuaded, upon the account of natural religion, that *there is a God, and that he is a rewarder of them that diligently seek him.*" Boyle's Christian Virtuoso, Part II. prop. 1.

of the Inductive Philosophy, thus breathes forth his pious meditation, " Thy creatures have been my books, but thy Scriptures much more. I have sought thee in the courts, fields, and gardens, but I have found thee in thy temples." *Bacon's Works*, V. 4. fol. p. 487.

The sentiment here quoted had been long familiar to him, for it pervades his writings; it is thus strikingly expressed in his immortal work. " Concludamus igitur theologiam sacram ex Verbo et Oraculis Dei, non ex lumine Naturæ aut Rationis dictamine hauriri debere. Scriptum est enim cœli enarrant Gloriam Dei, at nusquam scriptum invenitur, cœli enarrant Voluntatem Dei."* †

Having then this broad line marked out before us, and with a clear and perfect understanding, as to what we ought, and what we ought not to ex-

---

* Bacon De Augm. Scient. Lib. IX. ch. i.

† " Nothing," says Sir I. F. W. Herschel " can be more unfounded than the objection which has been taken *in limine*, by persons, well-meaning perhaps, certainly narrow minded, against the study of natural philosophy, and indeed against all science, —that it fosters in its cultivators an undue and overweening self-conceit, leads them to doubt the immortality of the soul, and to scoff at revealed religion. Its natural effect, we may confidently assert, on every well constituted mind, is and must be the direct contrary. No doubt, the testimony of natural reason, on whatever exercised, must of necessity stop short of those truths which it is the object of revelation to make known; but while it places the existence and principal attributes of a Deity on such grounds as to render doubt absurd and atheism ridiculous, it unquestion-

pect from the discoveries of Natural Philosophy, we may strenuously pursue our labours in the fruitful fields of Science, under the full assurance that we shall gather a rich and abundant harvest, fraught with endless evidences of the existence, and wisdom, and power, and goodness of the Creator.

"The Philosopher (says Professor Babbage) has conferred on the Moralist an obligation of surpassing weight; in unveiling to him the living miracles which teem in rich exuberance around the minutest atom, as well as through the largest masses of ever active matter, he has placed before him resistless evidence of immeasurable design."*

"See only (says Lord Brougham) in what contemplations the wisest of men end their most sublime enquiries!   Mark where it is that a Newton finally reposes after piercing the thickest veil that

ably opposes no natural or necessary obstacle to further progress; on the contrary, by cherishing as a vital principle an unbounded spirit of enquiry, and ardency of expectation, it unfetters the mind from prejudices of every kind, and leaves it open and free to every impression of a higher nature which it is susceptible of receiving, guarding only against enthusiasm and self-deception by a habit of strict investigation, but encouraging, rather than suppressing, every thing that can offer a prospect or a hope beyond the present obscure and unsatisfactory state.   The character of the true Philosopher is to hope all things not impossible, and to believe all things not unreasonable."   Discourse on the Study of Natural Philosophy, p. 7.

* Babbage on the Economy of Manufactures, 1 Ed. p. 319.

envelopes nature—grasping and arresting in their
course the most subtle of her elements and the
swiftest—traversing the regions of boundless space
—exploring worlds beyond the solar way—giving
out the law which binds the universe in eternal
order! He rests, as by an inevitable necessity,
upon the contemplation of the great First Cause,
and holds it his highest glory to have made the
evidence of his existence, and the dispensations
of his power and of his wisdom better understood
by men."*

If then it is admitted to be the high and peculiar privilege of our human nature, and a devo-
tional exercise of our most exalted faculties, to
extend our thoughts towards Immensity and into
Eternity, to gaze on the marvellous Beauty that
pervades the material world, and to comprehend
that Witness of himself, which the Author of the
Universe has set before us in the visible works of
his Creation; it is clear that next to the study of
those distant worlds which engage the contemplation of the Astronomer, the largest and most
sublime subject of physical enquiry which can
occupy the mind of Man, and by far the most interesting, from the personal concern we have in
it, is the history of the formation and structure of
the Planet on which we dwell, of the many and
wonderful revolutions through which it has

* Lord Brougham's Discourse of Natural Theology, 1 Ed.
p. 194.

passed, of the vast and various changes in organic life that have followed one another upon its surface, and of its multifarious adaptations to the support of its present inhabitants, and to the physical and moral condition of the Human race.

These and kindred branches of enquiry, coextensive with the very matter of the globe itself, form the proper subject of Geology, duly and cautiously pursued, as a legitimate branch of inductive science: the history of the Mineral kingdom is exclusively its own; and of the other two great departments of Nature, which form the Vegetable and Animal kingdoms, the foundations were laid in ages, whose records are entombed in the interior of the Earth, and are recovered only by the labours of the Geologist, who in the petrified organic remains of former conditions of our Planet, deciphers documents of the Wisdom in which the world was created.

Shall it any longer then be said, that a science, which unfolds such abundant evidence of the Being and Attributes of God, can reasonably be viewed in any other light than as the efficient Auxiliary and Handmaid of Religion? Some few there still may be, whom timidity or prejudice or want of opportunity allow not to examine its evidence; who are alarmed by the novelty, or surprised by the extent and magnitude of the views which Geology forces on their attention, and who would rather have kept closed

G.   Q Q

the volume of witness, which has been sealed
up for ages beneath the surface of the earth, than
impose on the student in Natural Theology the
duty of studying its contents; a duty, in which
for lack of experience they may anticipate a ha-
zardous or a laborious task, but which by those
engaged in it is found to afford a rational and
righteous and delightful exercise of their highest
faculties, in multiplying the evidences of the
Existence and Attributes and Providence of
God.*

The alarm however which was excited by the
novelty of its first discoveries has well nigh passed

* A study of the natural world teaches not the truths of revealed religion, nor do the truths of religion inform us of the
inductions of physical science.   Hence it is, that men whose
studies are too much confined to one branch of knowledge, often
learn to overrate themselves, and so become narrow-minded.
Bigotry is a besetting sin of our nature.   Too often it has been
the attendant of religious zeal : but it is perhaps most bitter and
unsparing when found with the irreligious.   A philosopher, understanding not one atom of their spirit, will sometimes scoff at
the labours of religious men ; and one who calls himself religious
will perhaps return a like harsh judgment, and thank God that
he is not as the philosophers,—forgetting all the while, that man
can ascend to no knowledge, except by faculties given to him by
his Creator's hand, and that all natural knowledge is but a re-
flexion of the will of God.   In harsh judgments such as these,
there is not only much folly, but much sin.   True wisdom consists in seeing how all the faculties of the mind, and all parts of
knowledge bear upon each other, so as to work together to a
common end ; ministering at once to the happiness of man, and
his Maker's glory.—Sedgwick's Discourse on the Studies of the
University, Cambridge, 1833, App. note F. p. 102, 103.

away, and those to whom it has been permitted
to be the humble instruments of their promulga-
tion, and who have steadily persevered, under the
firm assurance that " Truth can never be opposed to Truth," and that the works of God when
rightly understood, and viewed in their true relations, and from a right position, would at length
be found to be in perfect accordance with his
Word, are now receiving their high reward, in
finding difficulties vanish, objections gradually
withdrawn, and in seeing the evidences of Geology
admitted into the list of witnesses to the truth of
the great fundamental doctrines of Theology.*

The whole course of the enquiry which we have
now conducted to its close, has shown that the
physical history of our globe, in which some
have seen only Waste, Disorder, and Confusion,
teems with endless examples of Economy, and
Order, and Design ; and the result of all our
researches, carried back through the unwritten
records of past time, has been to fix more steadily

* One of the most distinguished and powerful Theological
writers of our time, who about 20 years ago devoted a chapter of
his work on the Evidence of the Christian Revelation, to the refuta-
tion of what he then called " the Scepticism of Geologists," has in
his recent publication on Natural Theology, commenced his considerations respecting the origin of the world, with what he now
terms " The Geological argument in behalf of a Deity."   Chalmers's Natural Theology, V. I. p. 229. Glasgow, 1835.

For Dr. Chalmers's interpretation of Genesis i. 1. et seq. see
Edinburgh Christian Instructor, April, 1814.

our assurance of the Existence of One supreme
Creator of all things, to exalt more highly our
conviction of the immensity of his Perfections,
of his Might, and Majesty, his Wisdom, and Goodness, and all sustaining Providence; and to penetrate our understanding with a profound and
sensible perception,* of the "high Veneration
man's intellect owes to God."†

The Earth from her deep foundations unites
with the celestial orbs that roll through boundless
space, to declare the glory and shew forth the
praise of their common Author and Preserver;
and the voice of Natural Religion accords harmoniously with the testimonies of Revelation, in
ascribing the origin of the Universe to the will of
One eternal, and dominant Intelligence, the Almighty Lord and supreme first cause of all things
that subsist—"the same yesterday, to-day and
for ever"—"before the Mountains were brought
forth, or ever the Earth and the World were
made, God from everlasting and world without
End."

* "Though I cannot with eyes of flesh behold the invisible
God; yet, I do in the strictest sense behold and perceive by all
my senses such signs and tokens, such effects and operations as
suggest, indicate, and demonstrate an invisible God."—Berk-
ley's Minute Philosopher. Dial. iv. c. 5.
† Boyle.

SUPPLEMENTARY NOTES.

P. 41. PROFESSOR KERSTEN has found distinctly formed crystals of prismatic Felspar on the walls of a furnace in which Copper slate and Copper Ores had been melted. Among these *pyro-chemically* formed crystals, some were simple, others twin. They are composed of Silica, Alumina, and Potash. This discovery is very important, in a geological point of view, from its bearing on the theory of the igneous origin of crystalline rocks, in which Felspar is usually so large an ingredient. Hitherto every attempt to make felspar crystals by artificial means has failed. See Poggendorf's Annalen, No. 22, 1834, and Jameson's Edin. New Phil. Journal.

P. 88. An account has recently been received from India of the discovery of an unknown and very curious fossil ruminating animal, nearly as large as an Elephant, which supplies a new and important link in the Order of Mammalia, between the Ruminantia and Pachydermata. A detailed description of this animal has been published by Dr. Falconer and Captain Cautley, who have given it the name of Sivatherium, from the Sivalic or Sub-Himalayan range of hills in which it was found, between the Jumna and the Ganges. In size it exceeded the largest Rhinoceros. The head has been discovered nearly entire. The front of the skull is remarkably wide, and retains the bony cores of two short thick and straight horns, similar in position to those of the four horned Antelope of Hindostan. The nasal bones are salient in a degree without example among Ruminants, and exceeding in this respect those of the Rhinoceros, Tapir, and Palæotherium, the only herbivorous animals that have this sort of structure. Hence there is no doubt that the Sivatherium was invested with a trunk like the Tapir. Its jaw is twice as large as that of a Buffalo, and larger than that of a Rhinoceros. The

remains of the Sivatherium were accompanied by those of the
Elephant, Mastodon, Rhinoceros, Hippopotamus, several Ruminantia, &c.

It is stated (p. 88) that there is a wider distance between the
living Genera of the Order Pachydermata than between those of
any other Order of Mammalia, and that many intervals in the
series of these animals have been filled up by extinct Genera and
Species, discovered in strata of the Tertiary series. The Sivatherium forms an important addition to the extinct Genera of
this intermediate and connecting character. The value of such
links with reference to considerations in Natural Theology has
been already alluded to, p. 114.

P. 106. Since this work was in the press, the author has seen
at Liège the very extensive collection of fossil Bones made by
M. Schmerling in the caverns of that neighbourhood, and has
visited some of the places where they were found. Many of
these bones appear to have been brought together like those in
the cave of Kirkdale, by the agency of Hyænas, and have evidently been gnawed by these animals; others, particularly those
of Bears, are not broken, or gnawed, but were probably collected
in the same manner as the bones of Bears in the cave of Gailen-
reuth, by the retreat of these animals into the recesses of caverns
on the approach of death; some may have been introduced by
the action of water.

The human bones found in these caverns are in a state of
less decay than those of the extinct species of beasts; they are
accompanied by rude flint knives and other instruments of flint
and bone, and are probably derived from uncivilized tribes that
inhabited the caves. Some of the human bones may also be the
remains of individuals who, in more recent times, may have been
buried in such convenient repositories.

M. Schmerling, in his Recherches sur les Ossemens Fossiles
des Cavernes de Liège, expresses his opinion that these human
bones are coeval with those of the quadrupeds, of extinct species,
found with them; an opinion from which the Author, after a
careful examination of M. Schmerling's collection, entirely dis-
sents.

P. 135. The Dinotherium has been spoken of as the largest of terrestrial Mammalia, and as presenting in its lower Jaw and Tusks a disposition of an extraordinary kind, adapted to the peculiar habits of a gigantic herbivorous aquatic Quadruped. The Author has recently been informed by Professor Kaup, of Darmstadt, that an entire head of this animal has been discovered at Epplesheim, measuring more than a yard in length and as much in breadth, and that he is preparing a description and figures of this head for immediate publication.

P. 446. In the conclusion of our chapter on the remains of animals of the lowest order, we noticed Ehrenberg's discoveries of the internal organization, and almost universal presence in the Air and Water, of microscopic living Infusoria, little expecting that before this work had issued from the press, they would also be found in a fossil state. In the London and Edin. Phil. Mag. Aug. 1, 1836, p. 158, there is an extract of a letter sent by M. Alexander Brongniart from Berlin to the Royal Academy of Sciences of Paris, announcing that Ehrenberg has also discovered the silicified remains of Infusoria in the stone called Tripoli (Polierschiefer of Werner), a substance which has been supposed to be formed from sediments of fine volcanic ashes in quiet waters. These petrified Infusoria form a large proportion of the substance of this kind of stone from four different localities, on which Ehrenberg has made his observations; they were probably living in the waters, at the time when they became charged with the volcanic dust, in which the Tripoli originated. It is added in this notice that the slimy Iron ore of certain marshes is loaded with Infusoria, of the genus Gallionella.

L'Institut, No. 166.

END OF VOL. I.

ERRATA.

VOL. I.

Page 48, Note, *for* Chapter *read* Cordier.
54, l. 4, *for* external *read* eternal.
55, l. 2, *for* organized *read* organic.
63, Note, l. 1, *for* Brogniart *read* Brongniart.
75, Note, l. 25, *for* upper *read* lower.
95, Note, l. 17, *for* reason *read* reasons.
123, l. 5, *for* Molte *read* Monte.
142, Note, *for* Cowper *read* Cooper.
155, l. 14, *for* Ilium *read* Ileum.
160, Note, *for* Weis *read* Weiss.
177, Note, l. 1, *for* (a) *read* (u.)
202, *for* peut-etre *read* est peut-être.
254, l. 18, *for* wherein *read* in which.
254, l. 21, *for* contains *read* contain.
264, l. 2, *for* Paleontology *read* Palæontology.
275, l. 6, *for* Megalicthys *read* Megalichthys.
281, l. 9, *for* Gyrodus *read* Microdon.
282, l. 14, *for* figs. 4. 5. *read* figs. 3. 4.
291, l. 19, *for* Myliobatis *read* Myliobates.
379, Note, l. 5, *for* calcerous *read* calcareous.
379, Note, l. 7, *for* Pl. 44 *read* 44′.
391, l. 2, *for* Brogniart *read* Brongniart.
420, l. 23, *for* fig. 2ª· *read* fig. 2, a.
431, l. 29, *for* Pl. 52, fig. 2, *read* Pl. 52, fig. 3.
432, l. 8, *for* Pentacrinite *read* Pentacrinites.
435, Note, l. 7, *for* Pentacrinite's *read* Pentacrinites.
439, Note, l. 3, *for* 13 *read* 14.
470, Note, l. 1, *for* Greswell *read* Creswell.
470, Note, l. 14, *for* in our Pl. 56, fig. 1, *read* in Count
 Sternberg's Tab. 7, fig. 5.
481, Note, l. 9, *for* dycotyledonous *read* dicotyledonous.
525, l. 5, *for* treasure *read* treasures.
528, Note, l. ult. *for* Herzothum *read* Herzogthum.

VOL. II.

Page 46, l. 19. *for* Myliobatis *read* Myliobates.
Plate 27ᵈ· B. 14, *for* Palata *read* Palate.
Plate 53, fig. 2, H. omitted at the Scapula.

C. Whittingham, Tooks Court, Chancery Lane.